SCHAUM'S OUTLINE OF

THEORY AND PROBLEMS

of

ANALYTICAL
CHEMISTRY

•

ADON A. GORDUS, Ph.D.
Professor of Chemistry
University of Michigan

•

SCHAUM'S OUTLINE SERIES
McGRAW-HILL BOOK COMPANY

New York   St. Louis   San Francisco   Auckland   Bogotá   Guatemala   Hamburg   Johannesburg
Lisbon   London   Madrid   Mexico   Montreal   New Delhi   Panama   Paris
San Juan   São Paulo   Singapore   Sydney   Tokyo   Toronto

ADON A. GORDUS received a B.S. in Chemistry in 1952 from the Illinois Institute of Technology and a Ph.D. in 1956 from The University of Wisconsin. He then joined the staff of The University of Michigan where he is Professor of Chemistry. He was a Guggenheim Fellow in 1973–1974 and recently was honored by students at The University of Michigan as the Phi Lambda Upsilon Outstanding Teacher. He has published over 70 research articles concerned with energy transfer processes, production of alternative fuels, and trace-element analysis of archaeological and historical artifacts, as well as of clinical, environmental, and criminalistic samples.

Schaum's Outline of Theory and Problems of
ANALYTICAL CHEMISTRY

1 2 3 4 5 6 7 8 9 10 11 12 13 14 15 16 17 18 19 20 SHP SHP 8 9 8 7 6 5

ISBN 0-07-023795-6

Sponsoring Editor, David Beckwith
Editing Supervisor, Marthe Grice
Production Manager, Nick Monti

Library of Congress Cataloging in Publication Data

Gordus, Adon A.
  Schaum's outline of analytical chemistry.

  Includes index.
  1. Chemistry, Analytic—Quantitative. I. Title.
QD101.2.G67 1985   545   84-27823
ISBN 0-07-023795-6

Preface

This Outline, which began as a study guide for students in quantitative analysis courses taught by the author, contains a number of original approaches introduced in these courses. They include the criteria for using various simplified equations in calculating the pH of weak-acid solutions and the solubilities of precipitates (including a rule for neglecting the water equilibrium), the use of a lever analogy in identifying principal species in polyprotic acid mixtures, the use of two different symbols to differentiate between half-cell and full-cell potentials, and a simple procedure for converting an equation based on ideal solutions to one in which non-ideal activity coefficients are needed. Numerous problems are concerned with non-ideal solutions. The precision of measurements and the resultant significant figures are emphasized throughout the Outline, and formulas for the propagation of errors are derived and applied. Although instrumental analysis is outside the scope of this book, introductory material is presented in Chapters 12 and 13 on phase equilibria, chromatography, and spectra. The last chapter treats the analytical uses of radioactivity, an important topic often omitted from standard texts and courses.

While directed primarily toward students of Analytical Chemistry, this Outline should also be of value in the more advanced (honors) introductory chemistry courses that include more rigorous treatments of ionic equilibrium.

It was the author's many students who defined the scope and depth of coverage of this Outline. Through their excellent questions, these students constantly helped to clarify the approaches used in the course: they were the real teachers.

Several students directly assisted in the preparation and reading of the text, and the editor, David Beckwith of McGraw-Hill, gave final shape to the Outline. The author is grateful to them all.

ADON GORDUS

Contents

CONTENTS

CONTENTS

Chapter 1

Analytical and Mathematical Review

1.1 UNITS, CONSTANTS, CONCENTRATIONS

Analytical chemistry by and large employs the International System of Units (SI), if it is agreed that the *gram* (used for masses) and the *liter* (used for fluid volumes) are to be taken as SI derived units. Table 1-1 collects important SI and non-SI units, with the latter marked by an asterisk. By definition, the SI *mole* is that amount of matter which consists of as many elementary units as are present in 0.012 kg of pure ^{12}C; it thus is identical with the traditional "gram mole." Remember that while the Kelvin and Celsius temperature scales employ the same degree, the zero of the Celsius scale is located at +273.15 K.

Decimal fractions and multiples of the SI units, as well as of the cal and eV, may be indicated by the prefixes given in Table 1-2. Thus, for instance,

$$1 \text{ L} = 1 \text{ dm}^3 \qquad 1 \text{ atm} = 101.325 \text{ kPa}$$

Table 1-1

Physical Quantity	Unit Name	Unit Symbol	Definition
Amount of Substance	mole	mol	
Density, Concentration	kilogram per cubic meter gram per liter	kg/m^3 g/L	 $1 \text{ g/L} = 1 \text{ kg/m}^3$
Electric Charge	coulomb	C	$1 \text{ C} = 1 \text{ A} \cdot \text{s}$
Electric Current	ampere	A	
Electric Potential	volt	V	$1 \text{ V} = 1 \text{ W/A}$ $= 1 \text{ J/C}$
Energy	joule calorie* electron volt*	J cal eV	$1 \text{ J} = 1 \text{ N} \cdot \text{m}$ $1 \text{ cal} = 4.184 \text{ J}$ (exactly) $1 \text{ eV} = 1.6022 \times 10^{-19} \text{ J}$
Force	newton	N	$1 \text{ N} = 1 \text{ kg} \cdot \text{m} \cdot \text{s}^{-2}$
Frequency	hertz	Hz	$1 \text{ Hz} = 1 \text{ s}^{-1}$
Length	meter centimeter angstrom* inch*	m cm Å in	 $1 \text{ cm} = 10^{-2} \text{ m}$ $1 \text{ Å} = 10^{-10} \text{ m}$ $1 \text{ in} = 2.54 \text{ cm}$ (exactly)

Table 1-1 *(cont.)*

Physical Quantity	Unit Name	Unit Symbol	Definition
Mass	kilogram gram pound mass* atomic mass unit*	kg g lbm u	$1\ g = 10^{-3}\ kg$ $1\ lbm = 453.6\ g$ $1\ u = 1.6606 \times 10^{-27}\ kg$
Power	watt	W	$1\ W = 1\ J/s$
Pressure	pascal atmosphere* torricelli*	Pa atm torr	$1\ Pa = 1\ N/m^2$ $1\ atm = 101\ 325\ Pa$ (exactly) $760\ torr = 1\ atm$
Radioactivity	becquerel Curie*	Bq Ci	$1\ Bq = 1\ decay/s$ $1\ Ci = 3.7 \times 10^{10}\ Bq$
Temperature	kelvin degree Celsius	K °C	$1\ °C = 1\ K$
Time	second	s	
Volume	cubic meter cubic centimeter liter fluid ounce*	m^3 cm^3 L oz	 $1\ cm^3 = 10^{-6}\ m^3$ $1\ L = 10^{-3}\ m^3$ $= 1000\ cm^3$ $1\ oz = 29.57\ mL$

Table 1-2

Fraction	Prefix	Symbol	Multiple	Prefix	Symbol
10^{-1}	deci	d	10^1	deka	da
10^{-2}	centi	c	10^2	hecto	h
10^{-3}	milli	m	10^3	kilo	k
10^{-6}	micro	μ	10^6	mega	M
10^{-9}	nano	n	10^9	giga	G
10^{-12}	pico	p	10^{12}	tera	T

Physical constants needed in analytical chemistry are listed in Table 1-3. The spectral energy factor will be of use in Chapter 13.

Table 1-4 gives the terms commonly used to state the concentration of a solution. In this book, *molarity* will be employed almost exclusively, and the symbol M will also be used as an abbreviation for the physical unit mol/L. It should be remembered that the molarity of a solution will vary (decrease) as its temperature increases, whereas *molality* is temperature independent. The two measures are related by the equation

$$m = \frac{1000M}{\rho - WM} \qquad (1.1)$$

in which ρ is the (temperature-dependent) density (in kg/m^3) of the solution and W is the molecular weight (in g/mol) of the solute. The *normality* of a solution depends on the reaction. For a solution that will be used in a redox reaction

$$\text{equivalent weight} = \frac{\text{molecular weight}}{\text{absolute change } n \text{ in oxidation state}}$$

For a nonredox reaction, the denominator is replaced by "number n of replaceable hydrogens, hydroxides, or their equivalent." In either case, n is reaction dependent. Normality and molarity are related via

$$N = nM \tag{1.2}$$

In any redox reaction at the equivalence point, the product of normality and volume—i.e., the number of equivalents—must be the same for the substance being oxidized and the substance being reduced. Using (1.2), this condition takes the form

$$n_o M_o V_o = n_r M_r V_r \tag{1.3}$$

Table 1-3

Avogadro constant	$N_A = 6.022 \times 10^{23}$ molecules·mol^{-1}
Faraday constant	$\mathscr{F} = 96.487$ kC·mol^{-1}
	$= 23.061$ kcal·V^{-1}·mol^{-1}
Gas constant	$R = 8.314$ J·K^{-1}·mol^{-1}
	$= 1.987$ cal·K^{-1}·mol^{-1}
	$= 82.05$ cm^3·atm·K^{-1}·mol^{-1}
	$= 0.08205$ L·atm·K^{-1}·mol^{-1}
Electronic charge magnitude	$e = 1.6022 \times 10^{-19}$ C
Planck constant	$h = 6.6262 \times 10^{-34}$ J·s
Speed of light	$c = 2.998 \times 10^8$ m·s^{-1}
Spectral energy factor	$hc = 1239.8$ nm·eV
	$= 1$ eV/8066 cm^{-1}

Table 1-4

Term	Definition
molarity (M)	= moles solute/liter solution
molality (m)	= moles solute/kg solvent
normality (N)	= equivalents solute/liter solution
formality (F)	= formula weights solute/liter solution
mole fraction (x)	= moles solute/(moles solute + moles solvent)
wt %	= (g solute/g solution)(100%)
ppm by wt	= μg solute/g solution
ppb by wt	= ng solute/g solution
meq/liter	= meq solute/liter solution
mg %	= mg solute/100 mL solution

Included in Table 1-4 are three terms used in clinical analysis. The first, wt %, is usually associated with the analysis of tissue samples, and is often qualified to indicate if it is based on a wet, dry, or ashed sample. The second, meq/liter, occurs in the analysis of ionic constituents in body fluids, such as Mg^{2+} in blood. The third, mg %, has a name that does not seem to fit its definition. Fortunately, this ambiguous term has fairly limited use in clinical analysis.

1.2 BALANCING REDOX EQUATIONS

STEP 1. Determine the oxidation states in both the reactant and product species for the element that undergoes oxidation and for the element that undergoes reduction. Fractional oxidation states are occasionally found, although they usually are a mixture of integer states. Assume values of oxidation states in the following sequence: (i) hydrogen is +1 except in H_2, where it is zero; (ii) oxygen is −2 except in peroxide, where it is −1, or in O_2, where it is zero. For organic compounds, recommended procedure is to assign H = +1, followed by O = −2, halogen = −1, and to calculate the oxidation state of carbon, if it is the remaining element, by difference. If the organic compound also contains nitrogen, then assign C = +4, and calculate the oxidation state of nitrogen by difference.

STEP 2. Calculate the *total increase in oxidation state* for the reactant ion or molecule that contains the element being *oxidized*, taking into account the number of atoms of the oxidized element per reactant ion or molecule. Similarly, calculate the *total decrease in oxidation state* for the reactant ion or molecule that contains the element being *reduced*. Balance these *reactants* by assigning them coefficients which make the two totals equal.

STEP 3. Many reactions require extra hydrogen- and/or oxygen-containing species to complete the balancing. If the reaction takes place in an acidic solution, then H^+ and/or H_2O may be added to the reactants and products as needed. If the reaction takes place in a basic solution, OH^- and/or H_2O may be added as needed. Of the three quantities H, O, and charge, at most two will be needed to complete the balancing. Select first that quantity which allows an unambiguous calculation. Thus, if the solution is acidic, the equation should *not* be first balanced in terms of hydrogen, since one cannot decide whether to add it as H^+ or as H_2O. Here, charge or oxygen balance is unambiguous; the former relies only on the addition of H^+, the latter only on the addition of H_2O. If the solution is basic, charge balance should be performed first, since it relies only on the addition of OH^-, whereas hydrogen or oxygen balance could use either OH^- or H_2O. Following the charge balance, either O or H balance can be performed using H_2O.

EXAMPLE 1.1 Complete and balance the redox reaction where ClO_3^- reacts with HNO_2 in an acidic solution to produce $Cl_2(g)$ and NO_3^-.

The chlorine is +5 in ClO_3^- and zero in $Cl_2(g)$, so that the reduction is $-5/Cl = -5/ClO_3^-$. The nitrogen is +3 in HNO_2 and +5 in NO_3^-, so that the oxidation is $+2/N = +2/HNO_2$. Therefore, balancing of the total oxidation and reduction changes requires that $2ClO_3^-$ react with $5HNO_2$, and the partially balanced equation is:

$$2ClO_3^- + 5HNO_2 \rightarrow Cl_2(g) + 5NO_3^-$$

The remaining balancing will be on charge and oxygen. The charge on the left is −2, whereas it is −5 on the right; therefore, $3H^+$ must be added to the right. There are a total of 16 oxygen atoms on the left, but only 15 on the right; hence, one H_2O molecule must be added to the right. The completely balanced equation is:

$$2ClO_3^- + 5HNO_2 \rightarrow Cl_2(g) + 5NO_3^- + 3H^+ + H_2O$$

The fact that the hydrogen atoms also balance serves as a check.

1.3 STOICHIOMETRY

Acid-base titrations are characterized by simple stoichiometry. If the reaction involves neutralization of a single hydrogen,

$$HX + MOH \rightarrow H_2O + MX$$

then mmols MOH = mmols HX. If the acid is diprotic and both hydrogens are neutralized,

$$H_2Y + 2MOH \rightarrow 2H_2O + M_2Y$$

then mmols MOH = 2(mmols H_2Y). If the acid is triprotic and all three hydrogens are neutralized,

$$H_3Z + 3MOH \rightarrow 3H_2O + M_3Z$$

then mmols MOH = 3(mmols H_3Z). The mmol relationships may be converted into volume relationships through the definition of molarity:

$$M = \frac{\text{mols}}{V(\text{L})} = \frac{\text{mmols}}{V(\text{mL})} \qquad \text{or} \qquad \text{mmols} = M \cdot V(\text{mL})$$

Other common stoichiometric relationships are based on oxidation-reduction processes.

EXAMPLE 1.2   The permanganate oxidation of ferrous ion proceeds in an acidic solution according to the reaction

$$MnO_4^- + 5Fe^{2+} + 8H^+ \rightarrow Mn^{2+} + 5Fe^{3+} + 4H_2O$$

How many mL of 0.0240 M MnO_4^- is required to oxidize 20.00 mL of 0.112 M Fe^{2+}?
    The stoichiometry requires that

$$\text{mmols Fe}^{2+} = 5(\text{mmols MnO}_4^-)$$
$$(MV)_{\text{Fe}^{2+}} = 5(MV)_{\text{MnO}_4^-}$$
$$(0.112)(20.00) = 5(0.0240)V$$
$$V = 18.7 \text{ mL}$$

If a series of chemical reactions are performed to obtain a final product which serves as the basis for the analysis, it is necessary to determine, in addition, the stoichiometric relationship from one reaction to the next.

EXAMPLE 1.3   A 0.250 g sample of a low-grade chromium ore was fused with sodium peroxide to convert the chromium to CrO_4^-. The sample was then dissolved in water, boiled, acidified (the CrO_4^- being converted to $Cr_2O_7^{2-}$), and titrated with 26.28 mL of Fe^{2+} solution. The products of the redox reaction were Fe^{3+} and Cr^{3+}. In a separate standardization titration, 46.48 mL of the Fe^{2+} was required to reduce 0.1172 g of $K_2Cr_2O_7$ standard. What is the wt % of Cr in the ore? What is the content in ppm by wt?
    The balanced redox reaction is

$$Cr_2O_7^{2-} + 6Fe^{2+} + 14H^+ \rightarrow 2Cr^{3+} + 6Fe^{3+} + 7H_2O$$

Since the molecular weight of $K_2Cr_2O_7$ is 294.19, the stoichiometry requires for the standardization:

$$\text{mmols Fe}^{2+} = 6(\text{mmols Cr}_2O_7^{2-})$$
$$M(46.48 \text{ mL}) = 6\left(\frac{117.2 \text{ mg}}{294.19 \text{ mg/mmol}}\right)$$
$$M = 0.05143$$

that is to say, the Fe^{2+} is 0.05143 M. The data for the analysis of the ore result in:

$$(0.05143)(26.28) \text{ mmol Fe}^{2+} = 6(\text{mmols Cr}_2O_7^{2-})$$

so that there was 0.2253 mmol $Cr_2O_7^{2-}$. Since there are two Cr per $Cr_2O_7^{2-}$, and since the atomic weight of Cr is 51.996, we have:

$$\frac{0.2253 \text{ mmol Cr}_2O_7^{2-}}{1} \cdot \frac{2(51.996) \text{ mg Cr}}{\text{mmol Cr}_2O_7^{2-}} \cdot \frac{100\%}{250 \text{ mg sample}} = 9.37 \text{ wt \% Cr}$$

or 9.37×10^4 ppm Cr by wt.

1.4   SOLVING EQUATIONS

The availability of small programmable calculators makes it possible to solve even fairly complicated equations of the sort that can arise from multiple equilibria systems. However, even in these cases, simplifications are sometimes useful.

Quadratic Equation

If an equation can be reduced to a quadratic, $ax^2 + bx + c = 0$, then:

$$x = \frac{-b \pm \sqrt{b^2 - 4ac}}{2a} \tag{1.4}$$

Successive Approximations

Especially for higher-order equations, successive approximations can be the most direct approach. A starting value for the unknown, x, is used in one of the x-terms, and the value of x is calculated on the basis of the remaining term or terms. This value becomes the new starting value, the calculation is repeated, ..., until a stable value is attained.

EXAMPLE 1.4   An equation for the hydrogen ion concentration, x, of a solution is:

$$x^3 + 0.0350\,x - 0.116 = 0$$

Find x.

First get an approximation for x by neglecting one of the x-terms in the equation (i.e., use the starting value $x_0 = 0$ in that term). For instance, if the x^3-term is neglected, then a first approximation for x is

$$x_1 = \frac{0.116}{0.0350} = 3.314$$

For the second approximation, substitute x_1 in the x^3-term and recalculate x:

$$x_2 = \frac{0.116 - 3.314^3}{0.0350} = -1037$$

The third approximation is

$$x_3 = \frac{0.116 - (-1037)^3}{0.0350} = 3.19 \times 10^{10}$$

It is clear by now that this set of successive approximations will lead nowhere. Therefore, let us instead start by neglecting the $0.0350\,x$-term; this results in

$$x_1 = (0.116)^{1/3} = 0.488$$

as the first approximation. The second approximation is

$$x_2 = [0.116 - (0.0350)(0.488)]^{1/3} = 0.463$$

The third approximation is

$$x_3 = [0.116 - (0.0350)(0.463)]^{1/3} = 0.464$$

which is within 1% of x_2. Thus, we can stop here, and write $x \approx 0.464$.

Graphical Intersection Method

On occasion, an approximate solution to a complicated equation is known. Let the equation be written in the form $A(x) = B(x)$. Values of each of A and B are calculated for two or more values of x in the immediate vicinity of the approximate solution. These values will plot as two straight lines (very nearly), whose point of intersection yields the solution x.

EXAMPLE 1.5   The theory of Chapter 3 gives for the solubility, S, of TlCl(s) the equation

$$\underbrace{2 \log S + 3.699}_{A(S)} = \underbrace{\frac{1.024\sqrt{S}}{1 + 0.984\sqrt{S}}}_{B(S)}$$

An approximate value of S, $S_0 = 1.40 \times 10^{-2}$, can be obtained from the solubility product for TlCl on the assumption that the solution is ideal. We compute the following values of A and B:

	A	B
$S_0 = 1.40 \times 10^{-2}$	-0.0087	$+0.1085$
$S_1 = 1.50 \times 10^{-2}$	$+0.0512$	$+0.1119$
$S_2 = 1.80 \times 10^{-2}$	$+0.2095$	$+0.1214$

S_1 was arbitrarily chosen as being within 10% of S_0. The fact that the corresponding A- and B-values are closer together than the values at S_0, but with the B-value still exceeding the A-value, shows that the desired solution must be larger than S_1. At S_2, however, the A-value is the greater; hence, the solution must lie between S_1 and S_2. Figure 1-1 shows the linear graphs and their intersection at $S \approx 1.62 \times 10^{-2}$.

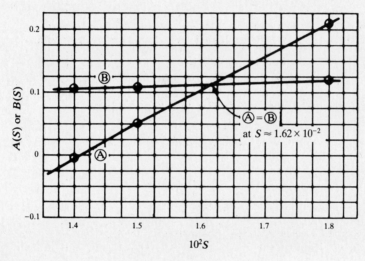

Fig. 1-1

1.5 POCKET CALCULATORS

A number of features of pocket calculators are especially useful in calculations for analytical chemistry.

Fractional Roots

The y^x function on a calculator can be used to calculate fractional roots. For example, $3.65^{1/3} = 3.65^{0.333}$. Therefore, keying in $y = 3.65$ and $x = 0.333$, followed by pressing the y^x button, will yield the cube root of 3.65, which is 1.54.

pH and pOH Calculations

If, for example, you are given $[\text{OH}^-] = 2.76 \times 10^{-4}$, simply key in that number and the $\log_{10}$ button will yield $-\text{pOH} = -3.559$. Now, $\text{pH} = 14.000 - \text{pOH}$, and $-\text{pOH}$ is already displayed; so all that remains is to add 14.000 to that value, obtaining $\text{pH} = 10.441$. [As indicated in Section 1.6, be certain that the number of decimal places in the pX value is equal to the number of digits in X.]

Range and Capacity

Most calculators have a maximum range of 10^{-99} to 10^{+99}. On those few occasions where calculations could exceed the range, carry the exponents in a separate calculation.

You should make use of the full calculator capacity (usually 8 to 10 digits) in all intermediate calculations. [Most calculators do this automatically, even though fewer digits might be displayed.] Round off the final value in accord with the rules for significant figures (Section 1.6).

Linear Regression

Many calculators have built-in programs for standard deviations and linear regression. Determine, for your calculator, how the values Σx, Σy, Σx^2, Σy^2, and Σxy are stored, since these summations can be useful in other statistical calculations described in Chapter 2.

1.6 SIGNIFICANT FIGURES

The rules that follow, while standard, must be used with the caution that *the actual uncertainty in a final value will always be greater than that given by these rules.*

RULE 1.   Values are reported so that they contain all digits known with certainty, plus a digit that is known only to within ± 1. If an actual uncertainty is *calculated*, for example, as a standard deviation, then the standard deviation automatically determines the number of significant figures. It is often useful to know the value of the standard deviation to two digits, and this requires that a second uncertain digit be shown in the original value.

EXAMPLE 1.6   A repeated series of measurements results in an average of 3.267 and a standard deviation of 0.038. This can be written as 3.267 ± 0.038, since it is clear from the value of the $\pm$-term that only the first two digits in the average value are certain. In such cases, the second uncertain digit is frequently displayed as a subscript: $3.26_7 \pm 0.03_8$. If only one digit of the standard deviation was to be kept, the rules of rounding would be applied and the result would be written as 3.27 ± 0.04.

RULE 2.   Zeros between nonzero digits are always significant. *Leading zeros* (as in 0.0081) merely locate the decimal point; they are never significant. *Trailing zeros* of integers may or may not be significant.

EXAMPLE 1.7   To indicate that a quantity notated as 30 000 has four significant figures, some authors would write $30\overline{\overline{0}}00$. The best practice in the case of trailing zeros—or in general to show which figures are significant—is to use scientific notation:
$$8.1 \times 10^{-3} \qquad 3.000 \times 10^4$$

RULE 3.   In the addition and/or subtraction of a set of values, the number of significant figures in the result is governed by the least certain value. For instance,
$$10.628 + 1.2 + 376.7 + 0.0039 = 388.5$$
since neither 1.2 nor 376.7 has digits past the first decimal place.

RULE 4.   In multiplying and/or dividing a set of values, the number of significant figures in the final value may not exceed the number of significant figures in the least precise multiplier or divisor. Thus,
$$\frac{(2.8)(108.7)}{(0.0373)(5298.3)} = 1.5$$
because the least precise value, 2.8, has only two digits.

EXAMPLE 1.8   An improvement on Rule 4 may be derived from the theory of Chapter 2 [see (2.9)], whereby the number of digits in the final value is determined by the *fractional* uncertainties in the multipliers or divisors. To see how this works, consider the evaluation of

$$X = \frac{(98)(379)}{(1305)(1.624)} \qquad Y = \frac{(103)(379)}{(1305)(1.624)}$$

based on (a) Rule 4, (b) the improvement.

(a) $X = 18$ (because 98 has only two digits) and $Y = 18.4$ (because 103 has only three digits).

(b)   Since the fractional uncertainties in the least significant values (98 and 103) are the same:

$$0.010 = 1.0\% = 1 \text{ part in } 100$$

the number of digits in either answer should be three, as this allows an uncertainty of 1 part in 100. Thus, $X = 17.5$ and $Y = 18.4$.

RULE 5.   If $Q = Kx^a$, where K is a constant and a is an accurately known constant exponent, then

$$\frac{\Delta Q}{Q} = a\,\frac{\Delta x}{x}$$

i.e., the fractional uncertainty in Q is equal to the fractional uncertainty in x multiplied by the exponent a.

EXAMPLE 1.9   If $Q = (0.0878)^{1/50} \approx 0.95251$ and if $\Delta x = \pm 0.0001$, then

$$\Delta Q = (0.95251)\left(\frac{1}{50}\right)\left(\frac{\pm 0.0001}{0.0878}\right) = \pm 2.2 \times 10^{-5}$$

and five digits are proper for Q. But if $Q = (0.0878)^{20} \approx 7.4112 \times 10^{-22}$, while $\Delta x = \pm 0.0001$, as before, then

$$\Delta Q = (7.4112 \times 10^{-22})(20)\left(\frac{\pm 0.0001}{0.0878}\right) = \pm 0.169 \times 10^{-22}$$

so that only two significant figures are justified: Q should be given as 7.4×10^{-22}.

RULE 6.   The number of *significant digits* in x determines the number of *decimal places* (the mantissa) in $\log x$, and conversely.

EXAMPLE 1.10   In a certain solution, $[H^+] = 3.75 \times 10^{-5}$ (relative to a standard state of 1 mol/L). Then,
$$pH = -\log [H^+] = -(0.574 - 5) = 4.426$$

Here, three decimal places are kept in the mantissa (0.574) because $[H^+]$ is given to three significant digits (3.75). On the other hand, if it is given that $pH = 10.7$, then

$$[H^+] = 10^{-pH} = 10^{-10}10^{-0.7} = \frac{1}{10^{0.7}} \times 10^{-10}$$

$$= \tfrac{1}{5} \times 10^{-10} = 2 \times 10^{-11}$$

where we evaluate the antilog of 0.7 to one significant digit (since 0.7 has only one decimal place).

If Rules 1–6 leave any doubt as to the proper number of significant figures to list for an answer, a good idea is to alter the last digit of the least significant value in the calculation. This will produce a change in the answer, which should then be truncated to include only the highest digit that changes.

1.7   IMPLICIT ASSUMPTIONS

Throughout this Outline:

1.   Unless otherwise limited by significant figures, any chemical assumptions underlying a calculation (e.g., that the ionization of X is negligible) must be borne out by that calculation to within 1%. If an assumption is not validated to better than 1%, it will be dropped and the calculation will be made without it.

2. Ionic solutions and gases will be assumed to be ideal unless otherwise stated. The activities of solids will be assumed equal to 1.00.

3. A temperature of 25°C will be assumed unless otherwise stated.

Solved Problems

CONCENTRATIONS

1.1 A blood serum sample is analyzed and found to contain 102.5 μg Ca^{2+}/mL serum. The density of the serum is 1053 kg/m³; AW Ca = 40.08. What is the Ca^{2+} concentration, in terms of (a) molarity?   (b) meq Ca^{2+}/L serum?   (c) ppm Ca^{2+} by wt?

(a)
$$\frac{102.5 \times 10^{-6} \text{ g}}{10^{-3} \text{ L}} \cdot \frac{1 \text{ mol}}{40.08 \text{ g}} = 2.557 \times 10^{-3} \ M$$

(b)
$$\frac{102.5 \text{ mg } Ca^{2+}}{1 \text{ L serum}} \cdot \frac{2 \text{ meq } Ca^{2+}}{40.08 \text{ mg } Ca^{2+}} = 5.115 \text{ meq } Ca^{2+}/\text{L serum}$$

(c)
$$\frac{102.5 \ \mu\text{g } Ca^{2+}}{1 \text{ mL serum}} \cdot \frac{1 \text{ mL serum}}{1.053 \text{ g serum}} = 97.34 \ \mu\text{g } Ca^{2+}/\text{g serum} = 97.34 \text{ ppm by wt}$$

1.2 Laboratory A reported the magnesium (AW = 24.31) content of a solution (density = 0.985 g/mL) as 194 ppm by wt, whereas laboratory B reported the magnesium content of another sample of the same solution as 8.15 millimolars. Which laboratory measurement was the larger?

Using data from A:

$$\frac{985 \text{ g soln}}{1 \text{ L soln}} \cdot \frac{194 \times 10^{-6} \text{ g Mg}}{\text{g soln}} \cdot \frac{1 \text{ mmol Mg}}{0.02431 \text{ g Mg}} = 7.86 \text{ mmol Mg/L soln}$$

Therefore the B measurement was the larger.

1.3 How many grams of $FeSO_4 \cdot (NH_4)_2SO_4 \cdot 6H_2O$ (MW = 392.14) must be dissolved and diluted to 250.0 mL to prepare an aqueous solution, of density 1.00 g/mL, that is 1.00 ppm Fe^{2+} by wt? (AW Fe = 55.85.)

$$\frac{250.0 \text{ mL}}{1} \cdot \frac{1.00 \text{ g soln}}{1 \text{ mL soln}} \cdot \frac{1.00 \times 10^{-6} \text{ g } Fe^{2+}}{1 \text{ g soln}} \cdot \frac{392.14 \text{ g cmpd}}{55.85 \text{ g } Fe^{2+}} = 1.76 \times 10^{-3} \text{ g cmpd}$$

1.4 How many mL of a 0.250 M KCl solution must be diluted to 1.000 L so that the diluted solution (density = 1.00 g/mL) is 400 ppm K^+ by wt? (MW KCl = 74.55; AW K = 39.10.)

$$\frac{1000 \text{ g dil soln}}{1 \text{ L dil soln}} \cdot \frac{400 \times 10^{-6} \text{ g } K^+}{1 \text{ g dil soln}} \cdot \frac{1 \text{ mol KCl}}{39.10 \text{ g } K^+} \cdot \frac{1000 \text{ mL KCl}}{0.250 \text{ mol KCl}} = 40.9 \text{ mL KCl/L dil soln}$$

1.5 How many mL of 0.250 M Fe^{3+} is needed to reach the equivalence point in the titration of 20.00 mL of 0.100 M Sn^{2+}? The redox reaction is

$$2Fe^{3+} + Sn^{2+} \rightarrow 2Fe^{2+} + Sn^{4+}$$

Sn^{2+} is oxidized, with $n_o = 2$, whereas Fe^{3+} is reduced, with $n_r = 1$. By (1.3),

$$2(0.100)(20.00) = 1(0.250)V_r, \qquad \text{or} \qquad V_r = 16.0 \text{ mL}$$

1.6 How many mL of $0.125\ M\ Cr^{3+}$ must be reacted with 12.00 mL of $0.200\ M\ MnO_4^-$ if the redox products are $Cr_2O_7^{2-}$ and Mn^{2+}?

Cr^{3+} is oxidized to $Cr_2O_7^{2-}$ (Cr oxidation state = +6/Cr), so that $n_o = 3$; MnO_4^- (oxidation state = +7) is reduced to Mn^{2+}, so that $n_r = 5$. By (1.3),

$$3(0.125)V_o = 5(0.200)(12.00) \qquad \text{or} \qquad V_o = 32.0\ \text{mL}$$

1.7 How many milliequivalents of Cr^{3+} and of MnO_4^- are involved in the reaction of Problem 1.6?

By (1.2), the Cr^{3+} normality is $N = 3(0.125) = 0.375$, so that

$$\text{meqs } Cr^{3+} = NV(\text{mL}) = (0.375)(32.0) = 12.0$$

By definition, meqs MnO_4^- must also equal 12.0, as may be checked by calculating the product of normality and volume for MnO_4^-.

BALANCING REDOX EQUATIONS

1.8 Balance the redox equation where H_2S reacts with UO_2^+ in an acidic solution to produce $S(s)$ and U^{4+}.

The oxidation state of sulfur is -2 in H_2S and zero in $S(s)$, so that there is an oxidation change of $+2/S = +2/H_2S$. The oxidation state of uranium is $+5$ in UO_2^+ and $+4$ in U^{4+}, so that there is a change in oxidation state of $-1/U = -1/UO_2^+$. Balancing the oxidation and reduction changes requires that $1H_2S$ react with $2UO_2^+$, giving the partially balanced equation

$$H_2S + 2UO_2^+ \rightarrow S(s) + 2U^{4+}$$

We now must balance any two of H, O, and charge, using H^+ and/or H_2O. The best choice is oxygen and charge, since hydrogen appears in both H^+ and H_2O. Since there are 4 oxygens on the left in the partially balanced equation and none on the right, add $4H_2O$ to the right to balance the oxygen. Since the charge is $+2$ on the left and $+8$ on the right, add $6H^+$ to the left to balance the charge. The completely balanced equation is then

$$6H^+ + H_2S + 2UO_2^+ \rightarrow S(s) + 2U^{4+} + 4H_2O$$

As a check, there are 8 hydrogens on either side of the equation.

1.9 Complete and balance the redox equation where MnO_4^- reacts with NH_4OH in a basic solution to produce NO_3^- and $MnO_2(s)$.

The manganese is $+7$ in MnO_4^- and $+4$ in MnO_2, so that the reduction is $-3/Mn = -3/MnO_4^-$. The nitrogen is -3 in NH_4OH and $+5$ in NO_3^-, so that the oxidation is $+8/N = +8/NH_4OH$. Balancing the oxidation and reduction changes requires that $8MnO_4^-$ react with $3NH_4OH$, and the partially balanced equation is:

$$8MnO_4^- + 3NH_4OH \rightarrow 8MnO_2(s) + 3NO_3^-$$

Two of the remaining three quantities, H, O, and charge, must be balanced using H_2O and/or OH^- as needed. Since H and O appear in both OH^- and H_2O, it is easiest to balance charge first. The partially balanced equation shows a total charge of -8 on the left and -3 on the right, so that adding $5OH^-$ to the right will balance the charge. We can now balance either H or O by adding H_2O as appropriate. Using H as the basis, we find 15H on the left and 5H (from the added $5OH^-$) on the right. Therefore, adding $5H_2O$ to the right will balance H, yielding the completely balanced equation

$$8MnO_4^- + 3NH_4OH \rightarrow 8MnO_2(s) + 3NO_3^- + 5OH^- + 5H_2O$$

The oxygen balance may be checked to confirm that the equation is correct.

1.10 Complete and balance the equation where $Cr_2O_7^{2-}$ reacts with H_2SO_3 in an acidic solution to produce Cr^{3+} and HSO_4^-.

Each chromium is +6 in $Cr_2O_7^{2-}$ and +3 in Cr^{3+}, so that the reduction is $-3/Cr = -6/Cr_2O_7^{2-}$ (since there are two chromium atoms per $Cr_2O_7^{2-}$). The sulfur is +4 in H_2SO_3 and +6 in HSO_4^-, so that the oxidation is $+2/S = +2/H_2SO_3$. Balancing the oxidation and reduction changes requires $3H_2SO_3$ reacting for every $Cr_2O_7^{2-}$; the partially balanced equation is:

$$Cr_2O_7^{2-} + 3H_2SO_3 \rightarrow 2Cr^{3+} + 3HSO_4^-$$

Of the remaining three quantities that must balance (H, O, charge), it is easiest to balance O and charge. There are 16 oxygen atoms on the left and 12 on the right in the partially balanced equation, so that $4H_2O$ must be added to the right to balance oxygen. The charge on the left is -2 and on the right $+3$, so that the addition of $5H^+$ on the left will balance the charge. The final equation,

$$Cr_2O_7^{2-} + 3H_2SO_3 + 5H^+ \rightarrow 2Cr^{3+} + 3HSO_4^- + 4H_2O$$

may be checked by a hydrogen balance.

1.11   Complete and balance the equation where MnO_4^- reacts with C_2H_5OH in an acidic solution to produce Mn^{2+} and $CO_2(g)$.

The manganese is +7 in MnO_4^- and +2 in Mn^{2+}, so that the total reduction change is $-5/Mn = -5/MnO_4^-$. Each carbon is -2 in C_2H_5OH and +4 in CO_2, so that the oxidation change is $+6/C = +12/C_2H_5OH$. Balancing the oxidation and reduction changes requires $12MnO_4^-$ reacting with $5C_2H_5OH$:

$$12MnO_4^- + 5C_2H_5OH \rightarrow 12Mn^{2+} + 10CO_2(g)$$

As the two remaining quantities to be balanced, it is best to choose oxygen and charge. Since there are 53 oxygen atoms on the left but only 20 on the right, $33H_2O$ must be added to the right to balance the oxygen. Since the charge is -12 on the left and $+24$ on the right, $36H^+$ must be added to the left to balance the charge. The completely balanced equation is:

$$12MnO_4^- + 5C_2H_5OH + 36H^+ \rightarrow 12Mn^{2+} + 10CO_2(g) + 33H_2O$$

The hydrogen balance will serve as a check.

1.12   Complete and balance the equation where $CO(NH_2)_2$ reacts with OBr^- in an alkaline solution to produce CO_2, N_2, and Br^-.

The nitrogen is -3 in $CO(NH_2)_2$ and zero in N_2, so that there is an oxidation change of $+3/N = +6/CO(NH_2)_2$. The bromine is +1 in OBr^- and -1 in Br^-, so that there is a reduction change of $-2/Br = -2/OBr^-$. Balancing these changes requires that $3OBr^-$ react with each $CO(NH_2)_2$, so that the partially balanced equation is:

$$CO(NH_2)_2 + 3OBr^- \rightarrow CO_2 + N_2 + 3Br^-$$

Two of the remaining three quantities (H, O, charge) must be balanced using OH^- and/or H_2O; it is easiest to balance charge first. In the partially balanced equation, both sides have a charge of -3, so that charge is already balanced and OH^- need not be added to either side. At most, we need to add H_2O. Considering H, there are four on the left but none on the right, so that adding $2H_2O$ to the right will balance the hydrogen and yield the completely balanced equation

$$CO(NH_2)_2 + 3OBr^- \rightarrow CO_2 + N_2 + 3Br^- + 2H_2O$$

The oxygen balance may be used to confirm that the equation is correct.

1.13   Complete and balance the equation where Cl_2 reacts with Ag^+ in an acidic solution to yield $AgCl(s)$ and ClO_3^-.

The silver is +1 in both Ag^+ and $AgCl$, so it does not undergo oxidation or reduction. Therefore, some of the Cl_2 must be oxidized and some reduced. Oxidation takes Cl_2 from the zero state to +5 in ClO_3^-, for a change of $+5/Cl = +10/Cl_2$. Reduction takes Cl_2 from the zero state to -1 in $AgCl$, for a reduction change of $-1/Cl = -2/Cl_2$. To balance the oxidation and reduction changes requires that we use $1Cl_2 \rightarrow 2ClO_3^-$ for every $5Cl_2 \rightarrow 10AgCl$. This results in the partially balanced equation

$$1Cl_2 + 5Cl_2 + 10Ag^+ \rightarrow 2ClO_3^- + 10AgCl(s)$$

where $10Ag^+$ has been added on the left to balance the silver. After division by 2, the equation reads:

$$3Cl_2 + 5Ag^+ \rightarrow ClO_3^- + 5AgCl(s)$$

Two of the remaining three quantities (H, O, charge) must be balanced using H^+ and/or H_2O. It is easiest to balance O and charge. There are no oxygens on the left in the simplified equation, but there are on the right; hence, add $3H_2O$ to the left. There is a charge of +5 on the left and −1 on the right, calling for the addition of $6H^+$ to the right. The balanced equation,

$$3Cl_2 + 5Ag^+ + 3H_2O \rightarrow ClO_3^- + 5AgCl(s) + 6H^+$$

may be checked by a hydrogen balance.

STOICHIOMETRY

1.14 Normal human plasma contains cations in the following concentrations: Na^+, 143 mmol/L; Ca^{2+}, 2.38 mmol/L; K^+, 4.51 mmol/L; Mg^{2+}, 1.32 mmol/L. If Cl^- is the only anion present in any appreciable concentration, what is the total weight (i.e., mass) of chloride salts of these four cations in 10.0 mL of human plasma? (MWs: NaCl = 58.44, KCl = 74.55, $CaCl_2$ = 110.99, $MgCl_2$ = 95.22.)

Since 10.0 mL is 0.01 L,

$$\frac{1.43 \text{ mmol } Na^+}{1} \cdot \frac{58.44 \text{ mg NaCl}}{1 \text{ mmol } Na^+} + \frac{0.0451 \text{ mmol } K^+}{1} \cdot \frac{74.55 \text{ mg KCl}}{1 \text{ mmol } K^+}$$

$$+ \frac{0.0238 \text{ mmol } Ca^{2+}}{1} \cdot \frac{110.99 \text{ mg } CaCl_2}{1 \text{ mmol } Ca^{2+}} + \frac{0.0132 \text{ mmol } Mg^{2+}}{1} \cdot \frac{95.22 \text{ mg } MgCl_2}{1 \text{ mmol } Mg^{2+}}$$

$$= 90.8 \text{ mg salts}$$

1.15 The iron, tin, and titanium in a sample were jointly precipitated as hydroxides, which were washed and ignited to produce Fe_2O_3, SnO_2, and TiO_2. This mixture of oxides, which amounted to 0.3768 g, was dissolved after fusion with potassium pyrosulfate. Analysis showed that 0.0322 g of Ti was present. The iron was determined separately as an oxide: 0.1837 g FeO was found. What was the weight (mass) of tin in the sample?

Molecular and atomic weights are: Ti = 47.90, TiO_2 = 79.90, Sn = 118.69, SnO_2 = 150.69, FeO = 71.85, Fe_2O_3 = 159.69. If x = g of Sn in sample, then:

$$\frac{0.0322 \text{ g Ti}}{1} \cdot \frac{79.90 \text{ g } TiO_2}{47.90 \text{ g Ti}} + \frac{x \text{ g Sn}}{1} \cdot \frac{150.69 \text{ g } SnO_2}{118.69 \text{ g Sn}} + \frac{0.1837 \text{ g FeO}}{1} \cdot \frac{159.69 \text{ g } Fe_2O_3}{2(71.85 \text{ g FeO})}$$

$$= 0.3768 \text{ g}$$

Solving, $x = 0.0937$ g Sn.

1.16 Silver arsenate, Ag_3AsO_4, is very insoluble. What weight of Ag_3AsO_4 precipitate forms when 27.00 mL of 0.180 M $AgNO_3$ is mixed with 15.00 mL of 0.130 M Na_3AsO_4?

The reaction is $3Ag^+ + AsO_4^{3-} \rightarrow Ag_3AsO_4(s)$. The molecular weight of Ag_3AsO_4 is 462.6. The mixture contains $(0.180)(27.00) = 4.86$ mmol Ag^+ and $(0.130)(15.00) = 1.95$ mmol AsO_4^{3-}. Since three times as much Ag^+ is needed as AsO_4^{3-}, the Ag^+ is the limiting reagent; thus,

$$\frac{4.86}{3} = 1.62 \text{ mmol } Ag_3AsO_4$$

is formed (and $1.95 - 1.62 = 0.33$ mmol AsO_4^{3-} is left over). Consequently,

$$(1.62 \text{ mmol } Ag_3AsO_4)(462.6 \text{ mg } Ag_3AsO_4/\text{mmol } Ag_3AsO_4) = 749 \text{ mg } Ag_3AsO_4$$

1.17 Ascorbic acid (vitamin C; MW = 176.13) is a reducing agent that reacts as follows:

$$C_6H_8O_6 \rightarrow C_6H_6O_6 + 2H^+ + 2e^-$$

The ascorbic acid can be determined by oxidation with a standard solution of I_2 (resulting in the reduction of I_2 to I^-), followed by analysis of the solution to determine the excess I_2. This latter reaction involves titration with thiosulfate, $Na_2S_2O_3$, yielding the redox products I^- and SO_4^{2-}.

A 200.0 mL sample of a citrus fruit drink was acidified with sulfuric acid, and 10.00 mL of 0.0250 M I_2 was added. Some of the I_2 was reduced by the ascorbic acid to I^-. The excess I_2 required 4.60 mL of 0.0100 M $Na_2S_2O_3$ for reduction. What was the vitamin C content of the drink, in "mg vitamin per mL drink"?

The two reactions are:

$$C_6H_8O_6 + I_2 \rightarrow C_6H_6O_6 + 2H^+ + 2I^-$$
$$5H_2O + S_2O_3^{2-} + 4I_2 \rightarrow 2SO_4^{2-} + 8I^- + 10H^+$$

We have:

$$\text{mmols excess } I_2 = 4(\text{mmols } S_2O_3^{2-}) = 4(0.0100)(4.60) = 0.184$$
$$\text{mmols } I_2 \text{ added} = (0.0250)(10.00) = 0.250$$

Therefore, $0.250 - 0.184 = 0.066$ mmol I_2 reacted with 0.066 mmol vitamin C, whence

$$\frac{0.066 \text{ mmol vitamin}}{200.0 \text{ mL drink}} \cdot \frac{176.13 \text{ mg vitamin}}{1 \text{ mmol vitamin}} = 0.058 \text{ mg vitamin/mL drink}$$

1.18 The sulfur in a 0.575 g sample of an organic thioamine was converted to H_2S, which was then absorbed in 15.00 mL of 0.0320 M I_2. This amount of I_2 was in excess of that required to oxidize the H_2S to elemental sulfur; I^- was the reduction product. The excess I_2 was determined by titration with thiosulfate, $S_2O_3^{2-}$, the reaction products being $S_4O_6^{2-}$ and I^-: 26.29 mL of 0.0175 M $Na_2S_2O_3$ was required to reduce the excess I_2. How much sulfur was in the original sample?

The reactions are:

$$H_2S + I_2 \rightarrow S + 2I^- + 2H^+$$
$$I_2 + 2S_2O_3^{2-} \rightarrow 2I^- + S_4O_6^{2-}$$

which imply

$$\text{mmols } I_2 = (\text{mmols S}) + \tfrac{1}{2}(\text{mmols } S_2O_3^{2-})$$
$$(0.0320)(15.00) = \frac{x \text{ mg S}}{32.064 \text{ mg/mmol}} + \tfrac{1}{2}(0.0175)(26.29)$$
$$x = 8.01 \text{ mg S}$$

1.19 The carbon in oxalic acid, $(COOH)_2$, is easily oxidized to CO_2 by acidic permanganate, MnO_4^-; the product is Mn^{2+}. 25.00 mL of an oxalic acid solution required 21.68 mL of $KMnO_4$ to oxidize the oxalic acid completely to CO_2. The $KMnO_4$ was standardized separately by reaction with Fe^{2+}; the redox product is Fe^{3+}. This standardization of 25.00 mL of the $KMnO_4$ required 32.85 mL of 0.1385 M Fe^{2+} to react completely. What was the concentration of the oxalic acid solution?

The reactions are:

$$MnO_4^- + 5Fe^{2+} + 8H^+ \rightarrow Mn^{2+} + 5Fe^{3+} + 4H_2O$$
$$5(COOH)_2 + 2MnO_4^- + 6H^+ \rightarrow 10CO_2 + 2Mn^{2+} + 8H_2O$$

For the standardization:

$$\text{mmols Fe}^{2+} = 5(\text{mmols MnO}_4^-)$$
$$(0.1385)(32.85) = 5(M)(25.00)$$

which gives the molarity of the permanganate as 0.03640 M. For the reaction with the oxalic acid:

$$\text{mmols acid} = \tfrac{5}{2}(\text{mmols MnO}_4^-)$$
$$(M)(25.00) = \tfrac{5}{2}(0.03640)(21.68)$$

which gives the molarity of the oxalic acid as 0.07892 M.

1.20 A 0.0600 g sample containing only 3-amino-1-propanol, C_3H_9NO (MW = 75.11), and 2-mercapto-2,3-thiazoline, $C_3H_5NS_2$ (MW = 119.21), was oxidized and the SO_2 formed in the reaction (as the only sulfur product) was absorbed in water containing excess H_2O_2. This resulted in the oxidation of the SO_2 to H_2SO_4 and the accompanying reduction of peroxide to H_2O. The solution then required 15.00 mL of 0.100 M NaOH for neutralization of the H_2SO_4. [H_2O_2 is too weak an acid to be titrated in aqueous solution.] What is the wt % of $C_3H_5NS_2$ in the sample?

$$\frac{(0.100)(15.00) \text{ mmol NaOH}}{1} \cdot \frac{1 \text{ mmol H}_2\text{SO}_4}{2 \text{ mmol NaOH}} \cdot \frac{1 \text{ mmol C}_3\text{H}_5\text{NS}_2}{2 \text{ mmol H}_2\text{SO}_4} \cdot \frac{119.21 \text{ mg C}_3\text{H}_5\text{NS}_2}{1 \text{ mmol C}_3\text{H}_5\text{NS}_2} \cdot \frac{100\%}{60.0 \text{ mg sample}}$$

$$= 74.5 \text{ wt \% C}_3\text{H}_5\text{NS}_2$$

1.21 Glycerol samples can be analyzed by reaction with an excess of HIO_4:

$$C_3H_8O_3 + 2HIO_4 \rightarrow 2HCHO + HCOOH + 2HIO_3 + 3H_2O \qquad (1)$$

The solution is then made basic and excess I^- is added to convert the remaining IO_4^- to IO_3^-, according to the reaction

$$2I^- + IO_4^- + H_2O \rightarrow IO_3^- + I_2 + 2OH^- \qquad (2)$$

The I_2 released in this reaction is then determined by reaction with H_3AsO_4:

$$H_3AsO_4 + I_2 + H_2O \rightarrow H_3AsO_4 + 2I^- + 2H^+ \qquad (3)$$

185.0 g of a sample containing an unknown amount of glycerol (MW = 92.10) in water was acidified and mixed with 50.00 mL of a $NaIO_4$ solution. (This amount of $NaIO_4$ is known to be in excess of that required for reaction.) Following oxidation of the glycerol, the solution was made basic and an excess of KI was added; the liberated I_2 required 12.64 mL of 0.0507 M H_3AsO_4 to reduce the I_2 to I^-. In a separate experiment, excess KI was added to 25.00 mL of $NaIO_4$, resulting in the formation of I_2; 14.17 mL of the 0.0507 M H_3AsO_4 was needed to convert the I_2 to I^-. What was the concentration of glycerol, in ppm by wt, in the sample?

In the standardization of the $NaIO_4$ according to reactions (2) and (3), 1.00 mmol of IO_4^- results in 1.00 mmol of I_2, which requires 1.00 mmol of H_3AsO_4. Therefore,

$$\text{mmols IO}_4^- = \text{mmols H}_3\text{AsO}_4 \qquad \text{or} \qquad (M)(25.00) = (0.0507)(14.17)$$

which gives the concentration of the $NaIO_4$ as 0.0287 M.

The analysis of the glycerol involves the IO_4^- in two reactions. Some of the IO_4^- is used in reacting with the glycerol in (1); for this reaction,

$$\text{mmols IO}_4^- = 2(\text{mmols glycerol})$$

The excess IO_4^- is used in the same way as in the standardization, so that

$$\text{mmols IO}_4^- = \text{mmols H}_3\text{AsO}_4$$

Therefore,

$$\text{total mmols } IO_4^- = 2(\text{mmols glycerol}) + (\text{mmols } H_3AsO_4)$$
$$(0.0287)(50.00) = 2(\text{mmols glycerol}) + (0.0507)(12.64)$$

from which mmols glycerol = 0.398. Then:

$$\frac{0.398 \text{ mmol glycerol}}{1} \cdot \frac{92.10 \text{ mg glycerol}}{1 \text{ mmol glycerol}} \cdot \frac{10^6}{1.850 \times 10^5 \text{ mg sample}} = 198 \text{ ppm by wt}$$

1.22 A 15.00 mL sample of a solution is 0.0400 M in Sn^{2+} and $x\,M$ in Fe^{2+}. Both ions are easily oxidized by $Cr_2O_7^{2-}$ in acidic solution; the redox products are Sn^{4+}, Fe^{3+}, and Cr^{3+}. If 18.00 mL of 0.1250 M $Cr_2O_7^{2-}$ is required to oxidize the Sn^{2+} and Fe^{2+} in this solution, what is the value of x?

The reactions are:

$$Cr_2O_7^{2-} + 3Sn^{2+} + 14H^+ \rightarrow 2Cr^{3+} + 3Sn^{4+} + 7H_2O$$
$$Cr_2O_7^{2-} + 6Fe^{2+} + 14H^+ \rightarrow 2Cr^{3+} + 6Fe^{3+} + 7H_2O$$

whence
$$\text{mmols } Cr_2O_7^{2-} = \tfrac{1}{3}(\text{mmols } Sn^{2+}) + \tfrac{1}{6}(\text{mmols } Fe^{2+})$$
$$(0.1250)(18.00) = \tfrac{1}{3}(0.0400)(15.00) + \tfrac{1}{6}(x)(15.00)$$
$$x = 0.820$$

1.23 A 0.2586 g sample contains only pure $Na_2C_2O_4$ (MW = 134.00) and pure KHC_2O_4 (MW = 128.13). It required 48.39 mL of 0.01660 M $KMnO_4$ for titration to the equivalence point in an acid solution where the MnO_4^- was reduced to Mn^{2+} and the $C_2O_4^{2-}$ oxidized to CO_2. How much KHC_2O_4 was present in the mixture?

The reaction is $2MnO_4^- + 5C_2O_4^{2-} + 16H^+ \rightarrow 2Mn^{2+} + 10CO_2 + 8H_2O$, and so

$$\text{mmols } C_2O_4^{2-} = \tfrac{5}{2}(\text{mmols } MnO_4^-) = \tfrac{5}{2}(0.01660)(48.39) = 2.008$$

Let x = g KHC_2O_4; then $0.2586 - x$ = g $Na_2C_2O_4$, and

$$\text{mols } KHC_2O_4 + \text{mols } Na_2C_2O_4 = 2.008 \times 10^{-3}$$
$$\frac{x}{128.13} + \frac{0.2586 - x}{134.00} = 2.008 \times 10^{-3}$$

Solving, $x = 0.229$ g KHC_2O_4.

SIGNIFICANT FIGURES

1.24 Evaluate $101.27 + 2.336 - 10.5 + 0.2973$.

The limiting term is 10.5, where only one decimal place is given. The answer must be stated as 93.4.

1.25 Evaluate $0.2724 - 0.0016 + 0.163 - 3.728$.

The limiting terms are the last two values, which are given to three decimal places. The answer is -3.294.

1.26 Evaluate

$$(a) \quad \frac{(0.168)(37.29)}{25.00} \qquad (b) \quad \frac{(0.093)(37.29)}{25.00} \qquad (c) \quad \frac{(0.014)(37.29)}{25.00}$$

using the ideas of Example 1.8.

(a) The least precise value is 0.168, which is assumed known to 1 part in 168. The answer can be given
     to three digits as 0.251, since the proportion

$$\frac{1}{168}(0.251) = 0.0015$$

places the uncertainty in the third decimal place.

(b) The least precise value is 0.093, which is assumed known to 1 part in 93. The answer can be given to
     three digits as 0.139, since the proportion

$$\frac{1}{93}(0.139) = 0.0015$$

places the uncertainty in the third decimal place.

(c) The least precise value is 0.014, which is assumed known to 1 part in 14. The answer can only be
     given to two digits as 0.021, since the proportion

$$\frac{1}{14}(0.021) = 0.0015$$

places the uncertainty in the third decimal place.

1.27   Evaluate

   (a) $(0.167)(32.58) - (0.101)(21.45)$ (c) $(0.183)(34.08) - (0.274)(22.68)$
   (b) $(0.167)(32.58) - (0.225)(21.45)$

(a) Each product can be given to three digits. The expression reduces to

$$5.44_1 - 2.16_6 = 3.27_5 = 3.28$$

(b) Each product can be given to three digits. The expression reduces to

$$5.44_1 - 4.82_6 = 0.61_5 = 0.62$$

and only two digits remain.

(c) Each product can be given to three digits. The expression reduces to

$$6.23_7 - 6.21_4 = 0.02_3 = 0.02$$

and only one digit remains.

1.28   Evaluate $Q = \sqrt[4]{0.0768}$.

   We have: $Q = (0.0768)^{0.25} = 0.5264$. Since (Rule 5)

$$\Delta Q = (0.5264)(0.25)\left(\frac{0.0001}{0.0768}\right) = 0.00017$$

four significant figures is appropriate for Q, the uncertainty being in the fourth digit.

1.29   Evaluate (a) $3.728 - \log(1.1 \times 10^{-7})$, (b) $-6.903 - \log(1.1 \times 10^{-7})$.

(a) By Rule 6, the number of decimal places in the log term is limited to two. Thus, the expression
     reduces to

$$3.728 + 6.95_9 = 10.68_7 = 10.69$$

(b) Now the expression reduces to $-6.903 + 6.95_9 = 0.05_6 = 0.06$; only one digit remains.

1.30   Evaluate f, given

$$-\log f = \frac{(0.5085)(2^2)\sqrt{0.00184}}{1 + (0.3281)(3)\sqrt{0.00184}}$$

where 2^2 and 1 are precise values, but 3 is known only to one digit.

Because $\sqrt{0.00184} = 0.0428_9$, the number of digits in the numerator is limited to three: its value is 0.0872_5. The value of the denominator is $1.000 \cdots + 0.04_2 = 1.04_2$. Therefore,

$$\log f = -\frac{0.0872_5}{1.04_2} = -0.0837_3$$

Since the value of $\log f$ is given to four decimal places, the antilog should be stated to four digits (Rule 6): $f = 0.8247$.

1.31   Evaluate f, given

$$-\log f = \frac{(0.5085)(3^2)\sqrt{0.00725}}{1 + (0.3281)(5)\sqrt{0.00725}}$$

where 3^2 and 1 are precise values, but 5 is known only to one digit.

As $\sqrt{0.00725} = 0.0851_4$, the value of the numerator is limited to three digits: 0.389_6. The denominator is

$$1 + 0.13_9 = 1.13_9$$

with only two digits in the fractional portion because the least reliable value that goes into that term is known to 1 part in 5, and

$$\tfrac{1}{5}(0.13_9) = 0.02_8$$

Therefore, the uncertainty is in the second digit of the value 0.13_9. We then have:

$$\log f = -\frac{0.389_6}{1.13_9} = -0.341_9$$

Since there are three decimal places in $\log f$, there should be three digits in f: $f = 0.455$.

Supplementary Problems

UNITS AND CONCENTRATIONS

1.32   How many mL of glycerol (MW = 92.1, density = 1.26 g/mL) must be used to prepare 250.0 mL of a 0.150 M glycerol solution?   *Ans.*   2.75 mL

1.33   How many mL of concentrated HCl (density = 1.18 g/mL, 36.0 wt % HCl, MW = 36.5) should be diluted to 1.00 L to produce a 0.100 M solution?   *Ans.*   8.59 mL

1.34   What is the calcium (AW = 40.1) concentration, in mol/L, of a blood serum sample (density = 1.05 g/mL) having a calcium content of 92.0 μg/mL?   *Ans.* 2.29×10^{-3} M

1.35   How many mL of 0.150 M MnO_4^- is required to react with 17.50 mL of 0.200 M Sn^{2+}, if the redox products are $MnO_2(s)$ and Sn^{4+}?   *Ans.*   15.6 mL

1.36   How many mL of 0.250 N Tl^{3+} is needed to react quantitatively with 10.00 mL of 0.175 M Sn^{2+}? The redox products are Tl^+ and Sn^{4+}.   *Ans.*   14.0 mL

BALANCING REDOX EQUATIONS

In Problems 1.37–1.47, complete and balance the redox equation, showing the total oxidation and the total reduction change for each reactant ion.

1.37 BrO_3^- reacts with Cu^+ in an acidic solution to produce Br^- and Cu^{2+}.

   Ans.

$$\overbrace{BrO_3^- + 6Cu^+}^{+1/Cu^+} + 6H^+ \rightarrow Br^- + 6Cu^{2+} + 3H_2O$$
$$\underbrace{\qquad\qquad\qquad}_{-6/Br^- = -6/BrO_3^-}$$

1.38 MnO_4^- reacts with I^- in an acidic solution to produce Mn^{2+} and I_2.

   Ans.

$$2MnO_4^- + \overbrace{10I^-}^{+1/I = +2/I_2} + 16H^+ \rightarrow 2Mn^{2+} + 5I_2 + 8H_2O$$
$$\underbrace{\qquad\qquad}_{-5/Mn = -5/MnO_4^-}$$

1.39 $S_2O_3^{2-}$ reacts with I_2 in a basic solution to produce SO_4^{2-} and I^-.

   Ans.

$$\overbrace{S_2O_3^{2-}}^{+4/S = +8/S_2O_3^{2-}} + 4I_2 + 10OH^- \rightarrow 2SO_4^{2-} + 8I^- + 5H_2O$$
$$\underbrace{\qquad\qquad}_{-1/I = -2/I_2}$$

1.40 NH_4OH reacts with Cl_2 in a basic solution to produce N_2 and Cl^-.

   Ans.

$$\overbrace{2NH_4OH}^{+3/N = +3/NH_4OH} + 3Cl_2 + 6OH^- \rightarrow N_2 + 6Cl^- + 8H_2O$$
$$\underbrace{\qquad\qquad}_{-1/Cl = -2/Cl_2}$$

1.41 MnO_4^- reacts with $As_2O_3(s)$ in an acidic solution to produce $Mn^{2+} + H_3AsO_4$.

   Ans.

$$4MnO_4^- + \overbrace{5As_2O_3(s)}^{+2/As = +4/As_2O_3} + 12H^+ + 9H_2O \rightarrow 4Mn^{2+} + 10H_3AsO_4$$
$$\underbrace{\qquad\qquad}_{-5/Mn = -5/MnO_4^-}$$

1.42 IO_3^- reacts with I^- in an acidic solution to produce I_2.

   Ans.

$$IO_3^- + \overbrace{5I^-}^{+1/I} + 6H^+ \rightarrow \tfrac{5}{2}I_2 + \tfrac{1}{2}I_2 + 3H_2O$$
$$\underbrace{\qquad\qquad}_{-5/IO_3^-}$$

1.43 NH_4OH reacts with OBr^- in a basic solution to produce $N_2(g)$ and Br^-.

   Ans.

$$\overbrace{2NH_4OH}^{+3/N = +3/NH_4OH} + 3OBr^- \rightarrow N_2(g) + 3Br^- + 5H_2O$$
$$\underbrace{\qquad\qquad}_{-2/Br = -2/OBr^-}$$

1.44 MnO_4^- reacts with N_2H_4 in a basic solution to produce $MnO_2(s)$ and $N_2(g)$.

$$\overset{+2/N\,=\,+4/N_2H_4}{4MnO_4^- + 3N_2H_4 \rightarrow 4MnO_2(s) + 3N_2(g) + 4OH^- + 4H_2O}$$

Ans.

$$-3/Mn = -3/MnO_4^-$$

1.45 $S_2O_3^{2-}$ reacts with MnO_4^- in an acidic solution to produce $S_4O_6^{2-}$ and MnO_4^{2-}.

$$\overset{+0.5/S\,=\,+1/S_2O_3^{2-}}{2S_2O_3^{2-} + 2MnO_4^- \rightarrow S_4O_6^{2-} + 2MnO_4^{2-}}$$

Ans.

$$-1/Mn = -1/MnO_4^-$$

1.46 PuO_2^{2+} reacts with Sn^{2+} in an acidic solution to produce PuO_2^+ and Sn^{4+}.

$$\overset{+2/Sn^{2+}}{2PuO_2^{2+} + Sn^{2+} \rightarrow 2PuO_2^+ + Sn^{4+}}$$

Ans.

$$-1/Pu = -1/PuO_2^{2+}$$

1.47 BH_4^- reacts with MnO_4^- in a basic solution to produce $H_2BO_3^-$ and $MnO_2(s)$.

$$\overset{+8/B\,=\,+8/BH_4^-}{3BH_4^- + 8MnO_4^- + H_2O \rightarrow 3H_2BO_3^- + 8MnO_2(s) + 8OH^-}$$

Ans.

$$-3/Mn = -3/MnO_4^-$$

STOICHIOMETRY

1.48 The nitrogen content of a 0.2476 g sample of a protein was determined by the Kjeldahl method. The NH_3 produced was distilled into 50.00 mL of HCl, which was in excess of the amount required for NH_3 neutralization. This solution required 18.76 mL of 0.1250 M NaOH for neutralization. A separate 25.00 mL aliquot of the HCl required 21.15 mL of the NaOH for neutralization. What was the wt % nitrogen in the protein sample?   *Ans.*   16.66%

1.49 A 1.762 g sample of impure $H_2C_2O_4$ (MW = 90.04) was dissolved in water and the impurities removed by filtration. 50.00 mL of 0.1028 M NaOH was added to the filtrate, this being in excess of the amount needed to neutralize both hydrogens of the $H_2C_2O_4$. The excess NaOH required 12.62 mL of 0.1251 M HCl to reach the equivalence point. What was the wt % $H_2C_2O_4$ in the sample?
Ans.   9.099%

1.50 When an excess of $Na_2S_2O_3$ was added to a 50.00 mL aliquot of hypoiodous acid, HOI, a redox reaction took place, resulting in SO_4^{2-} and I^- as products. The SO_4^{2-} was precipitated quantitatively as $BaSO_4(s)$ (MW = 233.4), filtered, dried, and found to weigh 1.209 g. What was the molarity of the HOI solution?   *Ans.*   0.2072 M

1.51 A 0.7538 g sample contains an unknown amount of As_2O_3 (MW = 197.84). The sample was treated with HCl and a reducing agent, resulting in the formation of $AsCl_3(g)$, which was distilled into a beaker of water. Following the hydrolysis reaction

$$AsCl_3 + 2H_2O \rightarrow HAsO_2 + 3H^+ + 3Cl^-$$

the amount of $HAsO_2$ was determined by titration with 0.05264 M I_2, requiring 33.64 mL to reach the equivalence point. The redox products in the titration were H_3AsO_4 and I^-. What was the wt % As_2O_3 in the sample?   *Ans.*   23.24%

SIGNIFICANT FIGURES

1.52   Evaluate $100 + 6 + 0.35 - 48$.   *Ans.*   58

1.53   Evaluate $38.2 + \sqrt{7.38} - (2.4)^3$.   *Ans.*   27

1.54   Evaluate $(0.1268)(13.29)/25.00$.   *Ans.* 6.741×10^{-2}

1.55   Evaluate $(0.1562)(28.27) - (0.1238)(35.64)$.   *Ans.* 4×10^{-3}

1.56   Evaluate $2.685 - \log(3.7 \times 10^4)$.   *Ans.* -1.88

1.57   Evaluate x, given $\log x = 6.25/(7.8 \times 10^2)$.   *Ans.*   1.019

1.58   Evaluate x, given $\log x = -(3.762 + 0.012)/(8 \times 10^4)$.   *Ans.*   0.99989

Chapter 2

Statistics and Probabilities

2.1 ACCURACY AND PRECISION

It is important in analytical chemistry to provide an indication of the accuracy and precision of a set of measurements. *Accuracy* is a measure of the difference between the average of a set of measurements and the true value. If the true value is known, as, for example, when a standard is being analyzed, then the difference between the known content of the standard and the average of the experimentally determined values determines the accuracy of those values.

EXAMPLE 2.1 If a standard containing exactly 20.00% Cl is analyzed by method A and there results an average value of 20.06% Cl, whereas analysis by method B results in an average value of 20.38% Cl, then the data from method A are more accurate than the data from method B.

Very often, however, the true value is *not* known and all that is possible is a statistical approximation of a range of values that includes the true value. In this case, evaluation of the *precision* of the measurements is first needed. Precision measures the degree of concentration of measurements about their mean; it is thus the inverse notion to *scatter*. Now, the usual index of scatter is the *standard deviation* of the set of measurements. The smaller the standard deviation, the smaller the scatter and the better the precision.

EXAMPLE 2.2 Suppose that analyses of the chloride standard of Example 2.1 gave:

$$\textit{method A} \qquad \% \ Cl = 20.06 \pm 0.25$$
$$\textit{method B} \qquad \% \ Cl = 20.38 \pm 0.07$$

Then the data resulting from method B are more precise, even though less accurate.

2.2 SYSTEMATIC AND RANDOM ERRORS

Two types of error can occur in measurements. *Systematic errors* are one-sided errors in that each is due to a specific positive or negative factor in the overall experimental procedure. For instance, an analytical balance that was not zeroed correctly might always read 0.1 mg too low, a gravimetric analysis might always result in less than 100% solid formation owing to partial solubility of the precipitate, a particular chemist might have a consistent tendency to titrate slightly past the equivalence point.

It is *random errors* that are the cause of scatter in the results of multiple measurements. When many repeated measurements are made, their frequency plot takes the form of a bell-shaped, *Gaussian curve*; random errors result both in positive and negative deviations from the true value, which is the average of the set of measurements. This is seen in Fig. 2-1, the observed frequency distribution for 6000 measurements of the same quantity. If a smaller number of repeated measurements had been made, the Gaussian shape would be less pronounced.

Except in a few initial problems in this chapter, we will assume that all systematic errors have been eliminated from an experiment and that only random errors remain. This will allow us to use standard statistical treatments to evaluate experimental data.

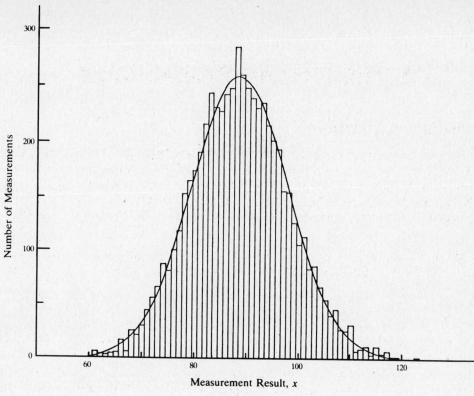

Fig. 2-1

2.3 STATISTICAL SYMBOLS AND DEFINITIONS

Out of all the possible probability distributions, we shall here be concerned with just two: the *Gaussian distribution* and another, allied to it, called *Student's t distribution*. [The *Poisson distribution* will be mentioned in Chapter 14.] The Gaussian distribution owes its special importance to a major result in probability theory, which, roughly paraphrased, says that the probability distribution describing any set of N repeated measurements goes over into a Gaussian distribution as $N \to \infty$. For this reason, the Gaussian distribution is also called the *normal distribution*.

Mathematically, a Gaussian distribution may be described by its *frequency function*, $y = \phi(x|\mu, \sigma_x)$, which plots as the bell-shaped Gaussian curve already mentioned. Here, x is the measurement result, and $y \, \Delta x$ gives the fraction of results that lie in the small interval from x to $x + \Delta x$. The two parameters, μ and σ_x, are, respectively, the mean (or average) and the standard deviation of the distribution. The Gaussian curve is centered on $x = \mu$, while the breadth of the central peak is governed by the value of σ_x. The smaller σ_x, the more tightly are the measured results clustered about the mean (i.e., the smaller the scatter). The Gaussian curve of Fig. 2-1 has the equation

$$y = 6000\phi(x|88.6, 9.2)$$

that is, it describes $N = 6000$ repeated measurements x, of which the arithmetic mean is $\mu = 88.6$ and the standard deviation is $\sigma_x = 9.2$.

Table 2-1 shows how to calculate the mean, standard deviation, and other statistics for finite and infinite samples of data. Observe, in lines 1 and 2, the use of Greek letters instead of Roman in the infinite case, and how sums are replaced by integrals. As $N \to \infty$, the sample average, $\bar{x}$, approaches the true (mean) value, μ, of the measurements, and the sample standard deviation, s_x, approaches the standard deviation of the actual measurement distribution (whether that distribution is Gaussian or not). Lines 3 and 4 of Table 2-1 involve *confidence coefficients* $t_{CL,\nu}$ which are obtained from areas under the frequency curve of Student's t distribution with parameter ν (see Appendix A). The use of these lines will be explained in Section 2.5; for now, note that $\bar{x}_N$ and s_{xN} in line 4 stand for the mean and standard deviation of the first N measurements.

Table 2-1

Statistic for N measurements of x	N finite	$N \to \infty$
Average (arithmetic mean)	$\bar{x} = \dfrac{\Sigma\, x_i}{N}$	$\mu = \int x\, \phi(x\|\mu, \sigma_x)\, dx$
Standard deviation	$s_x = \sqrt{\dfrac{\Sigma\,(x_i - \bar{x})^2}{N-1}}$	$\sigma_x = \sqrt{\int (x - \mu)^2\, \phi(x\|\mu, \sigma_x)\, dx}$
Estimate of true value, μ, at a confidence level CL	$\mu \approx \bar{x} \pm t_{CL,\nu}\, \dfrac{s_x}{\sqrt{N}}$ $(\nu = N - 1)$	$\mu = $ true value
Predicted range, at a given CL, for the next, or $(N+1)$st, measurement	$\bar{x}_N \pm t_{CL,\nu} \sqrt{\dfrac{N+1}{N}}\, s_{xN}$ $(\nu = N - 1)$	$\mu \pm t_{CL,\infty}\sigma_x$

EXAMPLE 2.3 Six portions of a well-mixed sample are analyzed separately for chloride. The data obtained are: 20.36, 19.98, 20.26, 20.18, 19.72, and 19.86% Cl. What are the average and the standard deviation of these data?

The average is

$$\bar{x} = \frac{\Sigma\, x_i}{N} = \frac{120.36}{6} = 20.06$$

The standard deviation can be calculated using an alternative form of the equation given in Table 2-1; this is the form that is used in pocket calculators that have a built-in standard deviation function.

$$s_x = \sqrt{\frac{\Sigma\, x_i^2 - N\bar{x}^2}{N-1}} = \sqrt{\frac{\Sigma\, x_i^2}{N-1} - \frac{(\Sigma\, x_i)^2}{N(N-1)}} \qquad (2.1)$$

For the data,

$$s_x = \sqrt{\frac{2414.73}{5} - \frac{(120.36)^2}{(6)(5)}} = \sqrt{482.9456 - 482.8843} = \sqrt{0.0613} = 0.25$$

As is seen in this example, the two terms under the square root are usually very similar in magnitude. As a result, it is necessary to calculate these terms to seven or eight digits in order to have two or three digits remaining in s_x.

2.4 THE GAUSSIAN ($N \to \infty$) DISTRIBUTION

The Gaussian distribution is a symmetrical distribution that, in theory, covers values of x from $-\infty$ to $+\infty$. Therefore, the *median* (middle value) is at the same time the *mode* (most likely value) and the mean (average value). As shown by Fig. 2-2, the Gaussian distribution (with mean 0 and standard deviation 1, which can always be arranged by suitably redefining the measurements x) is a limiting case of Student's t distribution as the parameter ν of that distribution approaches ∞. Hence, the last line of Appendix A pertains to the Gaussian distribution, from which we conclude that 68.3% of

x-data are included in the range $\mu - \sigma_x$ to $\mu + \sigma_x$ (i.e., $\mu \pm \sigma_x$), 95.5% of data are in the range $\mu \pm 2\sigma_x$, 99.7% of data are in the range $\mu \pm 3\sigma_x$. More generally, CL% of data are included in the range

$$\mu \pm t_{CL,\infty}\sigma_x$$

as implied by the last line of Table 2-1.

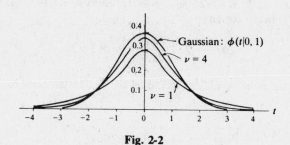

Fig. 2-2

The coefficients $t_{CL,\infty}$ can also be used to determine the probability that a future measurement will deviate from the true value by more than a certain amount. Since the frequency curve is symmetrical, the areas of the two tails are equal. Thus, for example, there are 95.5 chances in 100 that a value will fall in the range $\mu \pm 2\sigma_x$. In this case, each tail contains

$$\frac{100.0\% - 95.5\%}{2} = 2.25\%$$

of the total area ($=1$), which means that there are 2.25 chances in 100 that a value will be greater than $\mu + 2\sigma_x$, and also 2.25 chances in 100 that a value will be less than $\mu - 2\sigma_x$.

2.5 STUDENT'S t (N FINITE) DISTRIBUTION

Student's t distribution applies to a sample of N measurements when the parameter ν, called the *degrees of freedom*, is set equal to $N - 1$. As we can see from Fig. 2-2, the symmetrical frequency curve for Student's t is relatively broader than the Gaussian curve; and the breadth increases with decreasing ν. This is as it should be: smaller samples of data should show more scatter than do larger samples.

Line 4 of Table 2-1 indicates how the coefficients $t_{CL,\nu}$, tabulated in Appendix A, can be used to set up an interval which will include the next measured value x with a prescribed probability. The procedure is very much like that for the case $N \to \infty$, except that now the interval depends upon the *sample* mean $\bar{x}$ and the *sample* standard deviation s_x, as calculated from the first N measurements, rather than upon the true values μ and σ_x (which usually are unknown).

EXAMPLE 2.4 Based on the data of Example 2.3, determine a range of values such that there are 90 chances in 100 that the next (seventh) measurement will fall in this range.

From Example 2.3, $\bar{x}_6 = 20.06$ and $s_{x_6} = 0.25$. The degrees of freedom are $\nu = 6 - 1 = 5$. From Appendix A, $t_{90,5} = 2.015$. Thus, the desired range is

$$20.06 \pm (2.015)\left(\sqrt{\frac{7}{6}}\right)(0.25) = 20.06 \pm 0.54$$

Putting it a different way, there is only a 10% chance that the seventh measurement will turn out to be either greater than 20.60 or less than 19.52.

Line 3 of Table 2-1 has to do with a basic statistical question: How is the average, $\bar{x}$, of N measurements related to the true value, μ, of the quantity being measured? Measurements are subject to random errors, but there is nothing random about the true value. Hence, to talk about "the probability of μ being within so and so many units of an observed $\bar{x}$" would make no sense. We can, however, give a numerical estimate of our *confidence* that μ is within so and so of $\bar{x}$. This is what is done in line 3, but in inverse fashion: a degree of confidence (CL) is chosen first, and then "so and so" ($t_{CL,N-1}s_x/\sqrt{N}$) is calculated. Most scientists choose CL = 95%.

EXAMPLE 2.5 Give 95% confidence limits for the true chloride content in Example 2.3.

We have $v = N - 1 = 5$, $\bar{x} = 20.06$, $s_x = 0.25$, and, from Appendix A, $t_{95,5} = 2.571$. Thus, we are 95% confident that the true content μ falls between the limits

$$20.06 \pm (2.571)\frac{0.25}{\sqrt{6}} = 20.06 \pm 0.26$$

A report of the six measurements would cite as the result: % Cl = 20.06 ± 0.26 or 20.1 ± 0.3.

2.6 TESTS FOR SIGNIFICANCE

EXAMPLE 2.6 On Tuesday a certain acidic solution was determined to have a pH of 5.03; on Friday the same solution had a measured pH of 4.94. Is the difference to be attributed to random measuring errors, or is it to be considered *significant* (e.g., perhaps the solution became contaminated and its pH changed)?

Questions like that raised in Example 2.6 can be resolved only in terms of the notion of *confidence* (see Section 2.5); by convention, the 95% confidence level is used. In each case, we calculate the absolute value of Student's t corresponding to the observed difference. If $|t| > t_{95,v}$ (which would result *by pure chance* only 1 time in 20), the observed difference is *significant* (at the 95% confidence level). If $|t| < t_{95,v}$, the observed difference is *not significant* (at the 95% confidence level).

Formulas for $|t|$, in several important applications, follow.

Difference Between Measurement Average $\bar{x}$ and Known True Average μ

$$|t| = \frac{|\bar{x} - \mu|}{s_x}\sqrt{N} \qquad (2.2)$$

and $|t|$ is compared to $t_{95,N-1}$.

EXAMPLE 2.7 Formula (2.2) may be applied to *paired data*, in which case $\bar{x}$ and μ would represent average differences. Suppose that six different samples of a weak acid were each divided into two portions, a and b, and analyzed for their acid content by titration with the same standardized NaOH solution. For samples a, a color indicator was used to signal the end point in the titration; for samples b, the titration was followed with a pH meter. The paired measurements are displayed in Table 2-2.

Table 2-2

Sample number	% Acid by Method a	% Acid by Method b	Difference, x_i
1	32.57	32.43	+0.14
2	54.22	54.05	+0.17
3	38.71	38.66	+0.05
4	50.35	50.27	+0.08
5	41.63	41.64	−0.01
6	35.94	35.79	+0.15

The formulas of Table 2-1 yield $\bar{x} = 0.0967$, $s_x = 0.0692$. On the assumption that both methods of measurement are valid (called the *null hypothesis*, in statistical terminology), μ_a, the true average value under method a, must be equal to μ_b. Hence, for the differences x_i,

$$\mu = \mu_a - \mu_b = 0$$

and (2.2) gives

$$|t| = \frac{0.0967}{0.0692} \sqrt{6} = 3.42$$

which is larger than $t_{95,5} = 2.571$. We conclude, with 95% confidence, that the two methods show a significant difference in results. (In other words, we reject the null hypothesis at the 95% CL. It is interesting to note that method a gave the larger measurement 5 out of 6 times.)

Difference Between Measurement Average $\bar{x}$ and Known True Average μ, When True Standard Deviation σ_x Also Is Known

$$|t| = \frac{|\bar{x} - \mu|}{\sigma_x} \sqrt{N} \qquad (2.3)$$

and $|t|$ is compared to $t_{95,\infty}$ (Gaussian distribution).

Difference Between Average Values $\bar{x}$ and $\bar{y}$ of Two Sets of Measurements When True Standard Deviations σ_x and σ_y Are Known to Be Equal

$$|t| = \frac{|\bar{x} - \bar{y}|}{s_p} \sqrt{\frac{N_x N_y}{N_x + N_y}} \qquad (2.4)$$

where the *pooled standard deviation* is given by

$$s_p = \sqrt{\frac{s_x^2(N_x - 1) + s_y^2(N_y - 1)}{N_x + N_y - 2}} \qquad (2.5)$$

and $|t|$ is compared to $t_{95, N_x + N_y - 2}$.

EXAMPLE 2.8   In Problem 2.19, (2.4) is applied to two sets of data drawn from the same population (i.e., both chloride analyses were carried out on the same sample). It is also possible to use (2.4) when the two data sets may or may not come from the same population. For instance, an experimenter might have to decide (at a given CL) whether two samples of laboratory reagent came from the same batch. In this case, the null hypothesis would be that they did.

2.7 PROPAGATION OF ERROR

Sufficient past data are available to provide values of σ_x for certain laboratory operations. For example, $\sigma_x = 0.02$ mL for a typical buret reading. A typical analytical balance reading has $\sigma_x = 0.1$ mg, and a typical pH meter reading has $\sigma_x = 0.01$ pH units. Other values of the $(N \to \infty)$ standard deviation are given in Table 2-3.

Suppose now a quantity R depends on *statistically independent* measured quantities $x, y, z, \ldots$, via $R = f(x, y, z, \ldots)$; i.e.,

$$R_i = f(x_i, y_i, z_i, \ldots)$$

Then it is possible to estimate how errors in $x, y, z, \ldots$ are propagated to R, as follows. Make a Taylor expansion of R_i about the mean values:

$$R_i = f(\bar{x}, \bar{y}, \bar{z}, \ldots) + (x_i - \bar{x})\frac{\partial f}{\partial y}(\bar{x}, \bar{y}, \bar{z}, \ldots) + (y_i - \bar{y})\frac{\partial f}{\partial y}(\bar{x}, \bar{y}, \bar{z}, \ldots)$$

$$+ (z_i - \bar{z})\frac{\partial f}{\partial z}(\bar{x}, \bar{y}, \bar{z}, \ldots) + \cdots + (\text{higher-order terms})$$

or, identifying $f(\bar{x}, \bar{y}, \bar{z}, \ldots)$ with $\bar{R}$, the average value of R, and suppressing the arguments of the partial derivatives,

$$R_i - \bar{R} \approx (x_i - \bar{x})\frac{\partial f}{\partial x} + (y_i - \bar{y})\frac{\partial f}{\partial y} + (z_i - \bar{z})\frac{\partial f}{\partial z} + \cdots \qquad (2.6)$$

If now (2.6) is squared, divided through by $N-1$, and summed from $i = 1$ to N, the cross-product terms $[\Sigma (x_i - \bar{x})(y_i - \bar{y}),$ etc.] vanish, because of the statistical independence of $x, y, z, \ldots$. One is left with (see Table 2-1, line 2):

$$s_R{}^2 \approx \left(\frac{\partial f}{\partial x}\right)^2 s_x{}^2 + \left(\frac{\partial f}{\partial y}\right)^2 s_y{}^2 + \left(\frac{\partial f}{\partial z}\right)^2 s_z{}^2 + \cdots \qquad (2.7)$$

and this is the desired error-propagation formula. Usually we suppose N large and write σ's instead of s's in (2.7). Remember that the partial derivatives are to be evaluated at the average (or true) values.

Table 2-3

Volumetric Flasks		Pipets	
Volume, mL	σ_x, mL	Volume, mL	σ_x, mL
25	0.03	5	0.01
50	0.05	10	0.02
100	0.08	15	0.03
250	0.12	20	0.03
500	0.15	25	0.03
1000	0.50	50	0.05
2000	0.50	100	0.08

EXAMPLE 2.9 Let $R = K + ax + by + cz + \cdots$, where $K, a, b, c, \ldots$ are constants of either sign. Then (2.7) gives

$$\sigma_R = \sqrt{a^2\sigma_x{}^2 + b^2\sigma_y{}^2 + c^2\sigma_z{}^2 + \cdots} \qquad (2.8)$$

For instance, if $R = x - y$, with $\bar{x} = 5.3$, $\bar{y} = 2.6$, $\sigma_x = 0.8$, $\sigma_y = 0.4$, then

$$\bar{R} = 5.3 - 2.6 = 2.7$$
$$\sigma_R = \sqrt{(0.8)^2 + (0.4)^2} = 0.9$$

This result might be presented in shorthand fashion as:

$$(5.3 \pm 0.8) - (2.6 \pm 0.4) = 2.7 \pm 0.9$$

EXAMPLE 2.10 For $R = Kx^a y^b z^c \cdots$, (2.7) gives

$$\sigma_R = |\bar{R}|\sqrt{a^2\left(\frac{\sigma_x}{\bar{x}}\right)^2 + b^2\left(\frac{\sigma_y}{\bar{y}}\right)^2 + c^2\left(\frac{\sigma_z}{\bar{z}}\right)^2 + \cdots} \qquad (2.9)$$

For instance, if $R = x^3/y^2$, with $\bar{x} = 15.0$, $\bar{y} = 10.4$, $\sigma_x = 0.6$, $\sigma_y = 0.2$, then

$$\bar{R} = \frac{(15.0)^3}{(10.4)^2} = 31.2$$

$$\sigma_R = (31.2)\sqrt{3^2\left(\frac{0.6}{15.0}\right)^2 + (-2)^2\left(\frac{0.2}{10.4}\right)^2} = 3.9$$

and the value of R would be reported as 31.2 ± 3.9.

EXAMPLE 2.11 For

$$R = \frac{K}{2.3025 \cdots} \ln x = K \log_{10} x$$

$(x > 0)$, (2.7) gives

$$\sigma_R = (0.434 \cdots)|K| \frac{\sigma_x}{\bar{x}} \qquad (2.10)$$

For instance, suppose that a series of measurements has yielded 2.58×10^{-6} as the mean $[H^+]$, with a standard deviation of 0.03×10^{-6}; i.e.,

$$[H^+] = (2.58 \pm 0.03) \times 10^{-6}$$

Then, since $pH = -\log [H^+]$, we have (using Rule 6 of Section 1.6):

$$\overline{pH} = -\log (2.58 \times 10^{-6}) = 5.588$$

$$\sigma_{pH} = (0.434) \frac{0.03 \times 10^{-6}}{2.58 \times 10^{-6}} = 0.005$$

The calculated pH would be quoted as 5.588 ± 0.005.

2.8 REJECTION OF DATA

When a set of data contains an outlying value that appears to differ excessively from the average, the question will arise whether to retain or reject the outlying value. The value may be rejected at once if a systematic error is discovered. However, to discard data simply because they "do not look right" is invalid. The *Q-test* is a statistical criterion for rejection of data.

In applying the *Q*-test, you first identify the suspect value as x_1. The remaining data are arranged in ascending or descending order, according as x_1 seems too small or too large, respectively. For instance, if $N = 5$, then the following arrangement applies:

The *spread* is defined as $x_2 - x_1$. If the number of measurements is $N = 3$ to $N = 7$, then the *range* is defined as $x_N - x_1$; if $N = 8$, 9, or 10, the range is $x_{N-1} - x_1$. (We shall not pursue the theory beyond $N = 10$.) Now a *Q*-value $(0 < Q < 1)$ is calculated as

$$Q = \frac{\text{spread}}{\text{range}} \qquad (2.11)$$

Table 2-4.   Values of $Q_{CL,N}$

N \ CL	95% (1-tailed) = 90% (2-tailed)	99% (1-tailed) = 98% (2-tailed)
3	0.941	0.988
4	0.765	0.889
5	0.642	0.780
6	0.560	0.698
7	0.507	0.637
8	0.554	0.683
9	0.512	0.635
10	0.477	0.597

If Q exceeds $Q_{CL,N}$, as given in Table 2-4, then the suspect value, x_1, may be omitted. Ordinarily, the 90% (two-tailed) CL is used.

Occasionally, after one value has been rejected on the basis of the Q-test, a second value appears suspect. This second value is now labeled x_1 and the Q-test applied again to the $N - 1$ data. It is not recommended to apply the Q-test more than twice to a single set of data.

2.9 LINEAR LEAST-SQUARES DATA FIT

Consider a set of data points (x_i, y_i), such that the abscissas x_i are known precisely and all scatter is due to uncertainties in the ordinates y_i. The *method of least squares* gives as the equation of the "best" straight line through the data points:

$$y = mx + b \tag{2.12}$$

where [see (2.1) for s_x]

$$m = \frac{N \sum x_i y_i - (\sum x_i)(\sum y_i)}{N \sum x_i^2 - (\sum x_i)^2} = \frac{\overline{xy} - \bar{x}\bar{y}}{s_x^2(N-1)/N} \tag{2.13}$$

$$b = \bar{y} - m\bar{x} \tag{2.14}$$

Some pocket calculators have programs to evaluate m and b. The least-squares line is the "best" in the sense that the sum of the squared differences between the y-values it provides and the actual y_i is a minimum. For some purposes it is desirable to consider the slope and y-intercept of the least-squares line as experimental data. Their standard deviations are then given by the formulas

$$s_m = \frac{s_d}{s_x \sqrt{N-1}} \tag{2.15}$$

$$s_b = s_m \sqrt{\frac{\sum x_i^2}{N}} \tag{2.16}$$

where

$$s_d = \sqrt{\frac{\sum y_i^2 - b \sum y_i - m \sum x_i y_i}{N-2}} \tag{2.17}$$

EXAMPLE 2.12 The following flame-atomic absorption readings were obtained for a series of zinc standards:

x = ppm Zn^{2+}	0.50	1.00	1.50	2.00	2.50
y = absorbance	0.130	0.200	0.350	0.430	0.490

Find the equation of the least-squares fitted line through these data and give 95% confidence-level precision estimates for the slope and intercept.

Here, $N = 5$ and

$$\sum x_i = 5\bar{x} = 7.50 \qquad \sum y_i = 5\bar{y} = 1.600$$

$$\sum x_i^2 = 13.750 \qquad \sum y_i^2 = 0.6044 \qquad \sum x_i y_i = 5\overline{xy} = 2.875$$

By (2.13) and (2.14),

$$m = \frac{5(2.875) - (7.50)(1.600)}{5(13.750) - (7.50)^2} = 0.190$$

$$b = \frac{1.600}{5} - (0.190)\frac{7.50}{5} = 0.0350$$

so that the equation of the least-squares line is

$$y = 0.190x + 0.0350$$

Confidence intervals for the true values of m and b are given by expressions similar to line 3 of Table 2-1:

$$m \pm t_{CL,\nu} s_m \qquad \text{and} \qquad b \pm t_{CL,\nu} s_b \qquad\qquad (2.18)$$

From (2.1) and (2.17),

$$s_x = \sqrt{\frac{13.750}{4} - \frac{(7.50)^2}{(5)(4)}} = 0.7906$$

$$s_d = \sqrt{\frac{0.6044 - (0.0350)(1.600) - (0.190)(2.875)}{3}} = 0.0268$$

Then, from (2.15) and (2.16),

$$s_m = \frac{0.0268}{0.7906\sqrt{4}} = 0.0169$$

$$s_b = 0.0169 \sqrt{\frac{13.750}{5}} = 0.028$$

Using these standard deviations, along with

$$t_{CL,\nu} = t_{95,N-2} = t_{95,3} = 3.182$$

(from Appendix A), we obtain the intervals

> **for m** $0.190 \pm (3.182)(0.0169) = 0.190 \pm 0.054$
> **for b** $0.0350 \pm (3.182)(0.0281) = 0.035 \pm 0.089$

Linear Least Squares in Reverse

Suppose that N data points (x_i, y_i) have been fitted by the least-squares line. Now, for some fixed but unknown value x_u of x, let M measurements of y_u be made; denote the mean of these measurements by $\bar{y}_u$. We then have as an estimate of x_u the quantity

$$\hat{x}_u = \frac{\bar{y}_u - b}{m} \qquad\qquad (2.19)$$

the standard deviation of $\hat{x}_u$ being

$$s_{\hat{x}_u} = \frac{s_d}{m} \left[\frac{1}{M} + \frac{1}{N} + \frac{(\hat{x}_u - \bar{x})^2}{s_x^2(N-1)} \right]^{1/2} \qquad\qquad (2.20)$$

(If $\bar{y}_u$ is a theoretical, rather than a measured, value, (2.20) holds with $M = \infty$.)

EXAMPLE 2.13 A zinc solution of unknown concentration x_u was divided into three portions and each portion was analyzed using the procedure and apparatus of Example 2.12. The average absorbance was found to be $\bar{y}_u = 0.250$ ($M = 3$). (a) Find $\hat{x}_u$. (b) Give the 95% confidence interval for x_u.

(a) By (2.19) and Example 2.12,

$$\hat{x}_u = \frac{0.250 - 0.035}{0.190} = 1.13 \text{ ppm } Zn^{2+}$$

(b) By (2.20) and Example 2.12,

$$s_{\hat{x}_u} = \frac{0.0268}{0.190} \left[\frac{1}{3} + \frac{1}{5} + \frac{(1.13 - 1.50)^2}{(0.7906)^2(4)} \right]^{1/2} = 0.108 \text{ ppm } Zn^{2+}$$

Confidence intervals for x_u are given by $\hat{x}_u \pm t_{CL,\nu} s_{\hat{x}_u}$, where $\nu = N - 2$. Thus,

$$x_u \approx 1.13 \pm t_{95,3}(0.108) = 1.13 \pm (3.182)(0.108)$$
$$= 1.13 \pm 0.34 \quad \text{ppm } Zn^{2+}$$

Solved Problems

ACCURACY AND PRECISION

2.1   Two sets of measurements were made of the weight of active drug in a prescription pill preparation. Set A involved analysis of 25.0 mg pill samples, whereas set B involved 100.0 mg pill samples. Both sets of pills contained 80.0% (by weight) of the active drug. The data are:

	Set A ($N = 5$)	Set B ($N = 4$)
	19.8 mg	81.1 mg
	20.3	79.3
	20.6	80.4
	19.2	79.7
	19.7	
$\bar{x} =$	19.92	80.13
$s_x =$	0.55	0.79
$s_x/\sqrt{N} =$	0.24	0.40
$t_{95,\nu}s_x/\sqrt{N} =$	0.68	1.26
$\mu =$	20.00	80.00

Which set has: (*a*) the larger absolute error in the mean?   (*b*) the larger relative error in the mean?   (*c*) the larger relative range of values?   (*d*) the larger absolute precision at the 95% confidence level?   (*e*) the larger relative precision at the 95% confidence level?

(*a*)   Absolute errors are:

$$\text{set A}\quad |20.0 - 19.92| = 0.08 \qquad \text{set B}\quad |80.0 - 80.13| = 0.13$$

Set B has the larger absolute error.

(*b*)   Relative errors are:

$$\text{set A}\quad \frac{0.08}{20.0} = 0.0040 \qquad \text{set B}\quad \frac{0.13}{80.0} = 0.0016$$

Set A has the larger relative error.

(*c*)   Relative ranges are:

$$\text{set A}\quad \frac{20.6 - 19.2}{19.92} = 0.070 \qquad \text{set B}\quad \frac{81.1 - 79.3}{80.13} = 0.022$$

Set A has the larger relative range.

(*d*)   Half-widths of 95% confidence intervals are:

$$\text{set A}\quad t_{95,4}(0.24) = 0.68 \qquad \text{set B}\quad t_{95,3}(0.40) = 1.26$$

Set A has the narrower interval and therefore the larger absolute precision.

(*e*)   Relative amounts of scatter at the 95% confidence level are:

$$\text{set A}\quad \frac{0.68}{19.92} = 0.034 \qquad \text{set B}\quad \frac{1.26}{80.13} = 0.016$$

Set B has smaller relative scatter and, therefore, larger relative precision.

2.2   Two sets of analyses of the percent iron in a sample resulted in the following data:

$$\bar{x} \pm s_x = 36.27 \pm 0.16 \qquad N_x = 5$$
$$\bar{y} \pm s_y = 36.34 \pm 0.22 \qquad N_y = 8$$

The true value is $\mu = 36.32$. (a) Which set of data is more accurate? (b) Which set of data is more precise, at the 95% confidence level?

(a)   The accuracy of $\bar{x}$ is $|36.27 - 36.32| = 0.05$; of $\bar{y}$, $|36.34 - 36.32| = 0.02$. Therefore, $\bar{y}$ is more accurate.

(b)
$$t_{95,4} \frac{s_x}{\sqrt{5}} = (2.776) \frac{0.16}{2.236} = 0.20$$

$$t_{95,7} \frac{s_y}{\sqrt{8}} = (2.365) \frac{0.22}{2.828} = 0.18$$

The y-data are more precise, at the 95% CL.

RANDOM AND SYSTEMATIC ERRORS

2.3   An approximately 0.1 M NaOH solution is made by adding 4.0 g of solid NaOH to 1000 mL of water. The precise concentration of the solution is determined by titrating the NaOH against portions of a weak acid, potassium acid phthalate (KHP), obtained from the National Bureau of Standards and certified as being 99.99% pure. The KHP samples are weighed by difference on a balance. How would the calculated NaOH concentration be affected (in comparison with the true concentration), if (a) the KHP was not dried prior to weighing? (b) the balance was not zeroed properly and always read 1.00 mg too high? (c) the balance always read high, but in proportion to the weight recorded? (d) the NaOH solution was still warm when titrated against the KHP, whereas the NaOH is intended for use at 25 °C? (e) about 5.0 g, rather than 4.0 g, of NaOH was used to make the NaOH solution? (f) when the buret was filled prior to each titration, an air bubble became entrapped in the tip of the buret, but was dislodged during the titration? (g) an indicator that changes color in the pH range 3–5 was used rather than one that changes in the pH range 7–10, where the equivalence point occurs? (h) the person performing the titrations did not allow enough time for the remaining liquid adhering to the buret walls to drain down as the end point was reached?

The NaOH molarity is calculated from the equation

$$M = \frac{\text{mols KHP}}{\text{vol NaOH}} = \frac{\text{wt KHP}}{(\text{MW KHP})(\text{vol NaOH})}$$

(a)   Because of added H_2O, the measured weight of the KHP will be greater than the true weight; therefore, the calculated M will be greater than the true M.

(b)   The systematic error in the balance readings will cancel when weighing by difference, so that the measured weight of KHP will be the true weight. Therefore, the calculated M will be the true M.

(c)   The measured weight will be greater than the true weight, so the calculated M will be greater than the true M.

(d)   Since solutions expand upon heating, the measured volume of NaOH will be greater than the volume that would have been measured if the solution had been at room temperature. Therefore, the measured M will be less than the true M.

(e)   The difference does not affect the measurements; the true value will be obtained in either case (so long as no other errors exist). However, the true M found will be larger when 5.0 g is used than when 4.0 g is used.

(f)   As the bubble becomes dislodged, the level in the buret drops. Hence, the measured volume is incorrectly high, and the calculated M will be smaller than the true M.

(g)   The end point will be incorrectly identified as a volume that is less than the true volume. Therefore, the calculated M will be larger than the true M.

(h)   Because of liquid still adhering to the walls of the buret, the volume of NaOH that is measured is greater than the true volume. Therefore, the calculated M is less than the true M.

2.4   (*a*) In performing a titration, why is it better to use a solution of a concentration such that 30–45 mL is required, instead of a more concentrated solution that might only require about 5 mL to reach the end point?   (*b*) To make a standardized NaOH solution, why is it better to prepare a NaOH solution of approximate concentration and then standardize with KHP, rather than to weigh accurately some NaOH pellets, dissolve them, and dilute the solution in a volumetric flask?   (*c*) In analyzing a weak acid (e.g., 0.003 *M* acetic acid) by titration with 0.00300 *M* NaOH, why would you be able to obtain greater accuracy by using a pH meter (even if the meter always read 1.00 pH unit too low) instead of using a color-change indicator such as phenolphthalein?

(*a*)   The absolute uncertainty in the measurement is the same for both procedures, but the relative uncertainty is less when a larger volume is used.

(*b*)   Since NaOH pellets easily absorb water and CO_2, the pellets are not pure; whereas KHP can be weighed as a pure, dry solid.

(*c*)   The equivalence-point region for so dilute an acid and base does not involve a sufficiently sharp change in pH to span the pH range for a color indicator. Therefore, it is difficult to identify accurately the volume at the equivalence point using a color indicator. With a pH meter, it is possible to prepare first- and second-derivative plots of pH versus volume data, and these fix much more accurately the volume at the equivalence point. This is the case even if the pH meter readings are systematically low (or high), since differences in pH are required and any fixed systematic error in pH is canceled.

THE GAUSSIAN DISTRIBUTION

2.5   If $\mu \pm \sigma_x = 63.4 \pm 10.4$, what are the chances of an *x*-value being in the range 63.4 ± 26.8?

The half-width 26.8 is $26.8/10.4 = 2.58$ standard deviations. From the bottom row of Appendix A, this is the 99.0% confidence level. Therefore, there are 99 chances in 100 of a value being in this range.

2.6   If $\mu \pm \sigma_x = 33.57 \pm 0.25$, what are the chances that a measurement *x* will be greater than 34.06?

This value is $(34.06 - 33.57)/0.25 = 1.95$ standard deviations above the mean. Since 95.0% of the data are in the range $\mu \pm 1.96\sigma_x$, then 5.0% represents the sum of the areas of the two tails of the distribution, with 2.5% contained in each tail. Thus, there are 2.5 chances in 100 (1 chance in 40) that a measurement will be greater than 34.06.

2.7   Based on 1103 samples ($N \to \infty$), the Ca^{2+} content of human blood-serum is

$$\mu \pm \sigma_x = 94.0 \pm 5.5 \quad \mu g/mL$$

If a sample shows 110.5 μg/mL, (*a*) by how many standard deviations does it differ from the mean and (*b*) how often would you expect to find serum calcium values greater than this one?

(*a*)   This is $(110.5 - 94.0)/5.5 = 3.00$ standard deviations above the mean.

(*b*)   From the bottom row of Appendix A, $3.00\sigma_x$ is the 99.7% confidence level. Each tail of the curve contains $(100.0 - 99.7)/2 = 0.15\%$ of the total area (unity). Thus, there are only 0.15 chances in 100 (3 chances in 2000) of finding values greater than 110.5 μg/mL.

2.8   After analyzing a very large number of blood-serum samples from healthy adults, a clinical laboratory determined the blood-glucose level as $\mu \pm \sigma_x = 72.5 \pm 7.9$ mg% (see Table 1-4). What percentage of adults have glucose levels (*a*) less than 64.6?   (*b*) greater than 88.3?   (*c*) between 46.5 and 72.5?   (*d*) between 85.5 and 96.2?

(*a*)   64.6 is $(72.5 - 64.6)/7.9 = 1.00$ standard deviation below the mean. Since 68.3% of data are between $\mu \pm 1.00\sigma_x$,

$$\frac{100.0 - 68.3}{2} = 15.85\%$$

are less than $\mu - 1.00\sigma_x$.

(b)   88.3 is $(88.3 - 72.5)/7.9 = 2.0$ standard deviations above the mean. Since 95.5% of data are between $\mu \pm 2\sigma_x$,

$$\frac{100.0 - 95.5}{2} = 2.25\%$$

are greater than $\mu + 2\sigma_x$.

(c)   46.5 is $(72.5 - 46.5)/7.9 = 3.29$ standard deviations below the mean. Since 99.9% of data are between $\mu \pm 3.29\sigma_x$,

$$\frac{99.9\%}{2} = 49.95\%$$

are between $\mu - 3.29\sigma_x$ and μ.

(d)   85.5 is $(85.5 - 72.5)/7.9 = 1.65$ standard deviations above the mean (90% CL), whereas 96.2 is

$$\frac{96.2 - 72.5}{7.9} = 3.00$$

standard deviations above the mean (99.7% CL). Hence, 5.0% of data will be greater than 85.5 and 0.15% of data will be greater than 96.2. Therefore,

$$5.00\% - 0.15\% = 4.85\%$$

of data will be in the range 85.5–96.2.

2.9   Certain atomic-absorption spectral lamps are listed as having a life of 300 milliampere-hours. Assume that this is the average life, μ, and that $\sigma_x = 20$ mA·h. (a) To avoid making replacements prematurely, but to have only 5% probability of failure before replacement, how long should lamps be used? (b) What is the probability that a given lamp will last at least 340 mA·h?

(a)   By definition, the probability of a lifetime shorter than $\mu - t_{CL,\infty}\sigma_x$ is the area of the lower tail,

$$\frac{100\% - CL\%}{2}$$

Equating this area to 5% gives CL = 90, whence lamps should be changed after

$$\mu - t_{90,\infty}\sigma_x = 300 - (1.645)(20) = 267.1 \text{ mA·h}$$

(b)   340 mA·h is $300 + 2.00\sigma_x$. This is the 95.5% (two-tailed) confidence level. Since 2.25% of the area is in each tail, there are 2.25 chances in 100 (9 chances in 400) that a lamp will last this long.

2.10   Batteries in many new cars are warranted for 36 months. Assume that this duration of warranty was chosen by the battery manufacturers so that only 5 in 1000 batteries will fail during the warranty period, whereas 25 in 1000 will fail in 38 months. What is the average life of the batteries?

See Fig. 2-3. We have:

$$\mu - t_{99,\infty}\sigma_x = 36$$
$$\mu - t_{95,\infty}\sigma_x = 38$$

Solving for μ,

$$\mu = \frac{38t_{99,\infty} - 36t_{95,\infty}}{t_{99,\infty} - t_{95,\infty}}$$

$$= \frac{38(2.576) - 36(1.960)}{2.576 - 1.960} = 44.3_6 \text{ months}$$

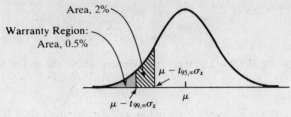

Fig. 2-3

2.11 Most medical test data are given in terms of a "normal range," which is the 95.5% (two-tailed) confidence interval. Suppose that a patient was tested for a disease that is known to exist in only 1 of every 1000 persons; those having the disease always register below the normal range on the test. What is the likelihood that the patient, if registering below the normal range, actually has the disease?

Since 2.25% (22.5 in 1000) will register below the normal range, but only 1 in 1000 has the disease, the likelihood of the patient having the disease is only 1 chance in 22.5 (2 chances in 45).

Equivalently, using the notion of *conditional probability*,

$$\text{Prob \{Sick|Below\}} = \frac{\text{Prob \{Sick and Below\}}}{\text{Prob \{Below\}}}$$

$$= \frac{\text{Prob \{Sick\}}}{\text{Prob \{Below\}}} = \frac{1/1000}{22.5/1000} = \frac{1}{22.5}$$

2.12 The combined Verbal + Math SAT scores of freshmen in a midwestern college followed a well-defined Gaussian distribution; of 3000 freshmen, 15 (0.50%) scored above 1500 and 750 (25.0%) scored below 1050. Find the mean, μ, and the standard deviation, σ_x, for this approximately infinite sample of scores.

We are given that

$$\mu + t_{99,\infty}\sigma_x = \mu + 2.576\sigma_x = 1500$$

$$\mu - t_{50,\infty}\sigma_x = \mu - 0.674\sigma_x = 1050$$

Solving simultaneously, $\sigma_x = 138.5$, $\mu = 1143$.

STUDENT'S t DISTRIBUTION

2.13 A set of seven measurements results in $\bar{x} \pm s_x = 15.74 \pm 0.38$. What is the range of values that includes the true value at a confidence level of 99.0%?

For $\nu = N - 1 = 6$, Appendix A gives $t_{99,6} = 3.707$. By Table 2-1, the desired range is

$$\bar{x} \pm t_{99,6}\frac{s_x}{\sqrt{N}} = 15.74 \pm (3.707)\frac{0.38}{\sqrt{7}} = 15.74 \pm 0.53$$

2.14 Seven measurements of the zinc content of a commercial vitamin-mineral supplement yielded 14.1, 15.2, 14.8, 15.5, 15.3, 14.6, and 14.9 mg Zn per capsule. Find (*a*) the sample average, (*b*) the sample standard deviation, and (*c*) the estimate of the true average value at the 95% confidence level.

(*a*)
$$\bar{x} = \frac{\Sigma x_i}{N} = \frac{104.4}{7} = 14.91 \text{ mg Zn}$$

(*b*) $\Sigma x_i = 104.4$ and $\Sigma x_i^2 = 1558.4$. From (*2.1*),

$$s_x = \sqrt{\frac{1558.4}{6} - \frac{(104.4)^2}{(7)(6)}} = \sqrt{259.7333 - 259.5086} = \sqrt{0.2247} = 0.47$$

(c) By Appendix A, $t_{95,6} = 2.447$, so that, from Table 2-1,

$$\mu \approx 14.91 \pm (2.447)\frac{0.47}{\sqrt{7}} = 14.91 \pm 0.44 \quad \text{mg Zn}$$

That is to say, we are 95% confident that μ is between 14.47 and 15.35 mg Zn.

2.15 Based on the data of Problem 2.14, what are the chances that the next (the eighth) measurement would be less than 13.05?

By line 4 of Table 2-1,

$$13.05 = \bar{x}_7 - t_{\text{CL},6}\sqrt{\frac{8}{7}}\, s_{x7} \quad \text{or} \quad 13.05 = 14.91 - t_{\text{CL},6}\sqrt{\frac{8}{7}}(0.47)$$

Solving, $t_{\text{CL},6} = 3.70$. From Appendix A, this is the 99.0% (two-tailed) confidence level, which is the same as the 99.5% (one-tailed) confidence level. Therefore, there are only 0.5 chances in 100 (1 chance in 200) that the eighth value would be less than 13.05.

2.16 A sample of impure glycol was analyzed in triplicate; the 95% confidence-level estimate of the true value of the glycol content (in percent) was calculated to be $\mu \approx 84.3 \pm 5.0$. Assume that the same standard deviation, s_x, is found for any set of N repeat measurements. How large must N be so that the 95% CL precision estimate of the mean will be less than 1.5 (rather than 5.0, as in the $N = 3$ analysis)?

It is given that

$$t_{95,2}\frac{s_x}{\sqrt{3}} = (4.303)\frac{s_x}{\sqrt{3}} = 5.0$$

from which $s_x = 2.0$. We now want to find the smallest N such that

$$t_{95,N-1}\frac{2.0}{\sqrt{N}} < 1.5 \quad \text{or} \quad \frac{t_{95,N-1}}{\sqrt{N}} < 0.75$$

Using the t_{95}-values in Appendix A, the ratio is calculated as: 1.05 for $N = 6$, 0.92 for $N = 7$, 0.84 for $N = 8$, 0.77 for $N = 9$, 0.72 for $N = 10$. Thus, N must be 10.

TESTS FOR SIGNIFICANCE

2.17 Six repeated measurements result in $\bar{x} \pm s_x = 20.06\% \pm 0.25\%$ for the percent chloride in a standard known to have a chloride content of $\mu = 20.00\%$. Is the difference statistically significant at the 95% CL?

By (2.2),

$$|t| = \frac{|20.06 - 20.00|}{0.25}\sqrt{6} = 0.59 < 2.571 = t_{95,5}$$

Therefore, the difference is not statistically significant at the 95% CL.

2.18 Six repeated measurements result in $\bar{x} = 20.06\%$ chloride for a chemical standard; prior analyses of this standard resulted in

$$\mu \pm \sigma_x = 20.00\% \pm 0.09\%$$

Is there a statistically significant difference between the average of the six measurements and the average for the infinite set, at the 95% CL?

By (2.3),
$$|t| = \frac{|20.06 - 20.00|}{0.09} \sqrt{6} = 1.63 < 1.960 = t_{95,\infty}$$

The measured difference, therefore, is not statistically significant at the 95% CL.

2.19 Two sets of chloride analyses of the same sample result in:
$$\bar{x} \pm s_x = 20.05 \pm 0.12 \qquad (N_x = 6)$$
$$\bar{y} \pm s_y = 20.21 \pm 0.07 \qquad (N_y = 5)$$

Is there a statistically significant difference between $\bar{x}$ and $\bar{y}$?

By (2.5),
$$s_p = \sqrt{\frac{(0.12)^2(5) + (0.07)^2(4)}{9}} = 0.10$$

and (2.4) then gives
$$|t| = \frac{|20.05 - 20.21|}{0.10} \sqrt{\frac{(6)(5)}{11}} = 2.6 > 2.262 = t_{95,9}$$

The difference in the means is statistically significant (at the 95% CL; but it is not significant at the 96.6% CL).

2.20 Personnel in a clinical laboratory routinely measure blood-serum calcium. Over a long period of time, they obtained data (in ppm Ca by wt) for healthy adults that averaged: $\mu \pm \sigma_x = 94.0 \pm 5.5$, in agreement with literature values. A trainee lab technician obtained for 16 measurements: $\bar{x} \pm s_x = 103.0 \pm 7.5$. Trainees are released for routine analyses when their data are in agreement with the true values at the 90% confidence level. Should this trainee be allowed to do routine analyses?

By (2.3),
$$|t| = \frac{|103.0 - 94.0|}{5.5} \sqrt{16} = 6.5 \gg 1.645 = t_{90,\infty}$$

The difference is (highly) significant at the 90% CL; in fact, it remains significant at the 99.9999% CL. With odds against him of one million to one, the trainee must not be allowed to do routine analysis; he is almost certainly introducing one (or more) systematic errors in his analytical procedure.

2.21 Over a long period of time, a quality control laboratory of a pharmaceutical company found that the content of the active ingredient in one of their products, a pill, was: $\mu \pm \sigma_x = 37.0 \pm 1.5$ mg. If the 95% confidence level (two-tailed) is chosen as the criterion of statistical significance, how *low* can the measured weight of ingredient be, if there is to be no significant difference between the long-term average and (a) the analysis of a single pill having content x mg? (b) the analysis of nine pills having average content $\bar{x}$ mg?

(a)   From Appendix A, $t_{95,\infty} = 1.960$. Therefore, a single value could be as low as
$$37.0 - (1.960)(1.5) = 34.1 \text{ mg}$$
and still be acceptable.

(b)   From (2.3) and $t_{95,\infty} = 1.960$, there results:
$$1.960 = \frac{|\bar{x} - 37.0|}{1.5} \sqrt{9}$$

Solving, $\bar{x} = 36.0$ or 38.0; the low value, therefore, is 36.0 mg.

2.22 It appeared that one of the spectrometers in a laboratory was giving erroneous absorption measurements. To test if this was the case, eight samples were divided into two portions and their absorptivities measured with the suspect spectrometer and with a spectrometer that was known to be reliable. The eight differences in absorptivities, x, yielded:

$$\bar{x} \pm s_x = 0.080 \pm 0.020$$

What conclusion should be drawn?

As in Example 2.7, we find:

$$|t| = \frac{0.080}{0.020}\sqrt{8} = 11$$

which greatly exceeds $t_{95.7} = 2.365$. With 95% confidence (indeed, with 99.999% confidence) we can state that the suspect instrument is no good.

PROPAGATION OF ERROR

2.23 Nine persons want to get on an elevator rated to hold 1500 lb. An old bathroom scale gives their weights as 180, 138, 224, 153, 177, 111, 169, 144, and 168 lb, for a total of 1464 lb. It is known that repeated measurements on this scale of any single weight between 100 and 250 lb obey a Gaussian distribution having $\sigma_x = 4.0$ lb. What is the probability that the combined weight of the nine persons is greater than 1500 lb?

For statistically independent variables $x, y, z, \ldots$, formula (2.8) is exact. (Recall that the parent relation (2.7) holds only to the first order; formulas (2.9) and (2.10) are *not* exact.) Moreover, if each variable has a Gaussian distribution, so does their linear combination R. For the present data, (2.8) gives

$$\sigma_R = \sqrt{9(4.0)^2} = 12.0 \text{ lb}$$

and the sample mean is $\bar{R} = 1464$ lb, which we take as the true mean μ of the Gaussian distribution of R. Now, 1500 lb will be a CL% right-hand confidence limit for this distribution (see Fig. 2-4) if

$$1464 + t_{CL,\infty}(12.0) = 1500 \qquad \text{or} \qquad t_{CL,\infty} = 3.00$$

whence, by Appendix A, CL = 99.7. Thus the area of the right-hand tail is 0.15% = 0.0015, and this is the required probability.

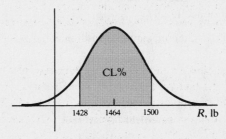

Fig. 2-4

2.24 Analysis of an amine using a back-titration Kjeldahl method resulted in the following equation for P, the percent nitrogen in the amine:

$$P = \frac{(\text{mmols acid}) - (\text{mL base})(\text{molarity base})}{(\text{mg sample})}(14.01)(100)$$

$$= \frac{(5.65 \pm 0.03) - (0.1250 \pm 0.0008)(15.53 \pm 0.04)}{750.0 \pm 0.3}(14.01)(100)$$

where the uncertainties are all standard deviations for infinite-sized samples. Calculate $\bar{P} \pm \sigma_P$.

By (2.9), the second term in the numerator is calculated as

$$1.941 \pm 1.941 \sqrt{\left(\frac{0.0008}{0.1250}\right)^2 + \left(\frac{0.04}{15.53}\right)^2} = 1.941 \pm 0.013$$

Based on (2.8), the numerator now is:

$$(5.65 \pm 0.03) - (1.941 \pm 0.013) = 3.71 \pm \sqrt{(0.03)^2 + (0.013)^2} = 3.71 \pm 0.03$$

Based on (2.9), the ratio is:

$$\frac{3.71 \pm 0.03}{750.0 \pm 0.3} = 0.00495 \pm 0.00495 \sqrt{\left(\frac{0.03}{3.71}\right)^2 + \left(\frac{0.3}{750.0}\right)^2} = 0.00495 \pm 0.00004$$

and finally,

$$\bar{P} + \sigma_P = (0.00495 \pm 0.00004)(14.01)(100) = 6.93 \pm 0.06$$

REJECTION OF DATA

2.25   The following five values were obtained for the wt % of an organic acid in a sample: 30.3, 31.1, 32.6, 36.7, 28.9. Determine if the value 36.7 may be rejected at the 90% (two-tailed) confidence level.

The data are first arranged in descending order so that the 36.7% value can be labeled x_1:

x_1	x_2	x_3	x_4	x_5
36.7	32.6	31.1	30.3	28.9

From (2.11), with $N = 5$,

$$Q = \frac{x_2 - x_1}{x_5 - x_1} = \frac{-4.1}{-7.8} = 0.53$$

which is less than $Q_{90.5} = 0.642$ (from Table 2-4). The value 36.7 may not be rejected.

2.26   The first three measurements in the standardization of a NaOH solution were: 0.1026, 0.1103, and 0.1085 M. A fourth and last measurement is about to be made. (a) How large, and (b) how small, can the fourth measurement be if it is not to be rejected under the Q-test at the 90% (two-tailed) confidence level?

(a)   By (2.11) and Table 2-4, the maximum measurement, x_1, satisfies

$$\frac{0.1103 - x_1}{0.1026 - x_1} = 0.765$$

Solving, $x_1 = 0.1354\ M$.

(b)   Similarly, the minimum measurement, x_1, satisfies

$$\frac{0.1026 - x_1}{0.1103 - x_1} = 0.765$$

or $x_1 = 0.0775\ M$.

LINEAR LEAST SQUARES

2.27   In the titration of a weak acid, let x denote the volume of base and y the second derivative of the volume of acid with respect to the pH. Then, in the vicinity of $y = 0$ (the equivalence point), y should be a linear function of x. Given the ten experimental points listed in Table 2-5, find 95% confidence limits for the equivalence-point volume of base.

Table 2-5

x_i, mL	30.64	30.68	30.75	30.79	30.82
y_i, mL	−1.486	−1.137	−0.781	−0.262	−0.180
x_i	30.89	30.96	31.00	31.07	31.13
y_i	0.080	0.383	0.623	0.828	1.202

For these $N = 10$ data,

$$\sum x_i = 308.7300 \qquad \sum y_i = -0.7300000$$

$$\sum x_i^2 = 9531.6665 \qquad \sum y_i^2 = 6.883576 \qquad \sum x_i y_i = -21.2584300$$

From (2.13), $m = 5.2153664$; from (2.14), $b = -161.08700$. Thus, $y = 0$ at

$$\hat{x} = \frac{-b}{m} = 30.88700$$

i.e., the equivalence-point volume is 30.89 mL.

From (2.17), $s_d = 0.141672$, and the standard deviation of $\hat{x}$ will be given by (2.20), in which $M = \infty$ because $y = 0$ is a theoretical value. Using $\bar{x} = 30.87300$ and $s_x = 0.165062$, we obtain $s_{\hat{x}} = 0.00862$. With $\nu = N - 2 = 8$, Appendix A gives $t_{95,8} = 2.306$. Hence the 95% confidence limits for $\hat{x}$ are

$$30.89 \pm (2.306)(0.00862) = 30.89 \pm 0.02 \quad \text{mL}$$

2.28 An organic halide is known to be either an iodide or a chloride. A sample of the halide was irradiated with neutrons, producing a radioactive element that must be either ^{128}I ($t_{1/2} = 25.0$ min) or ^{38}Cl ($t_{1/2} = 37.3$ min). A series of one-minute "counts" were taken and corrected for background, the ith "count" starting at time x_i (min). Results are shown in Table 2-6.

Table 2-6

x_i, min	1	6	11	16	21	26	31
Corrected counts in $(x_i, x_i + 1)$	70	68	56	44	41	40	32
$y_i = \ln$ (corr. cts.)	4.248	4.220	4.025	3.784	3.714	3.689	3.466

Since radioactive decay is described by an exponential equation, a plot of y versus x should be a straight line with slope $-(\ln 2)/t_{1/2}$. At the 95% CL, is the halogen Cl or I?

The sums are:

$$\sum x_i = 112.0000 \qquad \sum y_i = 27.14600$$

$$\sum x_i^2 = 2492.0000 \qquad \sum y_i^2 = 105.7889 \qquad \sum x_i y_i = 415.7410$$

Also ($N = 7$), $\bar{x} = 16.000000$ and $s_x = 10.801235$. By (2.13) and (2.14),

$$m = -0.026564 \qquad b = 4.30303$$

From (2.17), $s_d = 0.067390$; from (2.15), $s_m = 0.002547$. Since $t_{95.5} = 2.571$, the 95% confidence interval for m is

$$-0.0266 \pm (2.571)(0.00255) = -0.0266 \pm 0.0065$$

i.e., -0.0331 min^{-1} to -0.0201 min^{-1}. Since $t_{1/2} = -(\ln 2)/m$, the 95% confidence interval for the half-life is

$$\frac{-0.693}{-0.0331} = 20.9 \text{ min} \qquad \text{to} \qquad \frac{-0.693}{-0.0201} = 34.5 \text{ min}$$

The half-life of ^{38}Cl falls outside this range, while the half-life of ^{128}I is included; hence, it is much more probable that the halogen is ^{128}I.

2.29 In the potentiometric titration of Fe^{2+} with $Cr_2O_7^{2-}$, a plot of voltage versus volume of dichromate suggests that the equivalence-point volume is approximately 33.8 mL. Table 2-7 presents data from the first seven measurements at volumes greater than 33.9 mL; x (the more reliable data) is volume of dichromate, y is the derivative of x with respect to voltage. (a) Find the equivalence-point volume and the 95% CL precision estimate. (b) At the 95% CL, does the slope of the least-squares line agree with the theoretical value, which is 38.9 V^{-1}?

Table 2-7

x_i, mL	33.93	34.05	34.12	34.47	34.70	35.15	35.51
y_i, mL/V	5.8	10.1	15.3	24.6	40.7	53.3	64.6

(a) We have:

$$\sum x_i = 241.9300 \qquad \sum y_i = 214.4000$$

$$\sum x_i^2 = 8363.5753 \qquad \sum y_i^2 = 9645.4400 \qquad \sum x_i y_i = 7490.4280$$

$$\bar{x} = 34.5614 \qquad s_x = 0.59566290$$

From (2.13) and (2.14),
$$m = 37.793346 \text{ V}^{-1} \qquad b = -1275.5635 \text{ mL/V}$$

From (2.17), (2.15), and (2.16),
$$s_d = 2.753474 \qquad s_m = 1.887143 \qquad s_b = 65.230661$$

The least-squares line gives the equivalence-point volume as the abscissa corresponding to $y = 0$:

$$\hat{x} = \frac{-b}{m} = \frac{1275.5635}{37.793346} = 33.751 \text{ mL}$$

In (2.20), $M = \infty$ because $y = 0$ is a theoretical value; therefore, $s_{\hat{x}} = 0.04895$. From Appendix A, $t_{95.5} = 2.571$, and the 95% confidence interval for the equivalence point is

$$\hat{x} \pm t_{95.5} s_{\hat{x}} = 33.75 \pm (2.571)(0.04895) = 33.75 \pm 0.13 \quad \text{mL}$$

(b) The 95% confidence interval for the slope based on the experimental data is

$$m \pm t_{95.5} s_m = 37.79 \pm (2.571)(1.89) = 37.79 \pm 4.86 = 37.8 \pm 4.9 \quad \text{V}^{-1}$$

in agreement with the theoretical value.

Supplementary Problems

THE GAUSSIAN DISTRIBUTION

2.30 Analysis of 12.00 carat (50.00%) gold samples by a certain technique yields in the long term:

$$\mu \pm \sigma_x = 50.00\% \pm 0.14\%$$

How often would you expect to have an analysis above 50.42%?   *Ans.*   15 times in 10 000

2.31 Based on thousands of analyses, serum copper in healthy adults measures:

$$\mu \pm \sigma_x = 108.0 \pm 9.8 \quad \mu\text{g Cu/100 mL serum}$$

(a) What is the "normal range" (the 95.5% CL range)? (b) What percentage of healthy adults would have serum copper levels greater than 120.6? (c) What is the probability that the serum copper level of a healthy adult is between 117.8 and 124.1? (d) What percentage of healthy adults would register less than 60.0? (Persons having Wilson's disease generally register less than 60.0.)
Ans.   (a) 108.0 ± 19.6; (b) 10%; (c) 0.1085; (d) 0.00005%

2.32 Ten pounds of an impure sample of a solid weak acid was ground thoroughly to provide a homogeneous mixture. More than 10 000 portions of the mixture were analyzed for the wt % weak acid. It was found that 5.00% of the data indicated 72.00 wt% or greater, that 4.50% of the data were in the range 72.00 to 74.00 wt %, and that 5.00% of the data were in the range 71.22 to 72.00 wt %. Find $\mu \pm \sigma_x$.
Ans. 68.5 ± 2.1 wt %

2.33 The "normal range" (95.5% CL range) for [Na$^+$] in human whole blood is 0.1380 to 0.1460. (a) Find $\mu \pm \sigma_x$. (b) What range of [Na$^+$], centered on the mean, would encompass 50% of the data? (c) What percentage of healthy subjects would have [Na$^+$] greater than 0.1490? (d) Assume that a particular disease that occurs in one in 25 000 individuals is characterized by sodium levels greater than 0.1490. What are the chances that a person whose blood tests [Na$^+$] = 0.1500 has the disease?
Ans.   (a) 0.1420 ± 0.0020; (b) 0.1407 to 0.1433; (c) 0.035%; (d) 1 in 8.75

STUDENT'S *t* DISTRIBUTION

2.34 What is the 95% confidence interval for the value of μ if 21 analyses yield $\bar{x} \pm s_x = 101.0 \pm 11.3$?
Ans. 101.0 ± 5.1

2.35 Four measurements of the % KHP in a sample resulted in $\bar{x} \pm s_x = 62.3 \pm 1.2$. (a) What is the range for the true value, μ, at the 95% confidence level? (b) Find the value that would be exceeded by the next (fifth) measurement with probability 5/100 (1-tail).   *Ans.*   (a) 62.3 ± 1.9; (b) 65.5

2.36 Seven analyses for the phosphorus content of a fertilizer resulted in 16.2, 17.5, 15.4, 15.9, 16.8, 16.3, and 17.1%. Find (a) the standard deviation, s_x, and (b) the 95% confidence interval for the true value, μ.
Ans.   (a) 0.72; (b) 16.46 ± 0.67

TESTS FOR SIGNIFICANCE

2.37 The average [Na$^+$] in human whole blood is $\mu = 0.1420$. Analysis of ten whole-blood samples from healthy adults using a new analyzer gave $\bar{x} \pm s_x = 0.1400 \pm 0.0025$. Is $\bar{x}$ statistically consistent with the known μ?
Ans. $|t| = 2.5$, whereas $t_{95.9} = 2.262$. Therefore, there is a statistically significant difference.

2.38 A pooled blood sample is known to contain $\mu \pm \sigma_x = 85.00 \pm 0.20$ mg glucose per liter. A trained technologist wants to evaluate a new method of glucose analysis by performing 9 repeated tests of the blood sample. What is the widest range of measurements that would permit acceptance of the new method at the 95% confidence level?   *Ans.* $84.87 < \bar{x} < 85.13$ mg/L

2.39 A technician trainee in a clinical laboratory is released to do routine glucose analyses when the data obtained by the trainee agree at the 90% CL (two-tailed) with those of an experienced technologist. If the following results were obtained in two separate series of measurements:

$$\text{trainee} \qquad \bar{x} \pm s_x = 89.76 \pm 0.32 \quad \text{mg/L} \qquad (N_x = 7)$$
$$\text{technologist} \qquad \bar{y} \pm s_y = 89.43 \pm 0.25 \quad \text{mg/L} \qquad (N_y = 6)$$

should the trainee be allowed to do routine glucose analyses?
Ans. $|t| = 2.0 > t_{90,11} = 1.80$; the trainee should get more experience.

2.40 The technologist involved in the test described in Problem 2.39 wants to confirm the results, and repeats the test, this time using samples which are divided so that the technologist and trainee each analyze the same samples. A total of 10 such paired samples were analyzed and the average and standard deviation of the difference in glucose measurements was found to be:

$$\bar{x} \pm s_x = -0.3 \pm 0.4 \quad \text{mg/L}$$

Is this difference statistically significant? *Ans.* Yes: $|t| = 2.37 > 2.262 = t_{95,9}$.

PROPAGATION OF ERROR

2.41 The concentration of a NaOH solution is determined in a single experiment by titrating a weighed sample of pure, dried potassium acid phthalate (MW = 204.229) with the NaOH. The data are:

$$\text{weight KHP} = (11.6723 \pm 0.0001) - (10.8364 \pm 0.0001) \quad \text{g}$$
$$\text{volume NaOH} = (32.68 \pm 0.02) - (1.24 \pm 0.02) \quad \text{mL}$$

where the standard deviations are σ_x (infinite-sample) values. (a) What is the molarity of the NaOH? (b) Based on the propagation-of-error formula, what is the standard deviation of the molarity, σ_M?
Ans. (a) 0.1302 M; (b) $\sigma_M = 0.00012$ M

2.42 A mixture of five different chlorides is prepared for use as a chemical standard by weighing each chloride by difference and then combining the chlorides. The following weights, in grams, were obtained:

	NaCl	KCl	RbCl	$MgCl_2$	$CaCl_2$
$w_1 =$	3.6718	2.2774	3.1143	2.5682	1.6130
$w_2 =$	3.6203	2.2055	3.0759	2.5528	1.5866
$\Delta =$	0.0515	0.0719	0.0384	0.0154	0.0264

The balance used in the weighings is known to result in $\sigma_w = 0.2$ mg. (a) What is the value of σ_Δ for the weight of $MgCl_2$? (b) What is the value of $\sigma_{\Sigma\Delta}$ for the total weight of the sample? (c) What is the value of $\sigma_\%$ for the wt % $MgCl_2$ in the sample? *Ans.* (a) 0.2828 mg; (b) 0.632 mg; (c) 0.14 wt %

REJECTION OF DATA

2.43 Seven analyses were performed to determine the concentration of a solution of HCl. The data were: 0.1057, 0.1060, 0.1055, 0.1050, 0.1084, 0.1049, and 0.1040 M. (a) Can the 0.1084 M value be rejected on the basis of the Q-test at the 90% (two-tailed) confidence level? (b) If the 0.1084 M value can be rejected, can the 0.1040 M value be rejected next, using the same Q-test?
Ans. (a) spread/range = 0.546 > 0.507 = $Q_{90,7}$; therefore, reject. (b) spread/range = 0.450 < 0.560 = $Q_{90,6}$; therefore, do not reject.

LINEAR LEAST SQUARES

2.44 The absorbances (y) at 450 nm of five complexed copper solutions (of concentrations x) are fitted by a least-squares line having slope and vertical intercept

$$m \pm s_m = 0.0508 \pm 0.0086 \quad \text{ppm}^{-1} \qquad \text{and} \qquad b \pm s_b = 0.007 \pm 0.026$$

In addition, $s_d = 0.0272$ and $\bar{x} \pm s_x = 3.000 \pm 1.581$ ppm. A certain solution of complexed copper has an absorbance of 0.074. Find the 95% confidence interval for its concentration, if the absorbance is based on (a) a single measurement, and (b) the average of four measurements.

Ans.   (a) $M = 1$, $\hat{x} \pm t_{95,3}s_{\hat{x}} = 1.32 \pm 2.07$ ppm; (b) $M = 4$, $\hat{x} \pm t_{95,3}s_{\hat{x}} = 1.32 \pm 1.46$ ppm

2.45   The following data pertain to the region just past the equivalence point in the titration of a weak acid with NaOH:

x_i = mL NaOH	35.49	35.51	35.53	35.56	35.58	35.61	35.65	35.67	35.70
y_i = ΔmL/ΔpH	0.171	0.205	0.208	0.308	0.333	0.409	0.450	0.526	0.625

A plot of y versus x should, theoretically, be a straight line, with $y = 0$ at the equivalence point. (a) Find the 95% confidence interval for the equivalence point volume. (b) Is the 95% confidence interval for the slope of the least-squares line in accord with the theoretical value, 2.303?

Ans.   (a) $\hat{x} \pm t_{95,7}s_{\hat{x}} = 35.417 \pm 0.024$ mL; (b) Yes: $m \pm t_{95,7}s_m = 2.09 \pm 0.27$

Chapter 3

Free Energy and
Chemical Equilibrium

3.1 GIBBS FREE ENERGY

The *Gibbs free energy*, G, of a system is a thermodynamic state function, like the entropy, S, and the enthalpy, H. It is defined as

$$G = H - TS$$

where T is the (Kelvin) temperature of the system. It is useful to rewrite the definition in terms of the free-energy *change* accompanying some process. Letting

$$\Delta G \equiv G \text{ (final state)} - G \text{ (initial state)}$$

we have:

$$\Delta G = \Delta H - \Delta(TS)$$

If a thermodynamic quantity stays constant during the process, it is appended as a subscript. Thus, for instance,

$$\Delta G_T = \Delta H_T - T \, \Delta S_T$$

EXAMPLE 3.1 A simple, direct calculation of ΔG is possible only for a process involving an ideal gas. For any single substance, the first law of thermodynamics implies that

$$dG = V \, dP - S \, dT \tag{3.1}$$

or, at constant temperature,

$$dG_T = V \, dP \tag{3.2}$$

For n moles of an ideal gas, (3.2) may be integrated as:

$$\Delta G_T = \int_{P_1}^{P_2} V \, dP = \int_{P_1}^{P_2} \frac{nRT}{P} \, dP = nRT \ln \frac{P_2}{P_1}$$

Note that ΔG_T involves a pressure *ratio*. If P_1 is assigned the value 1.00 atm, defining the *standard state*, and if $P_2 = P$ is also measured in atm, we may write

$$\Delta G_T = nRT \ln (P/1.00) = RT \ln (P/1.00)^n \tag{3.3}$$

The importance of the Gibbs free energy resides in the following

free-energy principle $\Delta G_{P,T} \leq 0$

in which the *inequality* holds for a *spontaneous irreversible process* and the *equality* holds for a *reversible process*.

3.2 ACTIVITIES

In order to retain the simple form of (3.3) for substances other than ideal gases, a new term, *activity*, is defined via

$$\Delta G_T \equiv nRT \ln a = RT \ln a^n \tag{3.4}$$

46

The activity is dimensionless and is the ratio of a property in the actual state to that property in the standard state. Thus, $\Delta G_T = G_T - G_T^\circ$ is the change in free energy accompanying an isothermal transition from the standard state to the actual state.

Ideal Gases

Since the standard state for a gas is an ideal gas at 1.00 atm, $a = P$ (atm)/1.00 (atm), and (3.4) becomes (3.3). Most gases at pressures less than a few atmospheres may be considered ideal. At much higher pressures, the pressure in the actual state would be multiplied by a factor that corrects for non-ideality.

Pure Solids or Liquids

The standard state is $P^\circ = 1.00$ atm. For these materials, V can be taken as independent of P, so that (3.2) and (3.4) give

$$\Delta G_{T,V} = V(P - P^\circ) = nRT \ln a \tag{3.5}$$

By (3.5), the activity of the pure solid or liquid in the standard state is $a^\circ = 1.00$. For many substances (such as precipitates) at pressures smaller than 10 to 20 atm, $a \approx 1.00$.

Undissociated Species in Solution

Undissociated species X, such as weak acids, behave ideally at concentrations of about 5 M or less, and their activities can be approximated as the molarity divided by the standard-state concentration of an ideal 1.00 M solution: $a \approx M/(1.00 \text{ mol/L}) \equiv [X]$. (Here, as always, [ ] denotes *relative concentration*.)

Ions in Solution

Most ionic species $Z^\pm$ form ideal solutions at total ionic concentrations less than about 10^{-2} or 10^{-3} M. For such dilution, $a \approx [Z^\pm]$.

At higher concentrations, an *activity coefficient* is introduced to account for non-ideal effects:

$$a = f \cdot [Z^\pm] \tag{3.6}$$

In the literature, (3.6) is often stated in terms of the molality m; however, for aqueous solutions less concentrated than about 0.1 M, $m \approx M$—and we shall consider only such solutions. Values of f are well predicted from the *extended Debye-Hückel equation* (EDHE):

$$-\log f = \frac{Az^2\sqrt{I}}{1 + Bd\sqrt{I}} \tag{3.7}$$

Here, A and B are temperature-dependent constants, given in Table 3-1; I is the *ionic strength* of the solution, defined as

$$I = \frac{1}{2} \sum C_i z_i^2 \tag{3.8}$$

the summation including *all* ionic species in solution; C_i is the concentration (molarity or molality) of species i, whose ionic charge is z_i; and d is the "diameter," in angstroms, of the ion of interest, as given in Table 3-2. Table 3-3 is a listing of the EDHE f-values for ions of size d in aqueous solutions of ionic strength I at 25 °C.

Table 3-1

Temperature, °C	A	B
0	0.4883	0.3241
5	0.4921	0.3249
10	0.4960	0.3258
15	0.5000	0.3266
20	0.5042	0.3273
25	0.5085	0.3281
30	0.5130	0.3290
35	0.5175	0.3297
40	0.5221	0.3305
45	0.5270	0.3314
50	0.5319	0.3321
55	0.5371	0.3329
60	0.5425	0.3338

Table 3-2

d \ z	±1	±2	±3	±4
11				Th^{4+}, Ce^{4+}, Sn^{4+}
9	H^+		Al^{3+}, Fe^{3+}, Cr^{3+} Sc^{3+}, La^{3+}, Sm^{3+}	
8		Mg^{2+}, Be^{2+}		
6	Li^+, $C_6H_5COO^-$	Ca^{2+}, Cu^{2+}, Zn^{2+}, Sn^{2+} Mn^{2+}, Fe^{2+}, Ni^{2+}, Co^{2+}		
5	CCl_3COO^-	Sr^{2+}, Ba^{2+}, Cd^{2+}, Hg^{2+} Pb^{2+}, CO_3^{2-}, SO_3^{2-}, S^{2-} $(COO)_2^{2-}$, H $citrate^{2-}$	$citrate^{3-}$	$Fe(CN)_6^{4-}$
4	Na^+, IO_3^-, $H_2PO_4^-$, HCO_3^- HSO_3^-, $HAsO_4^-$, CH_3COO^-	Hg_2^{2+}, SO_4^{2-}, $S_2O_3^{2-}$ CrO_4^{2-}, HPO_4^{2-}, $S_2O_6^{2-}$	PO_4^{3-}, $Fe(CN)_6^{3-}$ $Co(NH_3)_6^{3+}$	
3	OH^-, F^-, CNS^-, HS^-, ClO_4^-, IO_4^- MnO_4^-, K^+, Cl^-, Br^-, I^-, CN^- NO_3^-, Rb^+, Cs^+, NH_4^+, Tl^+, Ag^+ $HCOO^-$, H_2 $citrate^-$, $CH_3NH_3^+$			

Table 3-3.   Activity Coefficients

d \ I	0.001	0.002	0.005	0.008	0.010	0.020	0.050	0.080	0.100
				Ionic charge $z = \pm 1$					
9	0.967	0.955	0.934	0.921	0.914	0.890	0.854	0.835	0.826
6	0.966	0.953	0.930	0.915	0.907	0.879	0.834	0.808	0.796
5	0.965	0.952	0.929	0.913	0.904	0.874	0.826	0.798	0.784
4	0.965	0.952	0.927	0.911	0.902	0.870	0.817	0.785	0.770
3	0.965	0.951	0.926	0.908	0.899	0.865	0.807	0.772	0.754
				Ionic charge $z = \pm 2$					
8	0.872	0.829	0.756	0.712	0.690	0.617	0.517	0.468	0.445
6	0.870	0.825	0.748	0.700	0.676	0.596	0.483	0.427	0.401
5	0.869	0.823	0.743	0.694	0.669	0.584	0.465	0.405	0.377
4	0.867	0.821	0.739	0.687	0.661	0.572	0.445	0.381	0.351
				Ionic charge $z = \pm 3$					
9	0.737	0.660	0.535	0.474	0.443	0.350	0.242	0.197	0.179
5	0.729	0.645	0.513	0.435	0.404	0.298	0.178	0.131	0.112
4	0.726	0.641	0.506	0.430	0.394	0.285	0.162	0.114	0.0949
				Ionic charge $z = \pm 4$					
11	0.588	0.486	0.348	0.282	0.252	0.173	0.0985	0.0727	0.0629
5	0.569	0.458	0.305	0.232	0.200	0.116	0.0467	0.0268	0.0202

EXAMPLE 3.2   Determine f_{H^+} and f_{OH^-} in a solution that is 0.0500 M HCl and 0.0600 M Ca(NO$_3$)$_2$ at 25 °C. Both HCl and Ca(NO$_3$)$_2$ dissociate completely in water; the ionic relative concentrations are:

$$[H^+] = 0.0500 \qquad [Cl^-] = 0.0500 \qquad [Ca^{++}] = 0.0600 \qquad [NO_3^-] = 0.1200$$

There will be negligible OH$^-$ in solution. The ionic strength is:

$$I = \tfrac{1}{2}\{[H^+](1^2) + [Cl^-](-1)^2 + [Ca^{++}](2)^2 + [NO_3^-](-1)^2\}$$
$$= \tfrac{1}{2}(0.0500 + 0.0500 + 0.2400 + 0.1200) = 0.2300$$

From Table 3-1, at 25 °C, $A = 0.5085$ and $B = 0.3281$. From Table 3-2, $d_{H^+} = 9$ and $d_{OH^-} = 3$. Therefore:

$$-\log f_{H^+} = \frac{(0.5085)(1^2)\sqrt{0.2300}}{1 + (0.3281)(9)\sqrt{0.2300}} = 0.10_1 \qquad \text{or} \qquad f_{H^+} = 0.79_3$$

$$-\log f_{OH^-} = \frac{(0.5085)(-1)^2\sqrt{0.2300}}{1 + (0.3281)(3)\sqrt{0.2300}} = 0.16_6 \qquad \text{or} \qquad f_{OH^-} = 0.68_3$$

3.3   CHEMICAL EQUILIBRIUM; EQUILIBRIUM CONSTANTS

Consider the general chemical reaction

$$\alpha A + \beta B + \cdots \rightarrow \lambda L + \mu M + \cdots \qquad (3.9)$$

which is supposed to occur at constant temperature and constant total pressure. The free-energy change,

$$\Delta G_{P,T} \equiv G_{P,T}(\text{products}) - G_{P,T}(\text{reactants})$$

in (3.9) is given by an expression analogous to (3.4):

$$\Delta(\Delta G_{P,T}) = \Delta G_{P,T} - \Delta G^\circ_{P,T} = RT \ln Q \qquad (3.10)$$

Here $\Delta G^\circ_{P,T}$ is the free-energy change in the hypothetical reaction (3.9) in which each reactant is in its standard state and each product is in its standard state; and Q, the *reaction quotient*, is given in terms of the actual activities by

$$Q = \frac{a_L{}^\lambda a_M{}^\mu \cdots}{\alpha_A{}^\alpha a_B{}^\beta \cdots} \qquad (3.11)$$

The free-energy principle (Section 3.1) tells us that if $\Delta G_{P,T}$ is negative, reaction (3.9) proceeds to the right, as written; if $\Delta G_{P,T}$ is positive, the reaction proceeds to the left; if $\Delta G_{P,T}$ is zero, the reaction proceeds to the right and to the left at equal rates. This last is the condition of (*dynamic*) *equilibrium*. At equilibrium, (3.10) gives

$$0 - \Delta G^\circ_{P,T} = RT \ln Q_{\text{eqm}} \qquad \text{or} \qquad Q_{\text{eqm}} = \exp\{-\Delta G^\circ_{P,T}/RT\} \equiv K$$

The number K is called the *thermodynamic equilibrium constant* because it depends only on thermodynamic quantities that stay fixed during the reaction. Rewriting (3.10) in terms of K,

$$\Delta G_{P,T} = RT \ln \frac{Q}{K} \qquad (3.12)$$

and (3.12) allows the quantitative prediction of whether a given combination of activities (i.e., starting concentrations or partial pressures) will result in the forward reaction, the backward reaction, or in chemical equilibrium.

Concentration-Based Equilibrium Constants

In the case of an ionic reaction, $a = fC$ [cf. (3.6)] for each ionic species. This factorization of a induces a like factorization of Q, and hence of K:

$$K = K_f K_C$$

whence $K_C = K/K_f$.

An ideal solution is characterized by $f = K_f = 1.00$ and $K = K_C$. It follows that any equilibrium expression derived for an ideal solution will be valid for a non-ideal solution if K is everywhere replaced by K/K_f.

EXAMPLE 3.3 Write the mass balance for a C_a M solution of the weak acid HA (with dissociation constant K_a) in terms of (a) [HA] and [A$^-$] and (b) [HA] and [H$^+$], if (i) the solution is ideal, and (ii) the solution is non-ideal.

(a) (i) $$[\text{HA}] + [\text{A}^-] = C_a$$

    (ii) Same as (i).

(b) (i) Since $K_a = [\text{H}^+][\text{A}^-]/[\text{HA}]$,

$$[\text{HA}] + \frac{K_a[\text{HA}]}{[\text{H}^+]} = C_a$$

(ii)
$$K_a = \frac{a_{\text{H}^+}\, a_{\text{A}^-}}{a_{\text{HA}}} = \frac{[\text{H}^+][\text{A}^-]}{[\text{HA}]} \frac{f_{\text{H}^+}\, f_{\text{A}^-}}{f_{\text{HA}}} = \frac{[\text{H}^+][\text{A}^-]}{[\text{HA}]} K_f$$

and so

$$[\text{HA}] + \frac{K_a[\text{HA}]}{K_f[\text{H}^+]} = C_a$$

which is the mass balance of (i) with K_a replaced by K_a/K_f.

COMBINING EQUILIBRIUM CONSTANTS

Equilibrium constants for individual reactions can be combined algebraically to yield the equilibrium constants for reactions which are combinations of the original reactions. The rules are: (1) If individual reactions are *added* together, then the overall equilibrium constant is the *product* of the equilibrium constants for the individual reactions. (2) If the *reverse* reaction is written, then the constant is the *reciprocal* of the constant for the forward reaction. (3) If the reaction is *doubled*, the equilibrium constant is *squared*; if the reaction is *tripled*, the equilibrium constant is *cubed*; etc. If the reaction is multiplied by one-half, then the new constant is the *square root* of the original constant, etc. More generally, if the reaction is multiplied by r (a rational number), then the equilibrium constant is raised to the rth power.

EXAMPLE 3.4 The equilibrium constant for the dissociation of hydrofluoric acid,

$$HF \rightarrow H^+ + F^-$$

is 6.76×10^{-4}; the equilibrium constant (solubility product) for lead fluoride,

$$PbF_2(s) \rightarrow Pb^{2+} + 2F^-$$

is 3.70×10^{-8}. Find the equilibrium constant for $Pb^{2+} + 2HF \rightarrow PbF_2(s) + 2H^+$.

This reaction can be written as the sum of twice the HF reaction and the reverse of the PbF_2 solubility:

$$2HF \rightarrow 2H^+ + 2F^- \qquad K = (6.76 \times 10^{-4})^2$$

$$Pb^{2+} + 2F^- \rightarrow PbF_2(s) \qquad K = \frac{1}{3.70 \times 10^{-8}}$$

$$\overline{Pb^{2+} + 2HF \rightarrow PbF_2(s) + 2H^+ \qquad K = \frac{(6.76 \times 10^{-4})^2}{3.70 \times 10^{-8}} = 12.4}$$

3.4 PRINCIPLE OF LE CHATELIER

The principle of Le Chatelier states that *when a stress is placed on a reaction mixture at equilibrium, the reaction will shift to relieve that stress.* Thus, starting with $Q = K$ in (*3.12*), if changes in the pressures or concentrations of any of the reactants or products make $Q > K$ ($Q < K$), the reaction will accelerate in the backward (forward) direction until the equilibrium state $Q = K$ is restored.

EXAMPLE 3.5 If 1.00 L of 0.100 M acetic acid is at equilibrium, how will the following changes affect the equilibrium *concentration* of H^+? (*a*) A few mL of 0.100 M HCl is added. (*b*) A small amount of solid sodium acetate is added. (*c*) The solution is diluted by a factor of three.

Acetic acid, HAc, is a weak acid, and a 0.1 M solution has a pH of about 2.9. The principal reaction taking place in solution is the dissociation of the acid: $HAc \rightarrow H^+ + Ac^-$.

(*a*) The addition of a product, H^+, causes the reaction to shift toward the reactants, resulting in a decrease in the temporary concentration of H^+. However, the shift toward the reactants does not consume all of the added H^+; the new equilibrium state has a higher H^+ concentration than prior to the addition of the HCl.

(*b*) The addition of product Ac^- also causes a shift toward the reactants. The new equilibrium concentration of H^+, therefore, will be lower.

(*c*) The dilution of an equilibrium mixture results in a temporary decrease in the concentrations of all species in solution and a resultant shift toward the side that contains the larger number of species (ions *and* undissociated molecules) in solution. [In the case of a gaseous reaction, an increase in volume (which is the same as diluting a solution) results in a shift to the side having the larger number of gaseous molecules.] In the present case, dilution will cause a shift toward the products, since there are two product species and only one reactant. This means that the *amount* (moles) of H^+ and the amount of Ac^- in solution will increase. However, the *concentration* of H^+ (and also the concentration of Ac^-) will decrease. This can be deduced intuitively by noting what would happen if continuing dilutions were made. In such a case, the solution must begin taking on, more and more, the properties of pure water. The pH must begin approaching 7.00, indicating that the H^+ concentration is decreasing.

The same conclusion can be reached by examining the reaction quotient. Prior to the three-fold dilution, the equilibrium concentrations were:

$$[H^+] = 1.34 \times 10^{-3} \qquad [Ac^-] = 1.34 \times 10^{-3} \qquad [HAc] = 0.0987$$

so that

$$K = \frac{(1.34 \times 10^{-3})(1.34 \times 10^{-3})}{0.0987}$$

Just after the dilution,

$$Q = \frac{[\frac{1}{3}(1.34 \times 10^{-3})][\frac{1}{3}(1.34 \times 10^{-3}]}{\frac{1}{3}(0.0987)} = \frac{K}{3}$$

Hence, $Q < K$, and the reaction shifts toward the products, H^+ and Ac^-. Let $x > 0$ be the number of moles of HAc that must dissociate to restore equilibrium, at which point

$$[HAc] = \frac{0.0987 - x}{3} \qquad [H^+] = [Ac^-] = \frac{(1.34 \times 10^{-3}) + x}{3}$$

Then the condition $Q = K$ takes the form

$$\frac{\left\{\dfrac{(1.34 \times 10^{-3}) + x}{3}\right\}^2}{\dfrac{0.0987 - x}{3}} = \frac{(1.34 \times 10^{-3})^2}{0.0987}$$

or

$$\left(1 + \frac{x}{1.34 \times 10^{-3}}\right)^2 = 3\left(1 - \frac{x}{0.0987}\right)$$

Solving the quadratic equation, $x = 0.97 \times 10^{-3}$ and

$$[H^+] = \frac{(1.34 + 0.97) \times 10^{-3}}{3} = 0.77 \times 10^{-3}$$

which is roughly one-half the original equilibrium concentration.

The principle of Le Chatelier may be extended to the case where the stress on the reaction is a change in temperature: one considers heat as a "reactant" in an endothermic reaction ($\Delta H > 0$), or as a "product" in an exothermic reaction ($\Delta H < 0$). Thus, increasing the "concentration" of heat (i.e., raising the temperature) will shift an endothermic reaction from reactants to products; whereas it will shift an exothermic reaction from products to reactants.

This effect can be described in terms of the equilibrium constant, whose temperature dependence is given by

$$\frac{d \ln K}{dT} = \frac{\Delta H^{\circ}_{P,T}}{RT^2} \tag{3.13}$$

If $\Delta H^{\circ}_{P,T}$, the standard-state heat of reaction, is assumed to be independent of temperature over the (small) temperature range T_1 to T_2, then integration of (3.13) yields

$$\ln \frac{K(T_2)}{K(T_1)} = -\frac{\Delta H^{\circ}_{P,T}}{R}\left(\frac{1}{T_2} - \frac{1}{T_1}\right) = \frac{\Delta H^{\circ}_{P,T}}{R}\left(\frac{T_2 - T_1}{T_1 T_2}\right) \tag{3.14}$$

If $\Delta H^{\circ}_{P,T} > 0$ (an endothermic standard reaction), (3.13) implies that $\ln K$—and hence K—is an increasing function of T. Therefore, a temperature rise will decrease Q/K, driving the reaction toward the products. On the other hand, if $\Delta H^{\circ}_{P,T} < 0$, a temperature rise will drive the reaction toward the reactants.

EXAMPLE 3.6   A reaction has $K(25°C) = 1.5 \times 10^{-6}$ and $\Delta H^\circ_{P,T} = 56.0$ kcal. Calculate $K(50°C)$.
From (3.14) and Table 1-3,

$$\ln \frac{K(50°C)}{1.5 \times 10^{-6}} = \frac{-56.0 \times 10^3 \text{ cal} \cdot \text{mol}^{-1}}{1.987 \text{ cal} \cdot \text{mol}^{-1} \cdot \text{K}^{-1}} \frac{25 \text{ K}}{(298 \text{ K})(323 \text{ K})} = -7.3_2$$

Thus, $K(50°C) = (1.5 \times 10^{-6})(6._6 \times 10^{-4}) = 9._9 \times 10^{-10}$.

Solved Problems

FREE ENERGY AND EQUILIBRIUM

3.1   The following reaction of ideal gases occurs at a temperature T for which $K = 4.00$:

$$A(g) + 2B(g) \rightarrow C(g) + D(g)$$

A vessel at temperature T contains a mixture of the gases, with $P_A = 3.00$ atm, $P_B = 2.00$ atm, $P_C = 5.00$ atm, and $P_D = 4.00$ atm. In which direction will the reaction proceed spontaneously?

$$Q = \frac{(5.00)(4.00)}{(3.00)(2.00)^2} = 1.67$$

which is less than K. Therefore, by (3.12), $\Delta G_{P,T} < 0$ and the reaction will proceed spontaneously from reactants toward products.

3.2   An ideal-gas reaction, $3A(g) + 2B(g) \rightarrow 2C(g) + D(g)$ has $K = 3.86 \times 10^3$ at $T = 300$ K. A gas mixture at 300 K has $P_A = 60.0$ torr, $P_B = 400$ torr, $P_C = 100$ torr, and $P_D = 550$ torr. In which direction will the reaction proceed spontaneously?

The activity of an ideal gas is the ratio of its pressure to the standard-state pressure, 1 atm = 760 torr. Thus,

$$Q = \frac{(100/760)^2(550/760)}{(60.0/760)^3(400/760)^2} = 92.6$$

Since Q is less than K, the reaction will proceed spontaneously from reactants toward products.

3.3   The equilibrium constant for the dissociation of acetic acid at 25 °C is 1.80×10^{-5}. If you add 0.100 mol of undissociated HAc and 0.500 mol of NaAc to 1.00 L of dissociated water ($[H^+] = [OH^-] = 1.00 \times 10^{-7}$), in which direction will the acetic acid reaction proceed spontaneously?

The initial activities (relative concentrations) are $[H^+] = 1.00 \times 10^{-7}$, $[Ac^-] = 0.500$, and $[HAc] = 0.100$. Therefore,

$$Q = \frac{(1.00 \times 10^{-7})(0.500)}{0.100} = 5.00 \times 10^{-7}$$

which is less than K. The reaction will proceed spontaneously from reactants toward products.

ACTIVITY COEFFICIENTS

3.4 $H_2O(l)$ is about as incompressible as steel; that is, its volume is essentially independent of pressure. What is the activity of $H_2O(l)$ at 25 °C and (a) 0.1 atm?   (b) 1.00 atm?   (c) 10 atm? (d) 100 atm?   (e) 1000 atm?

The constant density of H_2O is 1.0 g/mL and the molecular weight is 18.0; therefore, 1 mol of water occupies $V = 18.0$ mL. For a transition from the standard state $P° = 1.00$ atm to P(atm), (3.5) and Table 1-3 give

$$(18.0)(P - 1.00) = (1.00)(82.05)(298)(\ln a)$$

where both sides are in $mL \cdot mol^{-1} \cdot atm$. Substituting the given pressures, we find: (a) $a = 0.999264$, (b) $a = 1.00$ (the standard state), (c) $a = 1.0066$, (d) $a = 1.076$, (e) $a = 2.09$.

3.5   At 25 °C, the reaction $Br^- + AgCl(s) \rightarrow Cl^- + AgBr(s)$ has $K = 399$. The value of d in the EDHE is, from Table 3-2, 3 for both Br^- and Cl^-. Why is the concentration ratio $[Cl^-]/[Br^-]$ independent of the ionic strength of the solution and independent of whether the solution is considered ideal or non-ideal and obeying the EDHE?

The equilibrium constant for the reaction is

$$K = \frac{[Cl^-] f_{Cl^-} a_{AgBr}}{[Br^-] f_{Br^-} a_{AgCl}} = 399$$

At normal pressures, both a_{AgCl} and a_{AgBr} can be assumed to equal 1.00. If the solution is ideal, then

$$f_{Cl^-} = f_{Br^-} = 1.00$$

and $K = 399 = [Cl^-]/[Br^-]$. If the solution is non-ideal but EDHE applies, then, since the terms in EDHE are identical for f_{Cl^-} and f_{Br^-} (for any ionic strength),

$$f_{Cl^-} = f_{Br^-}$$

and, again, $K = 399 = [Cl^-]/[Br^-]$.

3.6   The reaction $CaF_2(s) + Sr^{2+} \rightarrow SrF_2(s) + Ca^{2+}$ has $K = 1.43 \times 10^{-2}$. The activities of $CaF_2(s)$ and $SrF_2(s)$ can each be considered equal to 1.00 and independent of the ionic strength. Consider two solutions undergoing the above reaction, one with an ionic strength $I_1 = 0.0100$ and another with $I_2 = 0.0500$, the ionic concentrations being determined primarily by added NaCl. Compare the values of $a_{Ca^{2+}}/a_{Sr^{2+}}$ for the two solutions.

Since

$$K = \frac{a_{SrF_2} a_{Ca^{2+}}}{a_{CaF_2} a_{Sr^{2+}}} = \frac{a_{Ca^{2+}}}{a_{Sr^{2+}}}$$

the activity ratio is the same for all ionic strengths.

3.7   Compare the values of the activity coefficient ratio $f_{Ca^{2+}}/f_{Sr^{2+}}$ for the two solutions of Problem 3.6.

According to the EDHE, the activity coefficient of an ion decreases with an increase in ionic strength. Moreover, the smaller the ionic size, d, the greater the decrease in f for a given increase in ionic strength. For this reaction, $d = 5$ for Sr^{2+} and $d = 6$ for Ca^{2+}. At $I = 0.0100$,

$$\text{ratio} = \frac{0.676}{0.669} = 1.010$$

At $I = 0.0500$,

$$\text{ratio} = \frac{0.483}{0.465} = 1.039$$

The ratio increases with an increase in ionic strength.

3.8 Compare the values of the concentration ratio $[Ca^{2+}]/[Sr^{2+}]$ for the two solutions of Problem 3.6.

> Since (Problem 3.6) the activity ratios are the same for the two ionic strengths, the concentration ratios must differ in an inverse manner to the activity coefficient ratios. That is to say, the concentration ratio must decrease with increasing ionic strength, because (Problem 3.7) the activity coefficient ratio increases.

3.9 At 25 °C, the ion product for water is $K_w = 1.00 \times 10^{-14} = a_{H^+} a_{OH^-}$. In pure water,

$$f_{H^+} = f_{OH^-} = 1.00$$

since the ionic strength is so very low. What are the activity coefficients for H^+ and OH^- in a 0.0300 M aqueous solution of $MgCl_2$?

> The $[H^+]$ and $[OH^-]$ will be negligibly small and the ionic strength will be determined by the $MgCl_2$:
>
> $$I = \tfrac{1}{2}([Mg^{2+}]2^2 + [Cl^-]1^2) = \tfrac{1}{2}\{(0.0300)(4) + (0.0600)(1)\} = 0.0900$$
>
> The activity coefficients are now determined from (3.7) and Tables 3-1 and 3-2 as
>
> $$-\log f_{H^+} = \frac{(0.5085)(1^2)\sqrt{0.0900}}{1 + (0.3281)(9)\sqrt{0.0900}} = 0.080_9 \qquad \text{or} \qquad f_{H^+} = 0.830$$
>
> $$-\log f_{OH^-} = \frac{(0.5085)(-1)^2)\sqrt{0.0900}}{1 + (0.3281)(3)\sqrt{0.0900}} = 0.11_8 \qquad \text{or} \qquad f_{OH^-} = 0.76_3$$

3.10 Find $[H^+]$ and a_{H^+} for the solution of Problem 3.9.

> Charge balance requires $[H^+] = [OH^-]$; hence,
>
> $$K_w = [H^+]f_{H^+}[OH^-]f_{OH^-} = [H^+]^2 f_{H^+} f_{OH^-}$$
> $$1.00 \times 10^{-14} = [H^+]^2(0.8301)(0.7625)$$
>
> Solving, $[H^+] = 1.26 \times 10^{-7}$ and
>
> $$a_{H^+} = [H^+]f_{H^+} = (1.26 \times 10^{-7})(0.830) = 1.04 \times 10^{-7}$$

3.11 The principal uncertainties in the calculation of activity coefficients by EDHE lie in the ionic size, d, and the ionic strength, I. Assume that a particular EDHE calculation reduces to

$$-\log f = \frac{0.253 \pm 0.008}{3.50 \pm 0.03}$$

where the precision limits are σ_x-values. Use the method of Section 2.7 to determine $\bar{f} \pm \sigma_f$.

> We have
>
> $$-\log \bar{f} = \frac{0.253}{3.50} = 0.0723 \qquad \text{or} \qquad \bar{f} = 0.8467.$$
>
> By (2.9), the uncertainty in
>
> $$\sigma_F = (0.0723)\sqrt{\left(\frac{0.008}{0.253}\right)^2 + \left(\frac{0.03}{3.50}\right)^2} = 0.00237$$
>
> To evaluate σ_f, use (2.10) in reverse ($K = -1$):
>
> $$\sigma_f = \frac{\bar{f}}{0.434}\sigma_F = \frac{0.8467}{0.434}(0.00237) = 0.0046$$
>
> Thus, $\bar{f} \pm \sigma_f = 0.8467 \pm 0.0046 = 0.847 \pm 0.005$.

COMBINING EQUILIBRIUM CONSTANTS

3.12   The reactions

$$Ag_3PO_4(s) \rightarrow 3Ag^+ + PO_4^{3-} \qquad \text{and} \qquad AgCl(s) \rightarrow Ag^+ + Cl^-$$

have respective equilibrium constants 1.30×10^{-20} and 1.78×10^{-10}. Find K for

$$Ag_3PO_4(s) + 3Cl^- \rightarrow 3AgCl(s) + PO_4^{3-}$$

The reaction of interest can be obtained by adding three times the reverse of the $AgCl(s)$ solubility reaction to the $Ag_3PO_4(s)$ solubility equation. Therefore,

$$K = \frac{(K_{sp} \text{ for } Ag_3PO_4)}{(K_{sp} \text{ for } AgCl)^3} = \frac{1.30 \times 10^{-20}}{(1.78 \times 10^{-10})^3} = 2.31 \times 10^9$$

3.13   Given

$$\begin{aligned}
Ca(OH)_2(s) &\rightarrow Ca^{2+} + 2OH^- & K &= 5.50 \times 10^{-6} \\
CaF_2(s) &\rightarrow Ca^{2+} + 2F^- & K &= 4.00 \times 10^{-11} \\
HF &\rightarrow H^+ + F^- & K &= 6.76 \times 10^{-4}
\end{aligned}$$

find K for $Ca(OH)_2(s) + 2HF \rightarrow CaF_2(s) + 2H_2O$.

The reaction can be formed as the sum of

$$\begin{aligned}
Ca(OH)_2 &\rightarrow Ca^{2+} + 2OH^- & K &= 5.50 \times 10^{-6} \\
2HF &\rightarrow 2H^+ + 2F^- & K &= (6.76 \times 10^{-4})^2 \\
Ca^{2+} + 2F^- &\rightarrow CaF_2(s) & K &= 1/(4.00 \times 10^{-11}) \\
2H^+ + 2OH^- &\rightarrow 2H_2O & K &= 1/(1.00 \times 10^{-14})^2
\end{aligned}$$

Hence, for the sum,

$$K = \frac{(5.50 \times 10^{-6})(6.76 \times 10^{-4})^2}{(4.00 \times 10^{-11})(1.00 \times 10^{-14})^2} = 6.28 \times 10^{26}$$

3.14   What is the equilibrium constant for the reaction

$$3Ag_2S(s) + Ba_3(AsO_4)_2(s) + 3H^+ \rightarrow 2Ag_3AsO_4(s) + 3HS^- + 3Ba^{2+}$$

given that the second dissociation constant for H_2S is 1.10×10^{-15} and solubility products for the solids are: $Ba_3(AsO_4)_2$, 7.70×10^{-51}; Ag_2S, 2.00×10^{-49}; Ag_3AsO_4, 1.00×10^{-22}?

The reaction can be obtained by adding three times the solubility equation for Ag_2S, the solubility equation for $Ba_3(AsO_4)_2$, three times the reverse of the second dissociation for H_2S, and twice the reverse of the solubility expression for Ag_3AsO_4. This results in:

$$\begin{aligned}
3Ag_2S(s) &\rightarrow 6Ag^+ + 3S^{2-} & K &= (2.00 \times 10^{-49})^3 \\
Ba_3(AsO_4)_2(s) &\rightarrow 3Ba^{2+} + 2AsO_4^{3-} & K &= 7.70 \times 10^{-51} \\
3H^+ + 3S^{2-} &\rightarrow 3HS^- & K &= 1/(1.10 \times 10^{-15})^3 \\
6Ag^+ + 2AsO_4^{3-} &\rightarrow 2Ag_3AsO_4(s) & K &= 1/(1.00 \times 10^{-22})^2
\end{aligned}$$

$$\overline{3Ag_2S(s) + Ba_3(AsO_4)_2(s) + 3H^+ \rightarrow 2Ag_3AsO_4(s) + 3HS^- + 3Ba^{2+}}$$

and

$$K = \frac{(2.00 \times 10^{-49})^3(7.70 \times 10^{-51})}{(1.00 \times 10^{-22})^2(1.10 \times 10^{-15})^3} = 4.63 \times 10^{-108}$$

(You must be careful in solving for K. Depending on your method of handling the very small exponents, you could exceed the 10^{-99} to 10^{99} range of your pocket calculator.)

PRINCIPLE OF LE CHATELIER

3.15 One liter of the following reaction mixture is at equilibrium at 25 °C:

$$Cu(s, 3.5\ g) + UO_2^{2+}\ (0.090\ M) + 4H^+\ (0.25\ M) \rightarrow Cu^{2+}\ (0.070\ M) + U^{4+}\ (0.0040\ M) + 2H_2O$$

Describe the effect on $[Cu^{2+}]$ of adding to the equilibrium mixture (a) a small amount of soluble $UO_2(NO_3)_2$; (b) a small amount of solid NaOH; (c) a few drops of concentrated acetic acid; (d) 1.0 g of metallic Cu; (e) a small amount of solid NaCl; (f) sufficient water to bring the volume to 2 L.

(a) UO_2^{2+} is a reactant, and the addition of it to the solution would cause the reaction to shift toward the products. The Cu^{2+} concentration would increase.

(b) The NaOH will react with some of the reactant H^+, causing the reaction to shift toward the reactants. The Cu^{2+} concentration will decrease.

(c) The acetic acid is a source of H^+ and will increase the concentration of this reactant. This will cause the concentration of Cu^{2+} to increase.

(d) The only requirement for equilibrium is that there be *some* solid Cu present; the amount is immaterial. The addition of more Cu metal has no effect on the equilibrium concentration of Cu^{2+}.

(e) Neither Na^+ nor Cl^- participates in the equilibrium reaction. Therefore there would be no effect on the concentration of Cu^{2+}. [If non-ideal effects are considered, then the addition of NaCl would change the ionic strength of the solution and so affect the activity coefficients of the ions. This might have a slight effect on the equilibrium concentration of Cu^{2+}.]

(f) The reaction will shift to the side having the larger number of species in solution. There are five reactant species in solution (UO_2^{2+} and four H^+) and two product species in solution (Cu^{2+} and U^{4+}). The reaction will shift toward the reactants and both the total amount and the concentration of Cu^{2+} will decrease.

3.16 At 25 °C, $K = 1.78 \times 10^{-10}$ and $\Delta H^\circ_{P,T} = +65.0\ kJ \cdot mol^{-1}$ for the reaction

$$AgCl(s) \rightarrow Ag^+ + Cl^-$$

What is the value of K for this reaction at 50 °C, if you assume that $\Delta H^\circ_{P,T}$ is independent of temperature between 25 °C and 50 °C?

From (3.14) and Table 1-3,

$$\ln \frac{K(50°C)}{1.78 \times 10^{-10}} = \frac{65.0 \times 10^3}{8.314} \left[\frac{25}{(298)(323)} \right]$$

whence $K(50°C) = 1.3_6 \times 10^{-9}$.

3.17 What is $\Delta G_{P,T}$ at 50 °C for $AgCl(s) \rightarrow Ag^+\ (0.250\ M) + Cl^-\ (0.465\ M)$, in a solution of ionic strength such that $f_{Ag^+} = 0.826$ and $f_{Cl^-} = 0.826$? Use any necessary information from Problem 3.16.

The reaction quotient is

$$Q = \frac{a_{Ag^+}\ a_{Cl^-}}{a_{AgCl}} = \frac{[(0.826)(0.250)][(0.826)(0.465)]}{1.00}$$

and (3.12) gives

$$\Delta G_{P,T} = (8.314)(323) \ln \frac{(0.826)^2(0.250)(0.465)}{1.36 \times 10^{-9}} = 4.80 \times 10^4\ J/mol$$

Observe that at 50 °C the reaction goes in the direction of recombination.

3.18   Explain why the solubility of $AgCl(s)$ at 25 °C increases upon the addition of a salt such as $NaNO_3$, which has nothing in common with the AgCl.

Moderate additions of a salt such as $NaNO_3$ increase the ionic strength of the solution, and therefore (see Table 3-3) decrease both activity coefficients f_{Ag^+} and f_{Cl^-}. Since

$$[Ag^+]\, f_{Ag^+}\, [Cl^-]\, f_{Cl^-} = K = \text{constant}$$

a decrease in f_{Ag^+} and f_{Cl^-} requires a corresponding increase in $[Ag^+]$ and $[Cl^-]$.

Supplementary Problems

FREE ENERGY AND EQUILIBRIUM

3.19   At 25 °C, the reaction

$$Ca_3(AsO_4)_2(s) + 6OH^- (0.100\ M) \rightarrow 3Ca(OH)_2(s) + 2AsO_4^{3-} (0.0200\ M)$$

has $K = 4.09 \times 10^{-3}$. Assume that the solution is ideal. In which direction will the reaction proceed spontaneously?
Ans. $Q = 400 > K$. Therefore, $\Delta G_{P,T} > 0$ and the reaction proceeds spontaneously in the reverse direction.

3.20   At 25 °C, the reaction

$$PbCl_2(s) + 2HF(0.300\ M) \rightarrow PbF_2(s) + 2H^+ (0.0150\ M) + 2Cl^- (0.0500\ M)$$

has $K = 1.98 \times 10^{-4}$. Assume that the solution is ideal. In which direction will the reaction proceed spontaneously?
Ans. $Q = 6.25 \times 10^{-6} < K$. Therefore, $\Delta G_{P,T} < 0$ and the reaction will proceed as written.

ACTIVITY COEFFICIENTS

3.21   Determine the activity coefficients and the value of Q for the reaction mixture of Problem 3.20, when the solution is considered non-ideal and its ionic strength is $I = 0.100$.
Ans. $f_{HF} = 1.00$, $f_{H^+} = 0.826$, $f_{Cl^-} = 0.754$; $Q = 2.42 \times 10^{-6}$

3.22   At 25 °C, find (*a*) the activity coefficients, (*b*) Q, and (*c*) the direction, of a spontaneous reaction for the mixture

$$AgCl(s) \rightarrow Ag^+ (1.50 \times 10^{-6}\ M) + Cl^- (1.30 \times 10^{-4}\ M)$$

The solution has ionic strength $I = 0.0500$, and $K(25°C) = 1.78 \times 10^{-10}$.
Ans.   (*a*) $f_{Ag^+} = 0.807$, $f_{Cl^-} = 0.807$; (*b*) $Q = 1.27 \times 10^{-10}$; (*c*) forward

COMBINING EQUILIBRIUM CONSTANTS

3.23   The solubility product for $Ca_3(AsO_4)_2(s)$ is $K = 6.80 \times 10^{-19}$. What is the solubility product for $Ca(OH)_2(s)$, given the equation and K-value of Problem 3.19?   *Ans.* 5.50×10^{-16}

3.24   The solubility product for $PbCl_2(s)$ is $K = 1.60 \times 10^{-5}$; the solubility product for $PbF_2(s)$ is $K = 3.70 \times 10^{-8}$. What is the dissociation constant for $HF \rightarrow H^+ + F^-$, given the equation and K-value of Problem 3.20?   *Ans.* 6.77×10^{-4}

PRINCIPLE OF LE CHATELIER

3.25 One liter of the reaction mixture

$$PbF_2(s) + C_2O_4^{2-}\ (3.00 \times 10^{-4}\ M) \rightarrow PbC_2O_4(s) + 2F^-\ (0.700\ M)$$

is at equilibrium at 25°C. What would be the effect on the concentration of $C_2O_4^{2-}$ if some NaF were added to the equilibrium mixture?   *Ans.*   The $C_2O_4^{2-}$ concentration would increase.

3.26 What would be the effect on the concentration of $C_2O_4^{2-}$ in the reaction of Problem 3.25 if 0.5 g of $PbF_2(s)$ were added to the equilibrium mixture?   *Ans.*   no effect

3.27 What would be the effects on the concentration and the amount of $C_2O_4^{2-}$ in the reaction of Problem 3.25 if the solution were diluted by a factor of two?
Ans.   Both the concentration and amount of $C_2O_4^{2-}$ would decrease.

3.28 What would be the effects on the concentration and the amount of F^- in the reaction of Problem 3.25 if the solution were diluted by a factor of two?
Ans.   The concentration of F^- would decrease; the amount of F^- would increase.

3.29 What would be the effect on the concentration of $C_2O_4^{2-}$ in the reaction of Problem 3.25 if some NaNO₃ were added to the equilibrium mixture? (Note that $Pb(NO_3)_2$ is soluble.)
Ans.   No effect, assuming that the solution is ideal and ionic-strength effects do not exist.

Chapter 4

Strong Acids–Strong Bases

4.1 THE WATER EQUILIBRIUM

Ion Product for Water

By definition, a *strong* acid or base is completely dissociated in aqueous solution (provided the concentration is not too great). Thus, to calculate the pH of such a solution, only one equilibrium, the dissociation of water, need be considered. The thermodynamic equilibrium constant for the dissociation of water,

$$H_2O \rightarrow H^+ + OH^-$$

(more precisely, $H_2O + H_2O \rightarrow H_3O^+ + OH^-$; but we may ignore the water of hydration) is given by

$$K = \frac{a_{H^+}\, a_{OH^-}}{a_{H_2O}} = \frac{[H^+]\, f_{H^+}\, [OH^-]\, f_{OH^-}}{[H_2O]\, f_{H_2O}} \tag{4.1}$$

The denominator of (4.1) can be considered a constant if the solute concentration is less than about 1 M and the pressure is less than about 10 atm. There are two reasons for this: the activity coefficient of H_2O can be approximated as 1.00 for pressures less than about 10 atm (Section 3.2) and the concentration of H_2O is insensitive to the addition of a limited amount of solute.

EXAMPLE 4.1 At 25 °C, the density of pure water is 997 kg/m^3 = 997 g/L, and so

$$M_w = \frac{997 \text{ g/L}}{18.0 \text{ g/mol}} = 55.4 \text{ mol/L}$$

A 0.102 M solution of NH_4Br in water at 25°C has density 1002.7 kg/m^3 and water concentration 55.2 mol/L.

We may therefore introduce the approximation that the denominator of (4.1) is a constant and may specify the water equilibrium in terms of the *ion product*

$$K_w \equiv K\,[H_2O]\, f_{H_2O} = a_{H^+} a_{OH^-} = [H^+]\, f_{H^+}\, [OH^-]\, f_{OH^-} \tag{4.2}$$

Temperature Effects

Energy is required to dissociate water into H^+ and OH^-. Therefore, as the temperature of water is increased, the following occurs: the reaction shifts toward the products H^+ and OH^-, increasing their concentration; the pH of pure water (a neutral solution), defined as

$$pH \equiv -\log a_{H^+} = -\log [H^+] - \log f_{H^+} \tag{4.3}$$

decreases, since $[H^+]$ increases and $f_{H^+} \approx 1.00$; the ion product for water increases, so that pK_w decreases. These temperature effects are illustrated in Table 4-1, due to H. S. Harned and W. J. Hamer. In this Outline, we take $K_w = 1.00 \times 10^{-14}$ and $pK_w = 14.000$ at 25 °C.

EXAMPLE 4.2 At 35 °C, what is the pH of (*a*) 0.100 M HCl? (*b*) 0.100 M NaOH? Assume that the activity coefficients are 1.00.

(*a*) $[H^+] = 0.100$, $a_{H^+} = 0.100$, and so, by (4.3), pH = 1.000.

(*b*) $[OH^-] = 0.100$, $a_{OH^-} = 0.100$, pOH = 1.000, and so, using Table 4-1,

$$pH = pK_w - pOH = 13.68 - 1.000 = 12.68$$

Table 4-1

Temperature °C	K_w	pK_w	pH = pOH of pure water	$[H^+] = [OH^-]$ of pure water
0	$1.1_4 \times 10^{-15}$	14.94	7.47	3.4×10^{-8}
10	$2.9_2 \times 10^{-15}$	14.53	7.27	5.4×10^{-8}
20	$6.8_1 \times 10^{-15}$	14.17	7.08	8.2×10^{-8}
25	$1.0_1 \times 10^{-14}$	14.00	7.00	1.0×10^{-7}
35	$2.0_9 \times 10^{-14}$	13.68	6.84	1.4×10^{-7}
50	$5.4_7 \times 10^{-14}$	13.26	6.63	2.3×10^{-7}
60	$9.6_1 \times 10^{-14}$	13.02	6.51	3.1×10^{-7}

4.2 CALCULATION OF pH

We shall derive a general equation for the hydrogen-ion concentration in a solution prepared by mixing V_a volume units of C_{sa}-molar strong acid, HA, with V_b volume units of C_{sb}-molar strong base, MOH. Assume that the solutions are ideal (unit activity coefficients). Determination of the unknown $[H^+]$, $[OH^-]$, $[M^+]$, and $[A^-]$ requires four equations; these are as follows.

1. Charge balance:

$$[H^+] + [M^+] = [OH^-] + [A^-]$$

2. The ion product for water:

$$K_w = [H^+][OH^-]$$

3. Mass balance for A:

$$[A^-](V_a + V_b) = C_{sa}V_a$$

4. Mass balance for M:

$$[M^+](V_a + V_b) = C_{sb}V_b$$

Elimination of $[OH^-]$, $[M^+]$, and $[A^-]$ among these equations produces a quadratic equation for $[H^+]$, which can be written in two convenient forms:

$$[H^+]^2 + \frac{V_bC_{sb} - V_aC_{sa}}{V_a + V_b}[H^+] - K_w = 0 \qquad (4.4a)$$

$$V_b\left(C_{sb} + [H^+] - \frac{K_w}{[H^+]}\right) = V_a\left(C_{sa} - [H^+] + \frac{K_w}{[H^+]}\right) \qquad (4.4b)$$

Observe that at the stoichiometric equivalence point, the coefficient of $[H^+]$ in (4.4a) vanishes; thus, the equation yields (at 25 °C)

$$[H^+] = \sqrt{K_w} = 1.00 \times 10^{-7} \qquad \text{or} \qquad pH = 7.000$$

as it must.

Equation (4.4b) is useful in plotting a titration curve. Having specified a pH (or hydrogen-ion concentration), you calculate $V_b \equiv V_T$ (if it is a titration of an acid with a base) or $V_a \equiv V_T$ (if it is a titration of a base with an acid). Figure 4-1 was constructed in this manner, using a small programmable calculator.

EXAMPLE 4.3 What volume of 0.200 M NaOH must be added to 40.00 mL of 0.100 M HCl to produce a solution having pH = 10.000?

If pH = 10.000, $[H^+] = 1.00 \times 10^{-10}$. Solving (4.4b), $V_b = 20.03$ mL. This is 0.03 mL (about one drop) past the equivalence point.

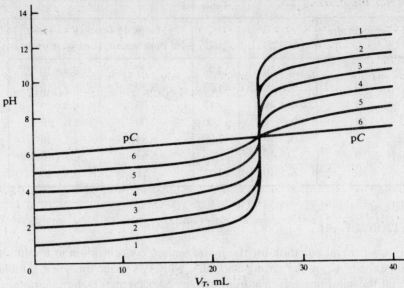

(a)   Titration of 25.00 mL of C M HCl with C M NaOH

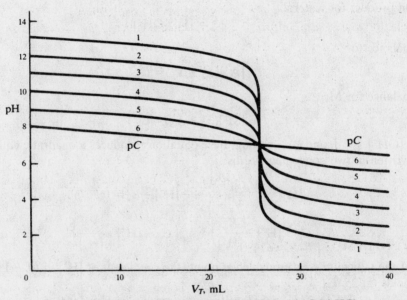

(b)   Titration of 25.00 mL of C M NaOH with C M HCl

Fig. 4-1

When the strong acid–base mixture is basic, the numerical solution of (4.4a) by the quadratic formula (1.4) requires more precision than is available with pocket calculators. In this case, it is better to use the analogous quadratic equation for $[OH^-]$:

$$[OH^-]^2 + \left[\frac{V_a C_{sa} - V_b C_{sb}}{V_a + V_b}\right][OH^-] - K_w = 0 \qquad (4.5)$$

Neglecting the Water Equilibrium

Most pH calculations are considerably simplified if the water equilibrium can be neglected as a source of H^+ or OH^-. The conditions (at $25\,°C$) for this important simplifying assumption are determined as follows.

If the overall solution is acidic, there must be a source of H^+ in addition to the dissociation of water. The water equilibrium and the starting and equilibrium concentrations of H^+ and OH^- may be indicated as:

$$\begin{array}{cccc} C\ start & A+10^{-7} & & 10^{-7} \\ H_2O \longrightarrow & H^+ & + & OH^- \\ C\ equil & A+10^{-7}-w & & 10^{-7}-w \end{array}$$

The presence of A moles of H^+ per liter causes the ion product for water to be exceeded; to restore equilibrium, the reaction shifts toward the reactant, consuming w moles each of H^+ and OH^- per liter. We see that the error in $[H^+]$ caused by neglecting the water equilibrium contribution, $10^{-7}-w$, will be less than 1%, provided

$$10^{-7}-w < 0.01[H^+]$$
$$[OH^-] < 0.01[H^+]$$
$$[H^+][OH^-] < 0.01[H^+]^2$$
$$1.00 \times 10^{-14} < 0.01[H^+]^2$$
$$1.00 \times 10^{-6} < [H^+]$$

that is, provided $pH < 6.000$.

If the overall solution is basic, a similar analysis gives the condition $pH > 8.000$.

If we neglect the water equilibrium, then, for acidic solutions, $[H^+]$ is determined only by the excess acid; (4.4a) becomes

$$[H^+] = \frac{V_a C_{sa} - V_b C_{sb}}{V_a + V_b} \tag{4.6}$$

For basic solutions, $[OH^-]$ is determined only by the excess base; (4.5) becomes

$$[OH^-] = \frac{V_b C_{sb} - V_a C_{sa}}{V_a + V_b} \tag{4.7}$$

One can use (4.6) and (4.7) instead of (4.4a) and (4.5) at all stages of a strong acid–strong base titration, except in the vicinity of the equivalence point, where the pH falls between 6.0 and 8.0.

4.3 TITRATION EQUIVALENCE POINT

As is seen in Fig. 4-1(a), there is a sharp increase in the pH for a small increase in the volume of added base near the equivalence point for solutions as dilute as $10^{-5}\,M$. This sharp increase, which allows locating the equivalence point on the graph, is due to the fact that the hydrogen-ion concentration has become very small and there is enough hydroxide in one drop of NaOH to neutralize the remaining hydrogen ion.

Color indicators can be used to signal this sharp change in pH. These indicators, which are weak acids, generally change color in a pH-interval of about two pH units. Therefore, if a color indicator is precisely to fix the equivalence point, it is necessary that the addition of one or two drops of titrant result in a change of at least two pH units. As Fig. 4-1(a) shows, the pH change at the equivalence point becomes less sharp as the acid–base concentrations decrease. Thus, in a titration of $10^{-4}\,M$ HCl with $10^{-4}\,M$ NaOH, even if the indicator changed color exactly in the range pH 6.0 to 8.0 (centered about the equivalence point), the indicator would begin changing color at $V_b = 24.51$ mL and would not complete the color change until $V_b = 25.50$ mL—a very wide spread. This does not mean that such dilute solutions cannot be successfully analyzed. What is required is that a pH meter be used and that the equivalence point be located, if possible, by other methods, such as inverse-derivative plots of the pH versus V_T data (see Problem 4.17).

Titration Error

If V_e is the volume of titrant at the true equivalence point and V_i is the volume of titrant corresponding to the color-indicator end point, then the *percent titration error*, $\%E$, is

$$\%E = \frac{V_i - V_e}{V_e}(100\%) \tag{4.8}$$

4.4 PRINCIPAL SPECIES

Principal species are those present in large concentrations, as compared with *minor species*, whose concentrations are less than 1% of the concentrations of the principal species.

EXAMPLE 4.4   In 0.10 M HCl, the concentrations of the species in solution would be: $[H^+] = 0.10$, $[Cl^-] = 0.10$, and $[OH^-] = 1.0 \times 10^{-13}$. Therefore, the hydrogen and chloride ions would be principal species, and the hydroxide ion would be a minor species.

In general, we will identify the principal species in a solution by a four-step procedure which consists in imagining that:

1. the compounds are added to water that is already at equilibrium and, therefore, at 25 °C, has $[H^+] = [OH^-] = 1.00 \times 10^{-7}$;
2. the compounds now present in solution dissociate into ions;
3. those species that can react with each other (such as H^+ from a strong acid and OH^- from a strong base) do so to the maximum extent;
4. the new solution contains these new species and the remaining ions and particles as the principal species.

We should then need to consider those shifts in ionic reaction that produce an equilibrium mixture. However, if we have identified the principal species correctly, then very little further reaction will usually occur. Sometimes, indeed, the identification of the principal species suffices to describe the equilibrium mixture.

Strong Acids and Bases

If a strong acid and a strong base are not in stoichiometric ratio, then H^+ *or* OH^- will be one of the principal species in the solution.

EXAMPLE 4.5   Determine the principal species in the following solutions:

$$(a)\ 0.1\ M\ \text{HCl}, 0.2\ M\ \text{NaOH} \qquad (b)\ 0.2\ M\ \text{HCl}, 0.2\ M\ \text{NaOH}$$

(a) The NaOH is partially neutralized and the principal species are 0.1 M Cl$^-$, 0.2 M Na$^+$, and 0.1 M OH$^-$.

(b) This is a stoichiometric mixture, with 0.2 M Na$^+$ and 0.2 M Cl$^-$ as the only principal species.

Soluble Salts

When soluble salts dissolve in water, they dissociate into ions, which constitute the principal species.

EXAMPLE 4.6 Find the concentrations of the principal species in the following solutions: (a) 0.1 M NaCl, (b) 0.2 M K_2SO_4, (c) 0.05 M NH_4Cl.

(a) $[Na^+] = 0.1$, $[Cl^-] = 0.1$; (b) $[K^+] = 0.4$, $[SO_4^{2-}] = 0.2$; (c) $[NH_4^+] = 0.05$, $[Cl^-] = 0.05$.

When mixtures of salts are dissolved in water, the various ions can interreact to form a weak acid or a weak base, a precipitate, or a complex. These possibilities are considered in subsequent chapters.

Solved Problems

pH OF AQUEOUS SOLUTIONS

4.1 At 25 °C, find $[H^+]$ if the pH is (a) 3.68, (b) 3.682, (c) 8.197, (d) −0.227, (e) 12.448. Assume ideal solutions.

(a) $[H^+] = a_{H^+} = $ antilog $(-pH) = 2.1 \times 10^{-4}$, (b) 2.08×10^{-4}, (c) 6.35×10^{-9}, (d) 1.69, (e) 3.56×10^{-13}.

4.2 At 25 °C, find $[H^+]$ if (a) pH = 3.68, $f_{H^+} = 0.928$; (b) pH = 3.682, $f_{H^+} = 0.928$; (c) pH = 8.197, $f_{H^+} = 0.866$; (d) pH = −0.227, $f_{H^+} = 0.467$; (e) pH = 12.448, $f_{H^+} = 0.825$.

In each case, $[H^+] = a_{H^+}/f_{H^+}$, with $a_{H^+} = $ antilog $(-pH)$ as in Problem 4.1.

(a) $\dfrac{2.1 \times 10^{-4}}{0.928} = 2.3 \times 10^{-4}$ (d) $\dfrac{1.69}{0.467} = 3.62$

(b) $\dfrac{2.08 \times 10^{-4}}{0.928} = 2.24 \times 10^{-4}$ (e) $\dfrac{3.56 \times 10^{-13}}{0.825} = 4.31 \times 10^{-13}$

(c) $\dfrac{6.35 \times 10^{-9}}{0.866} = 7.33 \times 10^{-9}$

4.3 What is the pH of a 0.020 M $CaCl_2$ solution at 25 °C, (a) assuming that the solution is ideal? (b) including activity effects?

(a) At 25 °C, $pK_w = 14.000$; therefore, pH = $\frac{1}{2}pK_w = 7.000$.

(b) The ionic strength is, by (3.8),

$$I = \tfrac{1}{2}[2^2(0.020) + (-1)^2(0.040)] = 0.060$$

From Tables 3-1 and 3-2, and (3.7), $f_{H^+} = 0.847$ and $f_{OH^-} = 0.793$; thus (4.2) becomes

$$1.00 \times 10^{-14} = [H^+](0.847)[OH^-](0.793) \tag{1}$$

However, as a result of the water dissociation, $[OH^-] = [H^+]$, so that (1) gives

$$[H^+] = \left[\frac{1.00 \times 10^{-14}}{(0.847)(0.793)} \right]^{1/2} = 1.22 \times 10^{-7}$$

Hence $a_{H^+} = [H^+]f_{H^+} = (1.22 \times 10^{-7})(0.847) = 1.03 \times 10^{-7}$

$$pH = -\log(1.03 \times 10^{-7}) = 6.987$$

4.4 At 50 °C, find the pH of (a) pure H_2O, (b) 0.100 M HCl, (c) 0.200 M NaOH, (d) a solution prepared by adding 10.00 mL of 0.500 M NaOH to 15.00 mL of 0.300 M HCl. Assume that solutions are ideal and that volumes add.

From Table 4-1, $K_w = 5.5 \times 10^{-14}$ at 50 °C.

(a) $pK_w = 13.26$; therefore, $pH = pK_w/2 = 13.26/2 = 6.63$.

(b) $[H^+] = 0.100$; $pH = 1.000$.

(c) $[OH^-] = 0.200$; $[H^+] = K_w/[OH^-] = 5.5 \times 10^{-14}/0.200 = 2.7_5 \times 10^{-13}$; $pH = 12.56$.

(d) The solution is 5.00 mmol NaOH + 4.50 mmol HCl in 25.00 mL. The NaOH will neutralize all of the HCl, so that the solution is the same as 25.00 mL containing 0.50 mmol NaOH and 4.50 mmol NaCl. Therefore,

$$[OH^-] = \frac{0.50}{25.00} = 0.020 \qquad [H^+] = \frac{5.5 \times 10^{-14}}{0.020} = 2.7_5 \times 10^{-12}$$

and $pH = 11.56$.

4.5 At 25 °C, what is the pH of the solution produced when 0.0700 mL of 36.0 wt % HCl (see Table 1-4) is diluted to 500 mL? The HCl has density 1.179×10^3 kg/m^3.

$$\left(\frac{0.0700 \times 10^{-3}\text{ L}}{1}\right)\left(\frac{1.179 \times 10^6\text{ g soln}}{10^3\text{ L}}\right)\left(\frac{36.0\text{ g HCl}}{100\text{ g soln}}\right)\left(\frac{1\text{ mol HCl}}{36.5\text{ g HCl}}\right)\left(\frac{1}{0.500\text{ L}}\right) = 1.63 \times 10^{-3}\ M$$

Thus, $[H^+] = 1.63 \times 10^{-3}$ and (ideal solution) $pH = -\log[H^+] = 2.788$.

4.6 At 25 °C, what is the pH of a saturated solution of $Ca(OH)_2$? The solubility of $Ca(OH)_2$ (s) is 1.50 kg/m^3 at 25 °C.

$$\left(\frac{1.50\text{ g Ca(OH)}_2}{1\text{ L}}\right)\left(\frac{1\text{ mol Ca(OH)}_2}{74.1\text{ g Ca(OH)}_2}\right)\left(\frac{2\text{ mol OH}^-}{1\text{ mol Ca(OH)}_2}\right) = 4.05 \times 10^{-2}\ M\ OH^-$$

whence $pOH = 1.393$ and $pH = 14.000 - 1.393 = 12.607$.

4.7 At 25 °C, find the pH of (a) an ideal, $3.7 \times 10^{-3}\ M$ HCl solution; (b) an ideal, $1.00 \times 10^{-8}\ M$ HCl solution.

(a) $[H^+] = 3.7 \times 10^{-3}$, as negligible H^+ arises from the H_2O dissociation; $pH = 2.43$.

(b) The hydrogen-ion concentration will *not* be 1.00×10^{-8} and the pH will *not* be 8.00, as this would be a basic solution. The solution must be acidic, since an acid is present; therefore the pH must be less than 7.00. Because the solution is so dilute, it will be necessary to solve a quadratic equation. The data are:

$$
\begin{array}{ccccc}
C \text{ start} & & 10^{-8}+10^{-7} & & 10^{-7} \\
& H_2O \longrightarrow & H^+ & + & OH^- \\
C \text{ equil} & & 10^{-8}+10^{-7}-w & & 10^{-7}-w
\end{array}
$$

from which $K_w = 1.00 \times 10^{-14} = (1.10 \times 10^{-7} - w)(10^{-7} - w)$, or

$$w^2 - 2.10 \times 10^{-7}w + 0.10 \times 10^{-14} = 0$$

Solving, we get $w = 2.05 \times 10^{-7}$ and 4.88×10^{-9}; the second root is the valid one, since $10^{-7} - w$ must be positive. Thus,

$$[H^+] = 1.10 \times 10^{-7} - 4.88 \times 10^{-9} = 1.05 \times 10^{-7} \qquad \text{and} \qquad pH = 6.979$$

An alternate method of solving the problem is to rewrite the $[OH^-]$ in the above equilibrium in terms of $[H^+]$, which is what we want to determine. Thus, $[OH^-] = [H^+] - 10^{-8}$, and

$$K_w = [H^+]([H^+] - 10^{-8}) \qquad \text{or} \qquad [H^+]^2 - 10^{-8}[H^+] - K_w = 0$$

Solving this quadratic gives $[H^+] = 1.05 \times 10^{-7}$ as the only positive root.

4.8 Let C_{sa} denote the concentration of a strong monoprotic acid, HA. Generalize Problem 4.7(b) by deriving an equation for $[H^+]$ as a function of C_{sa} in the case of an ideal aqueous solution of HA.

The general equation $(4.4a)$ holds in the limiting case $V_b = 0$, which is the case at hand. Thus,

$$[H^+]^2 - C_{sa}[H^+] - K_w = 0$$

is the desired equation.

4.9   At 25 °C, what changes in pH occur when

(a) H_2O (b) 0.0250 M HCl   (c) 0.0250 M NaOH

are diluted with water by a factor of ten? Assume solutions are ideal.

(a)   Initial pH = 7.000, final pH = 7.000; ΔpH = 0.000.

(b)   Initial pH = 1.602, final $[H^+]$ = 0.00250 and final pH = 2.602; ΔpH = 1.000.

(c)   Initial pH = 12.398, final $[OH^-]$ = 0.00250, final pOH = 2.602, final pH = 11.398; ΔpH = −1.000.

4.10   At 25 °C, what changes in pH occur when 0.100 mmol of a strong acid is added to 25.00 mL of

(a) H_2O (b) 0.0250 M HCl   (c) 0.0250 M NaOH

Assume ideality and no volume changes.

(a)   Initial pH = 7.000, final $[H^+]$ = 0.100/25.00 = 4.00×10^{-3} (the contribution from the water equilibrium can be neglected), final pH = 2.398; ΔpH = −4.602.

(b)   Initial pH = 1.602, final $[H^+]$ = 0.0250 + (0.100/25.00) = 0.0290, final pH = 1.538; ΔpH = −0.064.

(c)   Initial pH = 12.398, final $[OH^-]$ = 0.0250 − (0.100/25.00) = 0.0210, final pOH = 1.678, final pH = 12.322; ΔpH = −0.076.

4.11   Determine the pH at 25 °C of the solution that results from the addition of 20.00 mL of 0.0100 M Ca(OH)$_2$ to 30.00 mL of 0.0100 M HCl. Include activity effects.

We have initially

$$C_{sb}V_b = (0.0100)(20.00) = 0.200 \text{ mmol Ca(OH)}_2$$
$$C_{sa}V_a = (0.0100)(30.00) = 0.300 \text{ mmol HCl}$$

Since there are two OH^- per Ca(OH)$_2$, mmols OH^- = 0.400. After neutralization,

$$[OH^-] = \frac{0.400 - 0.300}{50.00} = 2.00 \times 10^{-3} \qquad [Ca^{2+}] = \frac{0.200}{50.00} = 4.00 \times 10^{-3} \qquad [Cl^-] = \frac{0.300}{50.00} = 6.00 \times 10^{-3}$$

and (3.8) gives as the ionic strength of the solution:

$$I = \tfrac{1}{2}[1^2(2.00 \times 10^{-3}) + 2^2(4.00 \times 10^{-3}) + 1^2(6.00 \times 10^{-3})] = 0.012$$

From Tables 3-1 and 3-2, and (3.7), we have f_{OH^-} = 0.891.

Then, by definition of pOH (cf. (4.3)),

$$pOH = -\log([OH^-]f_{OH^-})$$
$$-\log[(2.00 \times 10^{-3})(0.891)] = 2.749$$

and pH = pK_w − pOH = 14.000 − 2.749 = 11.251.

4.12   As might be expected, *heavy water* (deuterium oxide, D_2O) has K-values that differ from those for H_2O. At 25 °C, pure D_2O has pD = 7.450. Calculate, at 25 °C: (a) K_{hw}, (b) pD of a 0.100 M DCl solution, (c) pOD of a 0.500 M NaOD solution, (d) pD of a 0.300 M NaOD solution, (e) pD of a solution prepared by adding 10.00 mL of 0.200 M DCl to 15.00 mL of 0.250 M NaOD. Assume ideal solutions and additivity of volumes.

(a) pK_{hw} = pD + pOD = 7.450 + 7.450 = 14.900, K_{hw} = 1.26×10^{-15}.

(b) $[D^+]$ = 0.100, pD = 1.000.

(c) $[OD^-]$ = 0.500, pOD = 0.301.

(d) $[OD^-] = 0.300$, pOD = 0.523, pD = $14.900 - 0.523 = 14.377$.

(e) The mixture of $(10.00)(0.200) = 2.00$ mmol DCl and $(15.00)(0.250) = 3.75$ mmol NaOD is the same as 25.00 mL containing 2.00 mmol NaCl and 1.75 mmol NaOD. Thus,

$$[OD^-] = \frac{1.75}{25.00} = 0.0700$$

pOD = 1.155, pD = $14.900 - 1.155 = 13.745$.

TITRATION CURVES

4.13 In the titration of 15.00 mL of 0.200 M NaOH with 0.100 M HCl, calculate the pH for $V_a = 0, 2.00, 10.00, 20.00, 30.00, 40.00$ mL. Assume that the solutions are ideal.

The contribution from the water equilibrium can be neglected except at $V_a = 30.00$ mL, which is the equivalence point in the titration.

$V_a = 0.00$ mL. $[OH^-] = 0.200$, pOH = 0.699, pH = $14.000 - 0.699 = 13.301$.

$V_a = 2.00$ mL. $V_b C_{sb} = (15.00)(0.200) = 3.00$ mmol NaOH

$V_a C_{sa} = (2.00)(0.100) = 0.20$ mmol HCl

This is the same as 0.20 mmol NaCl and 2.80 mmol NaOH in 17.00 mL; thus,

$$[OH^-] = \frac{2.80}{17.00} = 0.165 \qquad pOH = 0.783 \qquad pH = 13.217$$

$V_a = 10.00$ mL. 3.00 mmol NaOH and $(10.00)(0.100) = 1.00$ mmol HCl is the same as 25.00 mL containing 1.00 mmol NaCl and 2.00 mmol NaOH:

$$[OH^-] = \frac{2.00}{25.00} = 0.0800 \qquad pOH = 1.097 \qquad pH = 12.903$$

$V_a = 20.00$ mL. 3.00 mmol NaOH and $(20.00)(0.100) = 2.00$ mmol HCl is the same as 35.00 mL containing 2.00 mmol NaCl and 1.00 mmol NaOH:

$$[OH^-] = \frac{1.00}{35.00} = 0.0286 \qquad pOH = 1.544 \qquad pH = 12.456$$

$V_a = 30.00$ mL. This is the equivalence point and the solution is the same as 3.00 mmol NaCl in 45.00 mL: pH = 7.000.

$V_a = 40.00$ mL. 3.00 mmol NaOH and $(40.00)(0.100) = 4.00$ mmol HCl is the same as 55.00 mL containing 3.00 mmol NaCl and 1.00 mmol HCl:

$$[H^+] = \frac{1.00}{55.00} = 0.0182 \qquad pH = 1.740$$

4.14 What is the pH in Problem 4.13 when $V_a = 20.00$ mL and activity effects are taken into account?

The solution is $2.00/35.00 = 0.0571$ M NaCl and 0.0286 M NaOH. The ionic strength is

$$I = \tfrac{1}{2}(1^2[Na^+] + 1^2[Cl^-] + 1^2[OH^-])$$
$$= \tfrac{1}{2}(0.0857 + 0.0571 + 0.0286) = 0.0857$$

From Tables 3-1 and 3-2, and (3.7), $f_{OH^-} = 0.766$. Therefore,

$$a_{OH^-} = [OH^-]f_{OH^-} = (0.0286)(0.766) = 0.0219$$

pOH = 1.659, and pH = $14.000 - 1.659 = 12.341$.

Note: Since OH^- is a principal species, its concentration is not affected by considerations of non-ideality. It is the minor species, H^+, that is affected. If the solution is considered ideal,

$$[H^+]_{ideal} = \frac{1.00 \times 10^{-14}}{0.0286} = 3.50 \times 10^{-13}$$

When it is considered non-ideal.

$$a_{H^+} = \frac{1.00 \times 10^{-14}}{0.0219} = 4.57 \times 10^{-13}$$

and we find that $f_{H^+} = 0.832$. Therefore,

$$[H^+] = \frac{4.57 \times 10^{-13}}{0.832} = 5.49 \times 10^{-13}$$

4.15 Calculate the pH when 24.9999 mL of 0.100 M NaOH was added in the titration of 25.0000 mL of 0.100 M HCl. Include corrections for non-ideal effects.

At this stage in the titration, the solution is the same as one with 2.49999 mmol of NaCl and 0.00001 mmol of HCl in a volume of 49.9999 mL. The ionic strength is due primarily to the principal species, NaCl; $I = 0.0500$. From this value, Tables 3-1 and 3-2, and (3.7), there result $f_{H^+} = 0.854$ and $f_{OH^-} = 0.807$. Because the concentration of H^+ from water is of the same (small) order of magnitude as that from the excess HCl,

$$\frac{0.00001}{49.9999} = 2 \times 10^{-7}$$

one may not neglect the water equilibrium. We have:

$$
\begin{array}{cccc}
C \ start & & 2 \times 10^{-7} + 10^{-7} & 10^{-7} \\
& H_2O \longrightarrow & H^+ & + \quad OH^- \\
C \ equil & & 3 \times 10^{-7} - w & 10^{-7} - w
\end{array}
$$

and (4.2) becomes

$$(0.854)(3 \times 10^{-7} - w)(0.807)(10^{-7} - w) = 1.00 \times 10^{-14}$$

or

$$w^2 - (4 \times 10^{-7})w + (1.5 \times 10^{-14}) = 0$$

The physically significant root is $w = 4 \times 10^{-8}$, whence

$$[H^+] = 2._6 \times 10^{-7} \qquad a_{H^+} = (2._6 \times 10^{-7})(0.854) = 2._2 \times 10^{-7} \qquad pH = 6.6_5$$

4.16 Find the pH at the equivalence point in the titration of 25.00 mL of 0.100 M HCl with 0.100 M NaOH. Include corrections for non-ideality.

The equivalence point is a solution having 0.0500 M Na^+ and 0.0500 M Cl^- as principal species. Both $[H^+]$ and $[OH^-]$ will be very small and approximately equal to 10^{-7}. However, charge balance requires that $[H^+] = [OH^-]$ in this solution.

The ionic strength is $I = 0.0500$, and, from Tables 3-1 and 3-2, and (3.7), $f_{H^+} = 0.854$ and $f_{OH^-} = 0.807$. Then, by (4.2),

$$[H^+](0.854)[OH^-](0.807) = [H^+](0.854)[H^+](0.807) = (0.689)[H^+]^2 = 1.00 \times 10^{-14}$$

Therefore, $[H^+] = 1.20 \times 10^{-7}$, $a_{H^+} = (1.20 \times 10^{-7})(0.854) = 1.02 \times 10^{-7}$, and

$$pH = -\log (1.02 \times 10^{-7}) = 6.989$$

4.17 If Fig. 4-1 is viewed sideways, it is seen that the equivalence point in a titration is characterized by a zero slope of the graph of V_T versus pH; that is,

$$\frac{dV_T}{d\text{pH}} \approx \frac{\Delta V_T}{\Delta \text{pH}} = 0$$

at the equivalence point. This zero value represents a minimum slope in the case of the titration of an acid with a base, and a maximum slope in the opposite case.

The data in Table 4-2 pertain to the titration of 20.00 mL of a dilute solution of NaOH with 3.252×10^{-4} M HCl, at 25 °C. Determine the concentration of the NaOH.

Table 4-2

V_T, mL	17.13	17.17	17.21	17.25	17.31	17.35
$\Delta V_T/\Delta\text{pH}$, mL	−0.591	−0.501	−0.388	−0.299	−0.175	−0.084
V_T, mL	17.39	17.42	17.47	17.53	17.58	17.63
$\Delta V_T/\Delta\text{pH}$, mL	−0.057	−0.102	−0.207	−0.336	−0.437	−0.545

The data are plotted in Fig. 4-2. If the one data point closest to the equivalence point (shown by the solid circle in Fig. 4-2) is omitted, the remaining data define two straight lines that intersect in $V_e = 17.38$ mL. Hence, the concentration of the NaOH is

$$\frac{(17.38)(3.252 \times 10^{-4})}{20.00} = 2.826 \times 10^{-4} \ M$$

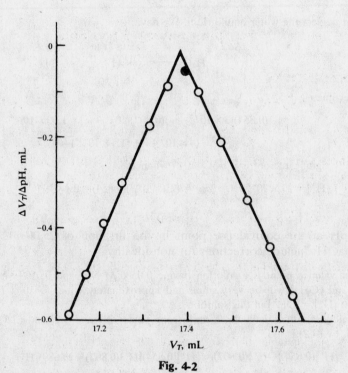

Fig. 4-2

4.18 What is the titration error if the true equivalence point is 25.00 mL but the change in a color indicator results in an end point of 24.78 mL?

From (4.8),

$$\%E = \frac{24.78 - 25.00}{25.00}(100\%) = -0.88\%$$

PRINCIPAL SPECIES

4.19   What are the concentrations of the principal species in a solution that is 0.2 M HCl, 0.1 M NaOH?

   The HCl is in excess and all of the NaOH will be neutralized. The concentrations of the principal species are, therefore, $[H^+] = 0.1$, $[Cl^-] = 0.2$, and $[Na^+] = 0.1$.

4.20   Find the concentrations of the principal species in a solution made by mixing 50.00 mL of 0.500 M HCl, 100.0 mL of 0.200 M NaOH, and 100.0 mL of H_2O.

   The mixture contains 25.0 mmol HCl and 20.0 mmol NaOH, all in 250.0 mL. The HCl is in excess. and 5.0 mmol will remain after reaction with NaOH. The concentrations of the principal species are:

$$[H^+] = \frac{5.0}{250.0} = 0.020 \qquad [Cl^-] = \frac{25.0}{250.0} = 0.100 \qquad [Na^+] = \frac{20.0}{250.0} = 0.0800$$

Supplementary Problems

pH OF AQUEOUS SOLUTIONS

4.21   At 25 °C, find $[H^+]$ if pOH is (a) 3.68, (b) 3.682, (c) 8.197, (d) −0.227, (e) 12.448. Assume that the solutions are ideal.
   Ans.   (a) 4.8×10^{-11}; (b) 4.81×10^{-11}; (c) 1.57×10^{-6}; (d) 5.93×10^{-15}; (e) 2.81×10^{-2}

4.22   At 25 °C, find $[OH^-]$ if (a) pH = 3.92, $f_{OH^-} = 0.727$; (b) pH = 5.28, $f_{OH^-} = 0.853$; (c) pOH = 4.31, $f_{OH^-} = 0.916$; (d) pOH = −0.628, $f_{OH^-} = 0.524$; (e) pH = 14.362, $f_{OH^-} = 0.685$.
   Ans.   (a) $1.1_4 \times 10^{-10}$; (b) $2.2_3 \times 10^{-9}$; (c) $5.3_5 \times 10^{-5}$; (d) 8.10; (e) 3.36

4.23   A solution at 25 °C is 0.080 M NaCl and 0.030 M CaCl$_2$. Calculate (a) the ionic strength of the solution, (b) f_{H^+} and f_{OH^-}, (c) the pH.   *Ans.*   (a) $I = 0.170$; (b) $f_{H^+} = 0.80$, $f_{OH^-} = 0.71$; (c) pH = 6.97

4.24   At 60 °C, determine the pH of (a) 0.0650 M HCl, (b) 0.125 M NaOH, (c) a solution prepared by adding 25.00 mL of 0.150 M NaOH to 15.00 mL of 0.200 M HCl. Assume that the solutions are ideal.
   Ans.   (a) 1.187; (b) 12.12; (c) 11.29

4.25   At 25 °C, 0.120 mL of a solution that is 50.0 wt % KOH (MW = 56.1) and has density 1.50×10^3 g/L is diluted to 250 mL. Assuming that the dilute solution is ideal, find its pH.   *Ans.*   11.807

4.26   At 25 °C, what is the pH of an ideal, 8.00×10^{-8} M NaOH solution?   *Ans.*   7.169

4.27   At 25 °C, calculate the pH *change* when 0.100 mmol of NaOH is added to 30.00 mL of (a) H_2O. (b) 0.0300 M HCl, (c) 0.0300 M NaOH. Assume that the solutions are ideal and that there is no change in volume when the NaOH is added.   *Ans.*   (a) 4.523; (b) 0.051; (c) 0.046

4.28   25.00 mL of 0.200 M HNO$_3$ was added to 30.00 mL of a NaOH solution of unknown concentration. At 25 °C, the pH of the mixture was 11.350. If the solutions were ideal, what was the original concentration of the NaOH?   *Ans.*   0.171 M

TITRATION CURVES

4.29 At 25 °C, 20.00 mL of 0.00150 M NaOH is titrated with 0.00100 M HCl. Calculate the pH for $V_a =$ 0, 5.00, 10.00, 20,00, 25.00, 29.00, 29.90, 29.99, 30.00, 30.01, 35.00, 40.00, 50.00 mL. Assume that the solutions are ideal. The $pC = 3$ curve in Fig. 4-1(b) indicates that, for most of the given volumes, either pH < 6 or pH > 8, making it possible to neglect the water equilibrium in most of the calculations.

    Ans. $V_a = 0$, pH = 11.176; $V_a = 5.00$, pH = 11.000; $V_a = 10.00$, pH = 10.824; $V_a = 20.00$, pH = 10.398; $V_a = 25.00$, pH = 10.046; $V_a = 29.00$, pH = 9.310; $V_a = 29.90$, pH = 8.301; $V_a = 29.99$, pH = 7.383; $V_a = 30.00$, pH = 7.000; $V_a = 30.01$, pH = 6.617; $V_a = 35.00$, pH = 4.041; $V_a = 40.00$, pH = 3.778; $V_a = 50.00$; pH = 3.544. [It is necessary to calculate the pH from the quadratic equation only for $V_a = 29.99, 30.00, 30.01$.]

PRINCIPAL SPECIES

4.30 What are the concentrations of the principal species in a solution that is 0.3 M NaOH and 0.2 M HCl?

    Ans. $[Na^+] = 0.3$, $[Cl^-] = 0.2$, $[OH^-] = 0.1$

4.31 What are the concentrations of the principal species in a solution that is prepared by mixing 50.0 mL of 0.200 M NaCl, 25.0 mL of 0.100 M NaOH, and 25.0 mL of 0.300 M HCl?

    Ans. $[Na^+] = 0.125$, $[Cl^-] = 0.175$, $[H^+] = 0.050$

Chapter 5

Simple Weak Acids
and Weak Bases

5.1 EQUILIBRIUM CONSTANTS

Weak Acids

A *weak acid*, HA, is an acid that is less than 100% dissociated in solutions of moderate concentration. For such an acid, the reaction

$$HA \rightarrow H^+ + A^- \tag{5.1}$$

has a measurable equilibrium constant, K_a, which in the case of an ideal solution is given by

$$K_a = \frac{[H^+][A^-]}{[HA]} \tag{5.2}$$

Weak Bases

Very few *weak bases* are described by the analogue of (5.1), $ROH \rightarrow R^+ + OH^-$. Instead, most weak bases are organic compounds, B, for which the reaction

$$B + H_2O \rightarrow BH^+ + OH^- \tag{5.3}$$

proceeds less than 100% in solutions of moderate concentration. The equilibrium constant, K_b, for an ideal solution is

$$K_b = \frac{[BH^+][OH^-]}{[B]} \tag{5.4}$$

(as usual, we have taken $a_{H_2O} = 1$).

Conjugate Bases

The product A^- of the dissociation of a weak acid is called a *conjugate base* because it can undergo hydrolysis according to the reaction

$$A^- + H_2O \rightarrow HA + OH^- \tag{5.5}$$

The equilibrium constant for (5.5) is denoted K_{cb}, or, sometimes, K_h (the *hydrolysis constant*). For an ideal solution,

$$K_{cb} = K_h = \frac{[HA][OH^-]}{[A^-]} = \frac{K_w}{K_a} \tag{5.6}$$

Reaction (5.5) explains why salts of weak acids are basic substances.

Conjugate Acids

Similarly, the reaction product BH^+ of a weak base is a *conjugate acid*, since it can undergo the reaction

$$BH^+ \rightarrow B + H^+ \tag{5.7}$$

For an ideal solution, the reaction equilibrium constant is

$$K_{ca} = \frac{[B][H^+]}{[BH^+]} = \frac{K_w}{K_b} \tag{5.8}$$

5.2   CALCULATION OF pH

General Equations

The following four equations hold for a $C_a\,M$ solution of a weak acid:

1.   Charge balance: $[H^+] = [A^-] + [OH^-]$
2.   Mass balance for HA: $C_a = [HA] + [A^-]$
3.   Water equilibrium: $K_w = [H^+][OH^-]$
4.   Acid equilibrium: $K_a = [H^+][A^-]/[HA]$

Elimination of [HA], [A$^-$], and [OH$^-$] yields the desired equation in [H$^+$] alone:

$$[H^+]^3 + K_a[H^+]^2 - (C_aK_a + K_w)[H^+] - K_aK_w = 0 \qquad (5.9)$$

As discussed in Section 3-3, the analogous equation for a non-ideal solution is

$$[H^+]^3 + \frac{K_a}{K_{af}}[H^+]^2 - \left(C_a\frac{K_a}{K_{af}} + \frac{K_w}{K_{wf}}\right)[H^+] - \frac{K_aK_w}{K_{af}K_{wf}} = 0 \qquad (5.10)$$

Equation (5.9) and its analogues as derived from Table 5-1 are first-degree in K and in C. Thus, these general equations can be used directly to calculate K, if pH and C are known, or to calculate C, if pH and K are known. However, being cubics in [H$^+$] (or [OH$^-$]), the equations are not convenient for the calculation of pH when K and C are given. We get around this difficulty, as in Chapter 4, by considering the reaction shifts that establish equilibrium, in the light of one or another simplifying assumption. This will lead to ranges of the parameters K and C over which the cubic equations can be replaced by linear or quadratic equations.

Table 5-1

To convert an equation for a $C_a\,M$ solution of a **weak acid** to an equation for:	Substitute:
A $C_b\,M$ solution of a **weak base**	C_b for C_a K_b for K_a [OH$^-$] for [H$^+$]
A $C_{cb}\,M$ solution of a **conjugate base** (same as a $C_s\,M$ solution of the salt of a weak acid)	C_{cb} (or C_s) for C_a K_{cb} for K_a [OH$^-$] for [H$^+$]
A $C_{ca}\,M$ solution of a **conjugate acid** (same as a $C_s\,M$ solution of the salt of a weak base)	C_{ca} (or C_s) for C_a K_{ca} for K_a

Simplified Equations

Imagine an initial $C_a\,M$ solution of *undissociated* weak acid, HA, in *dissociated* water. Concentration shifts a and w are necessary to bring about equilibrium:

$$
\begin{array}{lcccc}
initial & C_a & 10^{-7} & & 0 \\
& HA & \longrightarrow & H^+ & + & A^- \\
equilibrium & C_a - a & & 10^{-7} + a - w & & a
\end{array}
\qquad (5.11a)
$$

$$
\begin{array}{lccc}
initial & & 10^{-7} & 10^{-7} \\
& H_2O & \longrightarrow & H^+ & + & OH^- \\
equilibrium & & 10^{-7} + a - w & & 10^{-7} - w
\end{array}
\qquad (5.11b)
$$

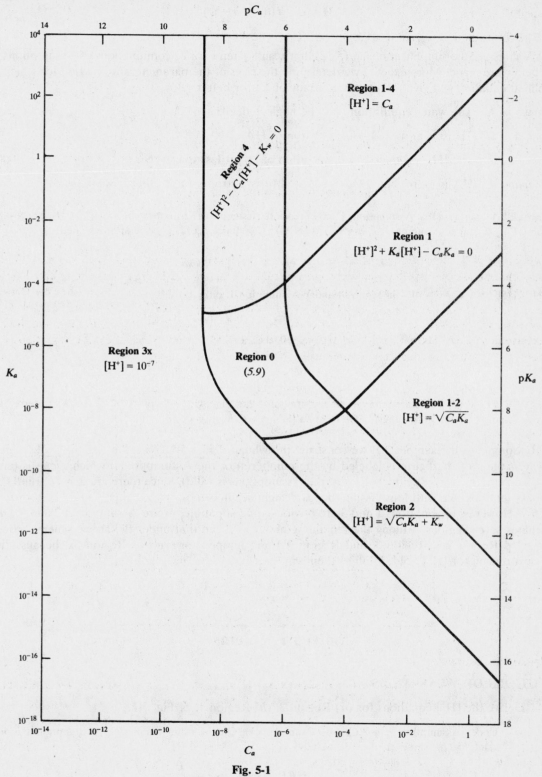

pC_a

K_a

Region 1-4
$[H^+] = C_a$

Region 4
$[H^+]^2 - C_a[H^+] - K_w = 0$

Region 1
$[H^+]^2 + K_a[H^+] - C_aK_a = 0$

Region 3x
$[H^+] = 10^{-7}$

Region 0
(5.9)

Region 1-2
$[H^+] = \sqrt{C_aK_a}$

pK_a

Region 2
$[H^+] = \sqrt{C_aK_a + K_w}$

C_a

Fig. 5-1

From $(5.11a)$, the equilibrium constant, (5.2), can be written as

$$K_a = \frac{[H^+]a}{C_a - a} = \frac{(10^{-7} + a - w)a}{C_a - a} \tag{5.12}$$

We consider the simplifications in (5.12) that result when various combinations of Assumptions 1 to 4 below are made. More exactly, we determine the sizes of the parameters K_a and C_a for which a simplified equation will give $[H^+]$ to an accuracy of 1% or better.

Assumption 1.   The water equilibrium may be neglected; $pH < 6.00$.

Assumption 2.   There is at most 1% dissociation of HA.

Assumption 3.   The $[H^+]$ produced by dissociation of HA is less than 1.99×10^{-9}, so that $pH = 7.000$.

Assumption 4.   HA is more than 99% dissociated (it is a strong, rather than a weak, acid).

EXAMPLE 5.1   Suppose that Assumptions 1 and 2 are valid. Assumption 1 is tantamount to $w \approx 10^{-7}$ (see page 63), so that $[H^+] \approx a$; under Assumption 2, $a \ll C_a$, or $[HA] \approx C_a$. Hence, (5.12) may be approximated by

$$K_a = \frac{[H^+]^2}{C_a} \qquad \text{or} \qquad [H^+] = \sqrt{C_a K_a}$$

Now, $[H^+] = \sqrt{C_a K_a} > 10^{-6}$, as required under Assumption 1, only if

$$C_a K_a > 10^{-12} \qquad \text{or} \qquad pC_a + pK_a < 12 \tag{1}$$

and Assumption 2 demands that $a = [H^+] = \sqrt{C_a K_a} < 0.01 C_a$, or

$$\frac{C_a}{K_a} > 10^4 \qquad \text{or} \qquad pC_a - pK_a < -4 \tag{2}$$

Inequalities (1) and (2) define a right-angle wedge, Region 1-2 of Fig. 5-1, in the plane of C_a and K_a (referred to logarithmic coordinates). In this region, the formula $[H^+] = \sqrt{C_a K_a}$ is good to within 1%.

Reasoning as in Example 5.1, we construct the whole of Fig. 5-1. With the exception of Region 3x, each region of the figure is labeled by the number(s) of the Assumption(s) which *alone* is (are) valid for that region (in Region 0, none of the Assumptions is valid). Since there are $2^4 = 16$ true/false combinations possible with four Assumptions, it might seem as though 16 regions should be shown in Fig. 5-1. However, assumptions 1 and 3 are obviously incompatible, as are Assumptions 2 and 4; this eliminates seven regions. Among the remaining nine, the three that under the above scheme would be labeled Region 2-3, Region 3, and Region 3-4, are lumped together as Region 3x, because the same expression for $[H^+]$ holds in all of them.

Solved Problems

SOLUTIONS OF WEAK ACIDS

5.1   Derive the $[H^+]$-equations for (a) Region 2, (b) Region 1, of Fig. 5-1.

    (a) Under Assumption 2, $a \ll C_a$ (see Example 5.1). On the other hand, since Assumption 1 is *not* valid, $(5.11b)$ shows that

$$a = [H^+] - [OH^-] = [H^+] - \frac{K_w}{[H^+]}$$

    Thus, (5.12) becomes approximately

$$K_a = \frac{[H^+]\left\{[H^+] - \dfrac{K_w}{[H^+]}\right\}}{C_a} = \frac{[H^+]^2 - K_w}{C_a}$$

whence

$$[H^+] = \sqrt{C_a K_a + K_w}$$

in Region 2.

(b) Under Assumption 1 $[H^+] \approx a$ (see Example 5.1), so that (5.12) becomes

$$K_a = \frac{[H^+]^2}{C_a - [H^+]} \qquad \text{or} \qquad [H^+]^2 + K_a[H^+] - C_a K_a = 0$$

5.2 Find the pH and the percent dissociation of a $C_a = 0.100\ M$ solution of hypochlorous acid, HClO, given $K_a = 2.95 \times 10^{-8}$. Justify any assumptions.

The point (C_a, K_a) lies in Region 1-2 of Fig. 5-1; hence,

$$[H^+] = \sqrt{(0.100)(2.95 \times 10^{-8})} = 5.43 \times 10^{-5}$$

and pH = 4.265. As pH < 6.00, Assumption 1 is justified.
The *percent dissociation* of a weak acid is defined via (5.11a):

$$\% \text{ dissoc} \equiv \frac{[A^-]}{[HA]_{\text{initial}}}(100\%) = \frac{a}{C_a}(100\%) \tag{5.13}$$

A useful alternative expression follows from (5.12):

$$\% \text{ dissoc} = \frac{K_a/[H^+]}{(K_a/[H^+]) + 1}(100\%) \tag{5.14}$$

For the present problem, Assumption 1 gives $a = [H^+] = 5.43 \times 10^{-5}$; hence, by (5.13),

$$\% \text{ dissociation} = \frac{5.43 \times 10^{-5}}{0.100}(100\%) = 0.0543\%$$

As this value is less than 1%, Assumption 2 is justified.

5.3 Repeat Problem 5.2 for formic acid, HCOOH, with $K_a = 1.78 \times 10^{-4}$.

The point (C_a, K_a) lies in Region 1 of Fig. 5-1; hence,

$$[H^+]^2 + (1.78 \times 10^{-4})[H^+] - (0.100)(1.78 \times 10^{-4}) = 0$$

Solving by the quadratic formula, (1.4), $[H^+] = 4.13 \times 10^{-3}$ and pH = 2.384 (Assumption 1 is justified).
By (5.13),

$$\% \text{ dissoc} = \frac{4.13 \times 10^{-3}}{0.100}(100\%) = 4.13\%$$

so that Assumption 2 is not justified. In Region 1, *only* Assumption 1 is valid.

5.4 Find the pH and the percent dissociation of a $C_a = 3.00 \times 10^{-4}\ M$ solution of HCN ($K_a = 6.17 \times 10^{-10}$).

The point (C_a, K_a) lies in Region 2 of Fig. 5-1; hence,

$$[H^+] = \sqrt{(3.00 \times 10^{-4})(6.17 \times 10^{-10}) + (1.00 \times 10^{-14})} = 4.42 \times 10^{-7}$$

and pH = 6.355 (Assumption 1 is not justified).
Since $K_a/[H^+] \ll 1$, (5.14) gives

$$\% \text{ dissoc} \approx \frac{K_a}{[\text{H}^+]} (100\%) = \frac{6.17 \times 10^{-10}}{4.42 \times 10^{-7}} (100\%) = 0.140\%$$

and Assumption 2 is justified.

5.5　Find the pH and the percent dissociation of a $C_a = 1.00 \times 10^{-6}\ M$ solution of HClO ($K_a = 2.95 \times 10^{-8}$).

　　The point (C_a, K_a) lies in Region 0 of Fig. 5-1. Although (5.9) could be solved algebraically, it is easier to use successive approximations or the graphical intersection method, as described in Section 1.4. Using the latter method requires two approximate values for $[\text{H}^+]$, which we can obtain from the simplified equations that hold in Regions 1-2 and 2: $[\text{H}^+] = 1.72 \times 10^{-7}$ and $[\text{H}^+] = 1.99 \times 10^{-7}$. Writing (5.9) in the form

$$\overbrace{[\text{H}^+]^3 + K_a[\text{H}^+]^2}^{A} = \overbrace{(C_aK_a + K_w)[\text{H}^+] + K_aK_w}^{B}$$

we compute:

	A	B
$[\text{H}^+] = 1.72 \times 10^{-7}$	5.96×10^{-21}	7.09×10^{-21}
$[\text{H}^+] = 1.99 \times 10^{-7}$	9.05×10^{-21}	8.16×10^{-21}

These four points define two straight lines that intersect in $[\text{H}^+] = 1.87 \times 10^{-7}$ (see Fig. 5-2), or pH = 6.728.

　　With $K_a/[\text{H}^+] = 0.158$, (5.14) gives:

$$\% \text{ dissoc} = \frac{0.158}{1.158} (100\%) = 13.6\%$$

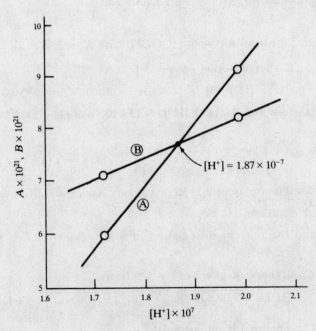

Fig. 5-2

5.6 Find the pH and the percent dissociation of $C_a = 2.00 \times 10^{-4} M$ iodic acid, HIO_3, with $K_a = 0.162$. Justify any assumptions.

The point (C_a, K_a) lies in Region 1-4 of Fig. 5-1, the strong-acid region. We have:

$$[H^+] = C_a = 2.00 \times 10^{-4} \qquad \text{and} \qquad pH = 3.699$$

Since pH < 6.00, Assumption 1 is justified.
With $K_a/[H^+] = 810$, (5.14) gives:

$$\% \text{ dissoc} = \frac{810}{811}(100\%) = 99.9\%$$

and Assumption 4 is justified.

5.7 Find the pH and the percent dissociation of a $C_a = 3.00 \times 10^{-7} M$ solution of phenol ($K_a = 1.05 \times 10^{-10}$).

The point (C_a, K_a) lies in Region 3x of Fig. 5-1; thus, pH = 7.000. Since $K_a/[H^+] = 1.05 \times 10^{-3} \ll 1$, (5.14) gives

$$\% \text{ dissoc} \approx \frac{K_a}{[H^+]}(100\%) = 0.105\%$$

(which justifies Assumption 2).

5.8 Find the pH and the percent dissociation of HAc in a solution that is made by mixing 25.00 mL of 0.200 M HCl with 25.00 mL of 0.200 M NaAc, given $K_a = 1.78 \times 10^{-5}$.

The HCl reacts completely with the NaAc to produce a solution that contains HAc and NaCl. The concentrations of the principal species are:

$$[HAc] = [Na^+] = [Cl^-] = \frac{5.00 \text{ mmol}}{50.00 \text{ mL}} = 0.100$$

Thus, the point $(C_a, K_a) = (0.100, 1.78 \times 10^{-5})$ lies in Region 1 of Fig. 5-1; the solution of the quadratic is $[H^+] = 1.32 \times 10^{-3}$, or pH = 2.879, since pH < 6.00, Assumption 1 is justified. Since $a = [H^+]$, (5.13) gives

$$\% \text{ dissoc} = \frac{1.32 \times 10^{-3}}{0.100}(100\%) = 1.32\%$$

(Assumption 2 is not valid.)

5.9 What is the pH of the solution of Problem 5.8 if non-ideal effects, based on the EDHE, (3.7), are included?

For this solution, the ionic strength is

$$I = \tfrac{1}{2}([H^+]1^2 + [Na^+]1^2 + [Cl^-](-1)^2 + [Ac^-](-1)^2 + [OH^-](-1)^2)$$

As $[OH^-]$ will be very small, it may be neglected; $[Na^+] = [Cl^-] = 0.100$. The precise values of $[H^+]$ and $[Ac^-]$ are not yet calculated, but we can use, as a first approximation, the values from Problem 5.8: $[H^+] = [Ac^-] = 0.00132$. Thus, $I \approx 0.101$. By (3.7),

$$f_{H^+} = 0.825 \qquad f_{OH^-} = f_{Cl^-} = 0.753 \qquad f_{Na^+} = f_{Ac^-} = 0.769$$

Also, since HAc is a neutral species, $f_{HAc} = 1.000$. Since Assumption 2 was invalid in Problem 5.8, we suppose it invalid here.
Substituting K_a/K_{af} for K_a (see Section 3.3) in the Region 1 equation results in

$$[H^+]^2 + \frac{K_a}{K_{af}}[H^+] - \frac{C_a K_a}{K_{af}} = 0$$

where
$$K_{af} = \frac{f_{H^+}f_{Ac^-}}{f_{HAc}} = \frac{(0.825)(0.769)}{1.000} = 0.634$$

Solving, $[H^+] = 1.66 \times 10^{-3}$, $a_{H^+} = (0.825)(1.66 \times 10^{-3}) = 1.37 \times 10^{-3}$, and

$$pH = -\log a_{H^+} = 2.863$$

Therefore Assumption 1 is valid. From (5.13), with $a = [H^+]$,

$$\% \text{ dissoc} = \frac{1.66 \times 10^{-3}}{0.100}(100\%) = 1.66\%$$

and Assumption 2 is not valid. We could now correct the ionic strength to $I \approx 0.102$, but this change would affect the calculated f-values by less than 1%.

5.10 A 2.56×10^{-3} M solution of salicylic acid, HSal, containing added NaCl has ionic strength $I = 0.100$ and pH = 2.943. For this solution, $f_{H^+} = 0.826$, $f_{OH^-} = 0.754$, $f_{Sal^-} = 0.796$, and $f_{HSal} = 1.000$. What is K_a for HSal?

When $[H^+]$ and C_a are known, it is easy to solve (5.9)—or, for a non-ideal solution, (5.10)—for K_a. However, a little additional simplicity accrues from use of the appropriate form of (5.12). For the present solution,

$$[H^+] = a_{H^+}/f_{H^+} = \frac{\text{antilog}(-2.943)}{0.826} = 1.38 \times 10^{-3}$$

so that Assumption 1 is valid and $[H^+] \approx a$. The non-ideal form of (5.12) is

$$K_a = \frac{K_{af}[H^+]a}{C_a - a} \qquad \text{where} \qquad K_{af} = (f_{H^+})(f_{Sal^-})/f_{HSal}$$

Substituting numerical values, we find, under Assumption 1,

$$K_a = \frac{(0.657)(1.38 \times 10^{-3})^2}{(2.56 \times 10^{-3}) - (1.38 \times 10^{-3})} = 1.06 \times 10^{-3}$$

SOLUTIONS OF WEAK BASES

5.11 Find the pH and the percent "dissociation" of a $C_b = 0.100$ M solution of the weak base hydroxylamine ($K_b = 9.55 \times 10^{-9}$). Justify any assumptions.

Weak-base problems are solved just like weak-acid problems, after the notation of Assumptions 1 through 4, Fig. 5-1, and all equations have been transformed according to Table 5-1.
Here, the point (C_b, K_b) lies in Region 1-2 of Fig. 5-1, so that

$$[OH^-] = \sqrt{C_b K_b} = 3.09 \times 10^{-5} \qquad \text{or} \qquad pOH = 4.510$$

and pH = 14.000 − 4.510 = 9.490. Since pOH < 6.00 (i.e., pH > 8.00), Assumption 1 is justified.
With $b = [OH^-] = 3.09 \times 10^{-5}$, (5.13) gives:

$$\% \text{ dissoc} = \frac{b}{C_b}(100\%) = \frac{3.09 \times 10^{-5}}{0.100}(100\%) = 0.0309\%$$

whereby Assumption 2 is justified.

5.12 Rework Problem 5.11 if the solution is non-ideal and contains sufficient NaCl to make the ionic strength unity. Assume the values

$$f_{BH^+} = 0.796 \qquad f_{H^+} = 0.826 \qquad f_{OH^-} = f_{Cl^-} = 0.754 \qquad f_{Na^+} = 0.770 \qquad f_B = 1.000$$

The non-ideal equation for Region 1-2 is

$$[OH^-] = \sqrt{\frac{C_b K_b}{K_{bf}}} \qquad \text{where} \qquad K_{bf} = (f_{BH^+})(f_{OH^-})/f_B$$

Substituting the numerical values, we obtain $[OH^-] = 3.99 \times 10^{-5}$; therefore,

$$a_{OH^-} = [OH^-]f_{OH^-} = (3.99 \times 10^{-5})(0.754) = 3.01 \times 10^{-5}$$
$$pOH = -\log a_{OH^-} = 4.522$$

and pH = $14.000 - 4.522 = 9.478$, justifying Assumption 1.

$$\% \text{ dissoc} = \frac{3.99 \times 10^{-5}}{0.100}(100\%) = 0.0399\%$$

and Assumption 2 is justified.

5.13 Find the pH and the percent dissociation of a $C_b = 0.0446\ M$ solution of the weak base urea ($K_b = 1.20 \times 10^{-14}$). Justify any assumptions.

The point (C_b, K_b) lies in Region 2 of Fig. 5-1; hence,

$$[OH^-] = \sqrt{C_b K_b + K_w} = 1.03 \times 10^{-7}$$

pOH = 6.989, and pH = $14.000 - 6.989 = 7.011$.
    By (5.14),

$$\% \text{ dissoc} = \frac{K_b/[OH^-]}{(K_b/[OH^-]) + 1}(100\%) \approx \frac{K_b}{[OH^-]}(100\%) = 1.17 \times 10^{-5}\%$$

whereby Assumption 2 is justified.

5.14 Find the pH and the percent dissociation of a $C_b = 0.0392\ M$ solution of the weak base piperidine ($K_b = 1.60 \times 10^{-3}$). Justify any assumptions.

The point (C_b, K_b) lies in Region 1 of Fig. 5-1; hence,

$$[OH^-]^2 + (1.60 \times 10^{-3})[OH^-] - (0.0392)(1.60 \times 10^{-3}) = 0$$

Solving, $[OH^-] = 7.16 \times 10^{-3}$, pOH = 2.145, and pH = 11.855 (Assumption 1 is justified).
    With $b = [OH^-]$, (5.13) gives

$$\% \text{ dissoc} = \frac{7.16 \times 10^{-3}}{0.0392}(100\%) = 18.3\%$$

5.15 For a weak base, B, determine (a) K_b, given that the pH of a 0.200 M solution of B is 12.554; (b) C_b, given that pH = 11.473 and $K_b = 5.25 \times 10^{-4}$.

(a) Assumption 1 holds for this solution, so that $b = [OH^-] = 0.0358$; (5.12) then gives:

$$K_b = \frac{[OH^-]b}{C_b - b} = \frac{(0.0358)^2}{0.200 - 0.0358} = 7.81 \times 10^{-3}$$

(b) Again, Assumption 1 is valid ($[OH^-] = 2.97 \times 10^{-3}$), so that (5.12) becomes

$$5.25 \times 10^{-4} = \frac{(2.97 \times 10^{-3})^2}{C_b - (2.97 \times 10^{-3})}$$

Solving $C_b = 1.98 \times 10^{-2}\ M$.

SOLUTIONS OF CONJUGATE ACIDS

5.16 Find the pH and the percent hydrolysis of a $C_{ca} = 0.0527\,M$ solution of NH_4Cl, if $K_b = 1.80 \times 10^{-5}$. Justify any assumptions.

We have

$$K_{ca} = \frac{K_w}{K_b} = 5.56 \times 10^{-10}$$

and the point (C_{ca}, K_{ca}) lies in Region 1-2 of Fig. 5-1. Thus,

$$[H^+] = \sqrt{C_{ca}K_{ca}} = 5.41 \times 10^{-6}$$

or pH = 5.267 (Assumption 1 is justified).
    With $a = [H^+] = 5.41 \times 10^{-6}$, (5.13) gives

$$\% \text{ hydrol} = \frac{5.41 \times 10^{-6}}{0.0527}(100\%) = 0.0103\%$$

whence Assumption 2 is justified.

5.17 Find the pH and the percent hydrolysis of a $C_{ca} = 0.0162\,M$ solution of piperidine hydrochloride (the salt of the weak base piperidine—see Problem 5.14). Justify any assumptions.

From Problem 5.14,

$$K_{ca} = \frac{K_w}{K_b} = \frac{10^{-14}}{1.60 \times 10^{-3}} = 6.25 \times 10^{-12}$$

Thus, the point (C_{ca}, K_{ca}) lies in Region 2 of Fig. 5-1.

$$[H^+] = \sqrt{(0.0162)(6.25 \times 10^{-12}) + 10^{-14}} = 3.33_5 \times 10^{-7}$$

whence pH = 6.477.
    Since $K_{ca}/[H^+] \ll 1$, (5.14) yields

$$\% \text{ hydrol} \approx \frac{K_{ca}}{[H^+]}(100\%) = 1.87 \times 10^{-3}\%$$

and Assumption 2 is justified.

5.18 Rework Problem 5.17 if non-ideal effects are considered; the solution contains enough NaCl to produce ionic strength $I = 0.100$. For this solution, from Table 3-3, $f_{H^+} = 0.826$, $f_{OH^-} = 0.754$, and, since B is a neutral species, $f_B = 1.000$. Assume that $f_{BH^+} = 0.796$.

The non-ideal equation holding in Region 2 is:

$$[H^+] = \sqrt{\frac{C_{ca}K_{ca}}{K_{caf}} + \frac{K_w}{K_{wf}}}$$

where

$$K_{caf} = f_B f_{H^+}/f_{BH^+} = 1.038 \qquad \text{and} \qquad K_{wf} = f_{H^+} f_{OH^-} = 0.623$$

Substituting, $[H^+] = 3.37 \times 10^{-7}$.
    The non-ideal version of (5.14) gives:

$$\% \text{ hydrol} \approx \frac{K_{ca}}{K_{caf}[H^+]}(100\%) = \frac{6.25 \times 10^{-12}}{(1.038)(3.37 \times 10^{-7})}(100\%) = 0.00179\%$$

so that Assumption 2 is justified.

5.19 A C_{ca} M solution of the salt of a weak base, BHCl, has pH = 4.502. If $K_b = 9.65 \times 10^{-7}$, find C_{ca}.

From (5.9),

$$C_{ca} = \frac{[H^+]^3 + K_{ca}[H^+]^2 - K_w[H^+] - K_{ca}K_w}{K_{ca}[H^+]}$$

Substitution of

$$[H^+] = \text{antilog}(-4.502) = 3.15 \times 10^{-5}$$
$$K_w = 10^{-14}$$
$$K_{ca} = \frac{K_w}{K_b} = 1.04 \times 10^{-8}$$

gives $C_{ca} = 0.0954$ M.

SOLUTIONS OF CONJUGATE BASES

5.20 Derive the $[OH^-]$ equation for Region 1-4 of Fig. 5-1.

Considering the concentration shifts in (5.5), we have

$$
\begin{array}{ccccccc}
\textit{initial} & C_{cb} & & & 0 & & 10^{-7} \\
& A^- & + & H_2O & \longrightarrow & HA & + & OH^- \\
\textit{equilibrium} & C_{cb}-b & & & & b & & 10^{-7}+b-w
\end{array}
$$

Neglect of the water equilibrium (Assumption 1) makes $[OH^-] \approx b$. Under Assumption 4, essentially all the A^- ends up as HA; i.e., $b \approx C_{cb}$. Thus, $[OH^-] \approx C_{cb}$ holds in Region 1-4.

5.21 Find the pH and percent hydrolysis of a $C_{cb} = 0.350$ M solution of NaAc ($K_a = 1.78 \times 10^{-5}$).

$K_{cb} = K_w/K_a = 5.62 \times 10^{-10}$; hence, the point (C_{cb}, K_{cb}) lies in Region 1-2 of Fig. 5-1, and

$$[OH^-] = \sqrt{(0.350)(5.62 \times 10^{-10})} = 1.40 \times 10^{-5}$$

pOH = 4.853, and pH = 14.000 − 4.853 = 9.147.
Since $K_{cb} \ll [OH^-]$, (5.14) gives

$$\% \text{ hydrol} \approx \frac{K_{cb}}{[OH^-]}(100\%) = 4.01 \times 10^{-3}\%$$

5.22 Find the pH and the percent hydrolysis of a $C_{cb} = 0.0568$ M solution of sodium dichloroacetate, NaA, given $K_a = 3.32 \times 10^{-2}$.

$K_{cb} = K_w/K_a = 3.01 \times 10^{-13}$, so that the point (C_{cb}, K_{cb}) lies in Region 2 of Fig. 5-1.

$$[OH^-] = \sqrt{(0.0568)(3.01 \times 10^{-13}) + 10^{-14}} = 1.65 \times 10^{-7}$$

pOH = 6.783, and pH = 7.217.
By (5.14), with $K_{cb} \ll [OH^-]$,

$$\% \text{ hydrol} \approx \frac{K_{cb}}{[OH^-]}(100\%) = 1.82 \times 10^{-4}\%$$

5.23 Find the pH and the percent hydrolysis of a $C_{cb} = 6.28 \times 10^{-4}$ M solution of the conjugate base NaA ($K_a = 3.54 \times 10^{-14}$), if the non-ideal solution contains sufficient NaCl to make the ionic strength $I = 0.100$. Assume the values

$$f_{H^+} = 0.826 \qquad f_{OH^-} = 0.754 \qquad f_{A^-} = 0.770 \qquad f_{HA} = 1.000$$

With $K_{cb} = K_w/K_a = 0.283$, the point (C_{cb}, K_{cb}) lies in Region 1-4 of Fig. 5-1. Supposing the $[OH^-]$ equation to hold under non-ideal conditions as well,

$$[OH^-] = C_{cb} = 6.28 \times 10^{-4}$$

$$pOH = -\log[OH^-] - \log f_{OH^-} = 3.325 \text{ (confirming Assumption 1), and}$$

$$pH = pK_w - pOH = 10.675$$

For a non-ideal solution, (5.14) becomes

$$\% \text{ hydrol} = \frac{K_{cb}/[OH^-]}{(K_{cb}/[OH^-]) + K_{cbf}}(100\%)$$

Substitution of the values

$$\frac{K_{cb}}{[OH^-]} = 451 \qquad K_{cbf} = \frac{f_{HA}\,f_{OH^-}}{f_{A^-}} = 0.979$$

yields

$$\% \text{ hydrol} = \frac{451}{452}(100\%) = 99.8\%$$

and Assumption 4 is justified.

5.24 A $0.00375\,M$ solution of the salt NaA of a weak acid has pH 7.554. Determine the acid dissociation constant K_a.

Solve the linear equation (5.9) for K_{cb} in terms of $[OH^-]$, C_{cb}, and K_w:

$$K_{cb} = \frac{[OH^-]\{[OH^-]^2 - K_w\}}{K_w + C_{cb}[OH^-] - [OH^-]^2}$$

Using the relations $[OH^-] = K_w/[H^+]$ and $K_a = K_w/K_{cb}$, obtain from this the formula

$$K_a = \frac{[H^+] + C_{cb} - (K_w/[H^+])}{(K_w/[H^+]^2) - 1} \tag{5.15}$$

From the data,

$$[H^+] = \text{antilog}(-7.554) = 2.79 \times 10^{-8}$$

$$\frac{K_w}{[H^+]} = \text{antilog}(7.554 - 14) = 3.58 \times 10^{-7}$$

$$\frac{K_w}{[H^+]^2} = \text{antilog}(15.108 - 14) = 12.8$$

$$C_{cb} = 3.75 \times 10^{-3}$$

Substitution in (5.15) yields

$$K_a \approx \frac{3.75 \times 10^{-3}}{11.8} = 3.18 \times 10^{-4}$$

Supplementary Problems

SOLUTIONS OF WEAK ACIDS

5.25 Find the pH and the percent dissociation of a $0.0375\,M$ solution of HCN ($K_a = 6.17 \times 10^{-10}$).
Ans. 5.318, 0.0128%

5.26 Find the pH and the percent dissociation of a $0.0684\,M$ solution of nitrous acid, HNO_2, if $K_a = 1.58 \times 10^{-3}$.
Ans. 2.016, 14.1%

5.27 Find the pH and the percent dissociation of a 4.25×10^{-3} M solution of HClO ($K_a = 2.95 \times 10^{-8}$).
Ans. 4.951, 0.264%

5.28 Find the pH and the percent dissociation of a 8.26×10^{-4} M solution of phenol ($K_a = 1.05 \times 10^{-10}$).
Ans. 6.507, 0.0338%

5.29 Find the pH and the percent dissociation of a 2.64×10^{-5} M solution of HClO ($K_a = 2.95 \times 10^{-8}$).
Ans. 6.058, 3.27%

5.30 Find the concentration of a solution of HNO_2 of pH 4.336, given $K_a = 1.58 \times 10^{-3}$.
Ans. $C_a = 4.30 \times 10^{-5}$ M

5.31 Find the concentration of a solution of HCN of pH 6.424, given $K_a = 6.17 \times 10^{-10}$.
Ans. $C_a = 2.14 \times 10^{-4}$ M

5.32 Find the dissociation constant of a 3.65×10^{-3} M solution of HA, if it has pH 3.535.
Ans. $K_a = 2.54 \times 10^{-5}$

5.33 Find the dissociation constant of a 6.75×10^{-5} M solution of HA, if it is 24.2% dissociated.
Ans. $K_a = 5.22 \times 10^{-6}$

SOLUTIONS OF WEAK BASES

5.34 Find the pH and the percent dissociation of a 0.260 M solution of the weak base pyridine ($K_b = 1.70 \times 10^{-9}$).
Ans. 9.323, 0.00809%

5.35 Find the pH and the percent dissociation of a 0.624 M solution of the weak base methylamine ($K_b = 5.25 \times 10^{-4}$).   Ans. 12.258, 2.90%

5.36 Find the concentration of a solution of pyridine of pH 7.588, given $K_b = 1.70 \times 10^{-9}$.
Ans. $C_b = 8.25 \times 10^{-5}$ M

SOLUTIONS OF CONJUGATE ACIDS

5.37 Find the pH and the percent hydrolysis of a 0.0335 M solution of pyridine hydrochloride (the salt of a weak base), if $K_b = 1.78 \times 10^{-9}$.   Ans. 3.366, 1.29%

5.38 Find the pH and the percent hydrolysis of a 0.0684 M solution of piperdine hydrochloride (the salt of a weak base), if $K_b = 1.60 \times 10^{-3}$.   Ans. 6.179, 9.46×10^{-4}%

SOLUTIONS OF CONJUGATE BASES

5.39 Find the pH and the percent hydrolysis of a 0.0816 M solution of sodium o-phthalate (the salt of a weak acid), if $K_a = 1.12 \times 10^{-3}$.   Ans. 7.934, 0.00104%

5.40 Find the pH and the percent hydrolysis of a 0.0415 M solution of sodium acetate (the salt of a weak acid), if $K_a = 1.78 \times 10^{-5}$.   Ans. 8.684, 0.0116%

5.41 Find the pH and the percent hydrolysis of a 0.0725 M solution of sodium cyanide (the salt of a weak acid), if $K_a = 4.93 \times 10^{-10}$.   Ans. 11.080, 1.66%

Chapter 6

Weak Acid or
Weak Base Titrations

6.1 INTRODUCTION

The present chapter may be considered an extension of Chapter 5: instead of the solution of a single acid (base, conjugate acid, conjugate base), we will study the mixture of two such reagents, usually in a titration. The general equation for $[H^+]$ or $[OH^-]$ turns out to be a cubic, like (5.9) when the titrant is a strong acid or strong base; but now the coefficients involve *three* parameters. The process of simplifying this equation is therefore even more complicated than in Chapter 5. Problem 6.15 will give some indication of why all these complications are necessary.

6.2 GENERAL EQUATION FOR TITRATION CURVES

Consider a mixture of V_a volume units of C_a° M *weak acid*, HA, having dissociation constant K_a, with V_b volume units of C_{sb}° M *strong base*, MOH (e.g., NaOH). The superscript on the concentrations is to emphasize that these are starting values, prior to any reaction; later we shall use the symbols C_a and C_{sb} to denote the concentrations just before equilibrium at an arbitrary stage of the reaction. The five concentration terms needed to define the mixture must satisfy the following five equations:

1. Charge balance: $\qquad\qquad [H^+] + [M^+] = [OH^-] + [A^-]$
2. Mass balance for acid: $\qquad C_a^\circ V_a = ([HA] + [A^-])(V_a + V_b)$
3. Mass balance for base: $\qquad C_{sb}^\circ V_b = [M^+](V_a + V_b)$
4. Water equilibrium: $\qquad\quad K_w = [H^+][OH^-]$
5. Acid equilibrium: $\qquad\quad\; K_a = [H^+][A^-]/[HA]$

Elimination of all unknowns except $[H^+]$ leads to

$$V_b\left(C_{sb}^\circ + [H^+] - \frac{K_w}{[H^+]}\right) = V_a\left(\frac{C_a^\circ K_a}{K_a + [H^+]} - [H^+] + \frac{K_w}{[H^+]}\right) \qquad (6.1)$$

a cubic in $[H^+]$ that in reality depends on just three parameters (because only three are involved in the system of five equations above); these parameters will be assigned in Section 6.3.

Equation (6.1) is the "acid form" of the general titration equation; under the conversions of Table 5-1, it will apply to a mixture of a conjugate acid and a strong base, a weak base and a strong acid, or a conjugate base and a strong acid. In the limit as $K_a \to \infty$, (6.1) goes over into (4.4b), the strong acid–strong base equation.

Our procedure in the remainder of this chapter will parallel that of Chapter 5. We shall replace (6.1)—more precisely, the system of five equations giving rise to (6.1)—by equivalent equations in the concentration shifts necessary to establish equilibrium. Solutions of these equations under a variety of simplifying assumptions will yield approximate solutions of (6.1) in various regions of parameter space. However, before carrying out this program, we wish to illustrate the use of (6.1) to construct titration curves.

EXAMPLE 6.1 Consider the titration of $V_a = 25.00$ mL of C° M HA ($K_a = 1.00 \times 10^{-5}$) with various volumes V_b of C° M NaOH. It is desired to obtain a titration curve for each of eight values of C°: 1.0, 10^{-1}, $10^{-2}, \ldots, 10^{-7}$.

The method is to employ (6.1) inversely. For a fixed value of $C°$, all quantities are known except V_b and $[H^+]$; and the equation is linear in V_b. Therefore, with the aid of a programmable calculator, one can easily find the values of V_b that correspond to assigned values of $[H^+]$. The resulting curves are shown in Fig. 6-1.

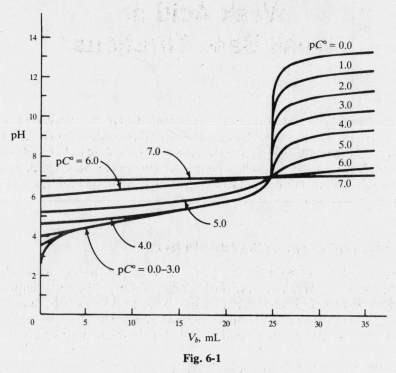

Fig. 6-1

Three characteristics of these titration curves are worth noting:

1. The pH halfway to the equivalence point (which is, of course, at $V_b = 25.00$ mL) is very nearly equal to the pK_a (5.000) for the more concentrated solutions ($pC° = 0, 1, 2, 3$); see Problem 6.3(b). The approximation becomes less valid as the solution becomes more dilute and/or as the strength of the acid increases. For very large values of K_a, where the acid is a strong acid, the approximation is totally invalid, as is shown by the curves of Fig. 4-1.

2. The pH exceeds 7.0 at the equivalence point in the titration of a weak acid with a strong base. This is due to the fact that the resultant salt solution involves some hydrolysis of the conjugate base, A^-, according to (5.5). It follows from (5.6) that the weaker the acid (the smaller the value of K_a), the stronger will be the conjugate base (the larger will be the value of K_{cb}) and the higher will be the pH at the equivalence point.

3. *Buffer solutions* are relatively concentrated (often 0.1 to 1.0 M) mixtures of a weak acid (or a weak base) and its salt; they have the property that they undergo very little change in pH when small amounts of H^+ or OH^- are added or when the solution is diluted, even by a factor of 10. Therefore, the best buffer solutions are those that correspond to the flattest portion of the weak-acid titration curve. Figure 6-1 shows this portion to lie about the half-equivalence point, which implies [see Problem 6.3(b)] that the optimal buffer should contain nearly equal amounts of weak acid and salt.

6.3 TITRATION REGIONS

Figure 6-2 presents a generalized curve for the titration of a weak acid with a strong base. In shape, the curve resembles those of Fig. 6-1, though $C_a° = C_{sb}°$ is not presumed in Fig. 6-2. The abscissa in Fig. 6-2 is not the volume V_b of added strong base, but the proportional quantity

$$\alpha = \frac{C_{sb}° V_b}{C_a° V_a} \equiv \text{fraction of HA titrated} \qquad (6.2)$$

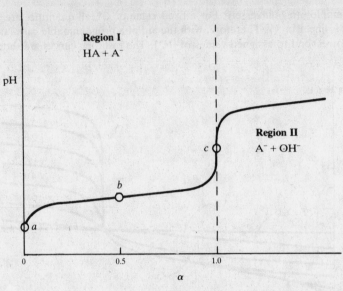

Fig. 6-2

By the usual notational changes, Fig. 6-2 may be made to apply to other titrations (e.g., for the titration of a weak base with a strong acid change the ordinate to pOH and interchange subscripts a and b in (6.2)).

Three points, a, b, and c, and two regions, I and II, are indicated in Fig. 6-2. Point a represents a $C_a^\circ M$ solution of the acid HA, whereas point c (the equivalence point) represents a $C_{cb} M$ solution of the conjugate base A^-. Thus, only a single principal species is present at a or at c, and the equation for $[H^+]$ involves only two parameters, C and K. Such equations have been treated in detail in Section 5.2.

Parameters in Regions I and II

At all stages of the titration other than point a ($\alpha = 0$) and point c ($\alpha = 1.0$), two principal species are present. For calculating equilibrium at an arbitrary stage, the pre-equilibrium concentrations of the two principle species prove to be more useful parameters than do the starting concentrations C_a° and C_{sb}°.

Region I ($0 < \alpha < 1$). In this region only part of the weak acid is neutralized. At any given stage, the principal species are HA and A^-, of respective concentrations

$$C_a = \frac{C_a^\circ V_a - C_{sb}^\circ V_b}{V_a + V_b} \qquad C_{cb} = \frac{C_{sb}^\circ V_b}{V_a + V_b} \tag{6.3}$$

As implied in Example 6.1, Region I falls into an acidic and a basic subregion. Clearing (6.1) of fractions, we obtain as the pH equation, useful in the acidic subregion,

$$[H^+]^3 + (C_{cb} + K_a)[H^+]^2 - (C_a K_a + K_w)[H^+] - K_a K_w = 0 \tag{6.4a}$$

or based on Table 5-1 as the pOH equation, useful in the basic subregion,

$$[OH^-]^3 + (C_a + K_{cb})[OH^-]^2 - (C_{cb} K_{cb} + K_w)[OH^-] - K_{cb} K_w = 0 \tag{6.4b}$$

Equations (6.4) involve three independent parameters: the two C's and a K. For the reaction parameter α, combination of (6.3) and (6.2) gives

$$\alpha = \frac{1}{1 + (C_a/C_{cb})} \tag{6.5}$$

Region II ($\alpha > 1$). In this region, which is past the equivalence point, all of the weak acid has been neutralized and excess strong base is present. The principal species are A^- and OH^-, with concentrations

$$C_{cb} = \frac{C_a^\circ V_a}{V_a + V_b} \qquad C_{sb} = \frac{C_{sb}^\circ V_b - C_a^\circ V_a}{V_a + V_b} \qquad (6.6)$$

The pOH equation (the region being entirely basic) takes the form

$$[OH^-]^3 + (K_{cb} - C_{sb})[OH^-]^2 - \{K_{cb}(C_{cb} + C_{sb}) + K_w\}[OH^-] - K_{cb}K_w = 0 \qquad (6.7)$$

and the reaction parameter may be written

$$\alpha = 1 + \frac{C_{sb}}{C_{cb}} \qquad (6.8)$$

6.4 CONCENTRATION SHIFTS TO EQUILIBRIUM

From Example 6.1, we know that in the titration of a weak acid the mixture is acidic at $\alpha = 0$ and basic for $\alpha \geq 1$. Therefore, Region I falls into acidic and basic subregions, whereas Region II is basic throughout.

Region I, Acidic Solution

The concentration shifts are:

$$
\begin{array}{ccccc}
\textit{initial} & C_a & & 10^{-7} & C_{cb} \\
& HA & \longrightarrow & H^+ \quad + & A^- \\
\textit{equilibrium} & C_a - a & & 10^{-7} + a - w & C_{cb} + a
\end{array} \qquad (6.9a)
$$

$$
\begin{array}{cccc}
\textit{initial} & & 10^{-7} & 10^{-7} \\
& H_2O & \longrightarrow & H^+ \quad + \quad OH^- \\
\textit{equilibrium} & & 10^{-7} + a - w & 10^{-7} - w
\end{array} \qquad (6.9b)
$$

where, as in Chapter 5, a mol/L of HA dissociates and w mol/L each of H^+ and OH^- back-react to restore the water equilibrium. From ($6.9a$),

$$K_a = \frac{[H^+](C_{cb} + a)}{C_a - a} \qquad (6.10)$$

Region I, Basic Solution

Here, b mol/L of the conjugate base reacts and w mol/L each of H^+ and OH^- back-react.

$$
\begin{array}{cccccc}
\textit{initial} & C_{cb} & & C_a & 10^{-7} \\
& A^- & + H_2O \longrightarrow & HA & + & OH^- \\
\textit{equilibrium} & C_{cb} - b & & C_a + b & 10^{-7} + b - w
\end{array} \qquad (6.11a)
$$

$$
\begin{array}{cccc}
\textit{initial} & & 10^{-7} & 10^{-7} \\
& H_2O & \longrightarrow & H^+ \quad + \quad OH^- \\
\textit{equilibrium} & & 10^{-7} - w & 10^{-7} + b - w
\end{array} \qquad (6.11b)
$$

From ($6.11a$),

$$K_{cb} = \frac{(C_a + b)[OH^-]}{C_{cb} - b} \qquad (6.12)$$

Region II (Basic Solution)

The concentration shifts are:

$$
\begin{array}{ccccc}
\textit{initial} & C_{cb} & & 0 & C_{sb} + 10^{-7} \\
& A^- & + H_2O \longrightarrow & HA & + OH^- \\
\textit{equilibrium} & C_{cb} - b & & b & C_{sb} + b + 10^{-7} - w
\end{array}
\qquad (6.13a)
$$

$$
\begin{array}{cccc}
\textit{initial} & & 10^{-7} & C_{sb} + 10^{-7} \\
& H_2O \longrightarrow & H^+ & + OH^- \\
\textit{equilibrium} & & 10^{-7} - w & C_{sb} + b + 10^{-7} - w
\end{array}
\qquad (6.13b)
$$

From (6.13a),

$$ K_{cb} = \frac{b[OH^-]}{C_{cb} - b} \qquad (6.14) $$

6.5 SIMPLIFIED EQUATIONS FOR THE TITRATION OF A WEAK ACID

The simplifying assumptions to be invoked are Assumptions 1 through 4 of Section 5.2, here retitled Assumptions 1a through 4a, plus:

Assumption 5a. The solution in Region I is acidic, and the concentration of additional conjugate base A^- arising from dissociation of C_a M HA is less than 1% of C_{cb}, the concentration of conjugate base already present.

Assumption 6a. The solution in Region I is basic, and the concentration of additional acid HA arising from hydrolysis of C_{cb} M conjugate base is less than 1% of C_a, the concentration of acid already present.

Assumption 7a. In Region II, the concentration b of additional hydroxide formed by reaction of the conjugate base is less than 1% of C_{sb}, the concentration of hydroxide from the excess strong base already present.

Assumption 8a. In Region II, more than 99% of the conjugate base reacts (it is thus a strong base), so that $b \approx C_{cb}$.

Table 6-1. Simplified Equations for Region I

Valid Assumptions	Simplified Equation	
1a, 4a	$[H^+] = C_a$	(6.15)
1a	$[H^+]^2 + (C_{cb} + K_a)[H^+] - C_a K_a = 0$	(6.16)
1a, 2a, 5a or 1b, 2b, 6a or 2a, 5a or 2b, 6a	$[H^+] = \dfrac{K_a C_a}{C_{cb}}$	(6.17)
1b	$[OH^-]^2 + (C_a + K_{cb})[OH^-] - C_{cb} K_{cb} = 0$	(6.18)
1b, 4b	$[OH^-] = C_{cb}$	(6.19)
4a	$[H^+]^2 - C_a[H^+] - K_w = 0$	(6.20)
4b	$[OH^-]^2 - C_{cb}[OH^-] - K_w = 0$	(6.21)
3a or 3b	$[H^+] = [OH^-] = 1.00 \times 10^{-7}$	(6.22)

EXAMPLE 6.2 The Assumptions are stated in "acid" form (hence the "a" after each number). The "base" form is obtained by the rules of Table 5-1; e.g.,

Assumption 3b. The $[OH^-]$ produced by hydrolysis of A^- is less than 1.99×10^{-9}, so that pOH = 7.000.

The simplified forms of (6.4) that obtain when various combinations of the Assumptions are applied to (6.10) or (6.12) are displayed in Table 6-1. Similarly, Table 6-2 shows the simplifications of (6.7) that result when certain of the Assumptions are applied to (6.14). Tables 6-1 and 6-2 reflect only the combinations of greatest practical importance. Obviously, if *none* of the Assumptions holds in a given case, the full equation (6.4) or (6.7) must be used.

Table 6-2. Simplified Equations for Region II

Valid Assumptions	Simplified Equation	
1b, 7a	$[OH^-] = C_{sb}$	(6.23)
1b	$[OH]^2 + (K_{cb} - C_{sb})[OH^-] - K_{cb}(C_{sb} + C_{cb}) = 0$	(6.24)
1b, 8a	$[OH^-] = C_{sb} + C_{cb}$	(6.25)
7a	$[OH^-]^2 - C_{sb}[OH^-] - K_w = 0$	(6.26)
8a	$[OH]^2 - (C_{sb} + C_{cb})[OH^-] - K_w = 0$	(6.27)
3b	$[OH^-] = 1.00 \times 10^{-7}$	(6.22)

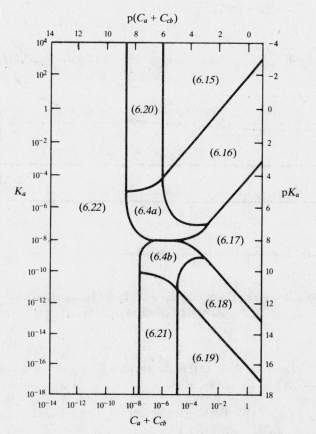

Fig. 6-3. The Case $\alpha = 0.100$

For practical applications of Tables 6-1 and 6-2, it is useful to have diagrams showing the parameter ranges over which the various simplified equations are valid. Since there are three free parameters, they must be subjected to a constraint to keep the diagrams planar (like Fig. 5-1).

Diagrams for Region I

Figure 6-3 shows which equations are valid in the specific case $\alpha = 0.100$; that is, it applies to the stage of titration at which 10% of the HA has been neutralized. Similar figures, also based on the free parameters $C_a + C_{cb}$ and K_a, would apply to other stages of the titration in Region I.

In Fig. 6-4, the free parameters are α and K_a. Under the constraint $C_a + C_{cb} = 0.1$, the first five equations of Table 6-1 are valid in the areas bounded by the solid curves; under $C_a + C_{cb} = 0.001$, they are valid in the areas bounded by the dashed curves (excepting the shaded areas, which require (6.4)). Equations (6.20), (6.21), and (6.22) do not apply to cases where $C_a + C_b > 0.001$ and, hence, do not appear in Fig. 6-4.

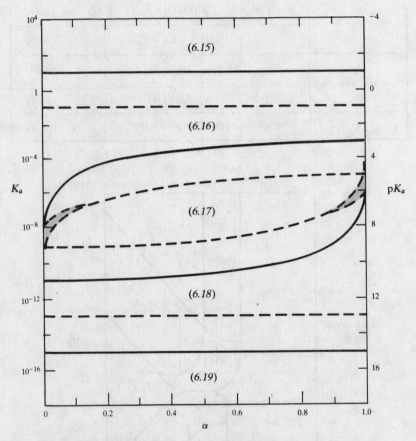

Fig. 6-4.   The Cases $C_a + C_{cb} = 0.1$ (Solid Boundaries) and
$C_a + C_{cb} = 0.001$ (Dashed Boundaries)

EXAMPLE 6.3 At the halfway point $(\alpha = 0.5)$ in the titration of a weak acid having $K_a = 1.0 \times 10^{-5}$, the appropriate equation is (6.17), if $C_a + C_{cb} \approx 0.1$—that is, if $C_a = C_{cb} \approx 0.05$. Thus, $[H^+] = K_a$, as predicted in Example 6.1.

If $C_a + C_{cb} \approx 0.001$—i.e., if $C_a = C_{cb} \approx 0.0005$—the proper equation is (6.16):

$$[H^+]^2 + (5.10 \times 10^{-4})[H^+] - (5.00 \times 10^{-9}) = 0$$

which has the solution

$$[H^+] \approx 0.96 \times 10^{-5} = 0.96 K_a$$

This is again in accord with Example 6.1, which stated that the approximation $[H^+] = K_a$ at the halfway point becomes worse as concentrations decrease.

Diagrams for Region II

Figure 6-5 is analogous to Fig. 6-3 and shows which equations are valid in the specific case $\alpha = 1.02$; that is, it applies to the stage of the titration at which the added strong base is 2% more than needed to reach the equivalence point, whence $C_{sb} = 0.02\,C_{cb}$. Similar figures, also based on the free parameters: C_{cb} and K_{cb}, would apply to other stages of the titration in Region II.

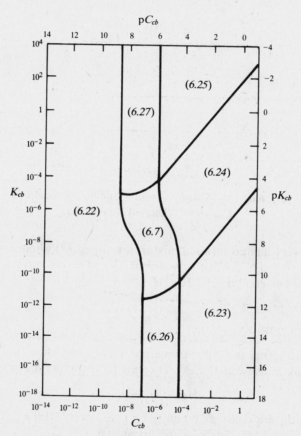

Fig. 6-5. The Case $C_{sb} = 0.02\,C_{cb}$

Figure 6-6 is analogous to Fig. 6-4. Solid curves are area boundaries under the constraint $C_{cb} = 0.1$; dashed curves, under $C_{cb} = 0.001$. The two free parameters are the *percent past the equivalence point,*

$$100(\alpha - 1) = 100(C_{sb}/C_{cb})$$

and K_{cb}. Equations (6.22), (6.26), and (6.27) do not apply to cases where $C_{cb} > 0.001$ and, hence, do not appear in Fig. 6-6.

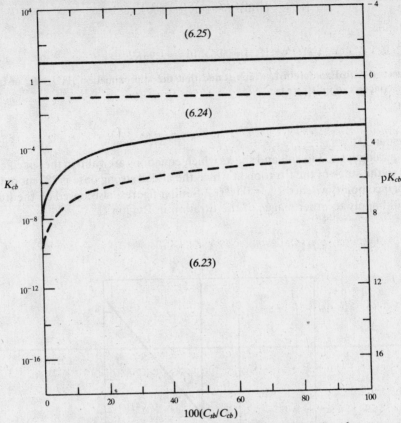

Fig. 6-6.   The Cases $C_{cb} = 0.1$ (Solid Boundaries) and
$C_{cb} = 0.001$ (Dashed Boundaries)

Most texts treat only very limited titration mixtures, using (6.17) whenever $0 < \alpha < 1.0$ and (6.23) whenever $\alpha > 1.0$. (See Problem 6.15.) The solved problems of this chapter consider the full range of possibilities for concentrations greater than $10^{-3}\ M$.

6.6   pH COLOR INDICATORS

Color indicators, which signal the equivalence point in the titration of a weak acid (or a weak base), are themselves weak acids (weak bases). The general formula for the indicator dissociation is:

$$HIn \rightarrow H^+ + In^-$$

where HIn or In⁻, or both, are colored. When $pH = pK_a$, then $[HIn] = [In^-]$; therefore, crossover from predominately HIn to predominately In⁻ occurs at that pH. As a result, an indicator to be used in a given titration is chosen so that its pK_a approximates as closely as possible the pH at the equivalence point. The actual color change takes place gradually, more or less in the pH-range $pK_a - 1.0$ to $pK_a + 1.0$. Therefore, pH color indicators are useful in titrations only when there is a change of at least two pH-units when the drop of titrant needed to reach the equivalence point is added. As is seen in Fig. 6-1, a change of two pH-units will occur only for solutions where, for $K_a = 10^{-5}$, the concentration is at least $10^{-3}\ M$.

Solved Problems

TITRATION CURVE EQUATION

6.1   It required 38.72 mL of 0.0678 M NaOH to reach the equivalence point in the titration of 25.00 mL of acetic acid ($K_a = 1.78 \times 10^{-5}$). What volume of NaOH was added when the pH was 5.000?

The original concentration of the HAc was $C_a^o = (38.72)(0.0678)/25.00 = 0.105$ M. With $[H^+] = 1.00 \times 10^{-5}$, all quantities in ($6.1$) are known except V_b. Solving the linear equation, $V_b = 24.8$ mL.

REGION I: DERIVATIONS

6.2   Derive (6.16) of Table 6-1.

Since Assumption 1a is valid, we may take $a = [H^+]$ in (6.9). Then, (6.10) becomes

$$K_a = \frac{[H^+](C_{cb} + [H^+])}{C_a - [H^+]}$$

which rearranges to (6.16).

6.3   (a) Derive (6.17) of Table 6-1 in the event that Assumptions 1b, 2b, and 6a are valid. (b) Show that wherever (6.17) holds, pH = pK_a at the halfway point in the titration of HA.

(a)   Under Assumption 1b, $b = [OH^-] = K_w/[H^+]$; under Assumption 2b, $[A^-] = C_{cb}$. Assumption 6a translates to $[HA] = C_a$. Then, (6.12) becomes

$$\frac{K_w}{K_a} = \frac{C_a(K_w/[H^+])}{C_{cb}}$$

which rearranges to (6.17).

(b)   By (6.5), $\alpha = 1/2$ implies $C_a/C_{cb} = 1$; and (6.17) then gives $[H^+] = K_a$, or pH = pK_a.

REGION I: PREDICTING pH

6.4   Approximately what will be the pH of a solution made by mixing 25.00 mL of 0.100 M HAc ($K_a = 1.78 \times 10^{-5}$) with 12.50 mL of 0.100 M NaOH?

The NaOH, which is present in a limited amount as compared with the HAc, will neutralize part of the HAc. The concentrations of the principal species will be:

$$[Na^+] = [Ac^-] = \frac{1.25 \text{ mmol}}{37.50 \text{ mL}} = 0.0333 \qquad [HAc] = \frac{1.25 \text{ mmol}}{37.50 \text{ mL}} = 0.0333$$

This solution is the same as that which exists halfway to the equivalence point in a titration of HAc with NaOH; it corresponds to point b in Fig. 6-2. From Problem 6.3(b), pH $\approx$ pK_a = 4.750.

6.5   Predict the pH of a solution that is made by mixing 25.00 mL of 0.100 M HAc ($K_a = 1.78 \times 10^{-5}$) with 10.00 mL of 0.100 M NaOH.

Since 1.00 mmol of NaOH is added to 2.50 mmol of HAc, $1.00/2.50 = 40\%$ of the acid will be neutralized, putting the mixture just to the left of point b in Fig. 6-2. Since pH $\approx$ pK_a = 4.75 at point b, and since the titration curve is relatively flat in this region, the pH of the mixture will be slightly less than pK_a, perhaps in the range 4.4–4.6.

6.6   Predict the pH of a solution that is made by mixing 25.00 mL of 0.100 M NH_4OH ($K_b = 1.80 \times 10^{-5}$) with 17.50 mL of 0.100 M HCl.

This is a mixture of 2.50 mmol NH_4OH and 1.75 mmol HCl; it corresponds to a titration of NH_4OH at the stage where $1.75/2.50 = 70\%$ of the base has been neutralized; that is, to a point to the right of point b in the "base" version of Fig. 6-2. At point b, $pOH \approx pK_b = 4.75$, or

$$pH \approx 14.00 - 4.75 = 9.25$$

Therefore, this mixture will have a pH that is slightly less, perhaps in the range 8.8–9.0.

6.7   Predict the pH of a mixture of 25.00 mL of 0.100 M HAc ($K_a = 1.78 \times 10^{-5}$), 50.00 mL of 0.100 M NaAc, and 25.00 mL of 0.100 M HCl.

The HCl will react with some of the NaAc to produce additional HAc. The concentrations of the principal species will be:

$$[Na^+] = \frac{5.00 \text{ mmol}}{100.00 \text{ mL}} = 0.0500 \qquad [Cl^-] = \frac{2.50}{100.00} = 0.0250$$

$$[Ac^-] = \frac{5.00 - 2.50}{100.00} = 0.0250 \qquad [HAc] = \frac{2.50 + 2.50}{100.00} = 0.0500$$

From (6.5),

$$\alpha = \frac{1}{1 + \dfrac{0.0500}{0.0250}} = 0.333$$

Therefore, the solution is to the left of point b (Fig. 6-2), where the $pH \approx pK_a = 4.75$, and so the pH should be approximately 4.3–4.5.

REGION I: CALCULATING pH

6.8   What is the pH of the solution of Problem 6.4? Justify any assumptions.

Because $C_a + C_{cb} = 0.0666 \approx 0.1$, the solid curves of Fig. 6-4 can be used to identify the appropriate equation. From the K_a-value and $\alpha = 0.5$, (6.17) should apply, so that

$$[H^+] = K_a = 1.78 \times 10^{-5} \qquad \text{or} \qquad pH = 4.750$$

Since the pH is less than 7.00, Assumption 1a is valid, making $a = [H^+]$. Since this value of a is

$$\frac{1.78 \times 10^{-5}}{0.0333}(100\%) = 0.0535\%$$

of either C_a or C_{cb}, Assumptions 2a and 5a are also valid.

6.9   Repeat Problem 6.8, including corrections for non-ideal effects. For this solution,

$$f_{H^+} = 0.870 \qquad f_{OH^-} = 0.834 \qquad f_{Na^+} = f_{Ac^-} = 0.842 \qquad f_{HAc} = 1.000$$

Substituting K_a/K_{af} for K_a in (6.17) results in the non-ideal equivalent

$$[H^+] = \frac{C_a K_a}{C_{cb} K_{af}} \qquad \text{where} \qquad K_{af} = \frac{f_{H^+} f_{Ac^-}}{f_{HAc}} = 0.733$$

Evaluating: $[H^+] = 2.43 \times 10^{-5}$, $a_{H^+} = (2.43 \times 10^{-5})(0.870) = 2.11 \times 10^{-5}$, $pH = -\log a_{H^+} = 4.676$ (Assumption 1a is valid).

Because the amount of HAc dissociation is $a = [H^+]$, (5.13) gives:

$$\% \text{ dissoc of HAc} = \frac{a}{C_a}(100\%) = 0.0730\%$$

$$\% \text{ extra Ac}^- \text{ from dissoc} = \frac{a}{C_s}(100\%) = 0.0730\%$$

and Assumptions 2a and 5a are also valid.

6.10 What is the pH of the solution of Problem 6.6? Justify any assumptions.

We have: $C_b = 0.75 \text{ mmol}/42.50 \text{ mL} = 0.018$, and $C_{ca} = 1.75/42.50 = 0.0412$. Thus, $C_b + C_{ca} = 0.059$ and

$$\alpha = \frac{1}{1 + \frac{0.018}{0.0412}} = 0.70$$

(as found in Problem 6.6). Locating the point (α, K_b) in the "base" version of Fig. 6-4, we find that, because $C_b + C_{ca} \approx 0.1$, the appropriate equation is the "base" analog of (6.17):

$$[\text{OH}^-] = \frac{K_b C_b}{C_{ca}} = \frac{(1.80 \times 10^{-5})(0.018)}{0.0412} = 7.9 \times 10^{-6} \qquad \text{or} \qquad \text{pOH} = 5.10$$

Since pOH < 6.00, Assumption 1b is valid and $b = [\text{OH}^-]$. Therefore,

$$\% \text{ dissoc of NH}_4\text{OH} = \frac{7.9 \times 10^{-6}}{0.018}(100\%) = 0.44\%$$

$$\% \text{ extra NH}_4^+ \text{ from dissoc} = \frac{7.9 \times 10^{-6}}{0.0412}(100\%) = 0.012\%$$

Thus, Assumptions 2b and 5b are also valid.

6.11 What is the pH of a solution made by mixing 20.00 mL of 0.100 M HClO ($K_a = 2.96 \times 10^{-8}$) and 12.00 mL of 0.100 M NaOH? Justify any assumptions.

The NaOH partially neutralizes the HClO, and the concentrations of the principal species are:

$$C_{cb} = [\text{ClO}^-] = \frac{1.20 \text{ mmol}}{32.00 \text{ mL}} = 0.0375 \qquad C_a = [\text{HClO}] = \frac{2.00 - 1.20}{32.00} = 0.025$$

Thus, $C_a + C_{cb} = 0.063 \approx 0.1$ and

$$\alpha = \frac{1}{1 + \frac{0.025}{0.0375}} = 0.60$$

The coordinates α, K_a in Fig. 6-4 indicate that (6.17) should be valid:

$$[\text{H}^+] = \frac{(2.96 \times 10^{-8})(0.025)}{0.0375} = 2.0 \times 10^{-8} \qquad \text{or} \qquad \text{pH} = 7.70$$

Since the solution is basic but pH < 8.00 (pOH > 6.00), Assumption 1b is *not* valid. Therefore,

$$b = [\text{OH}^-] - [\text{H}^+] = 4.8 \times 10^{-7}$$

$$\% \text{ hydrol of ClO}^- = \frac{4.8 \times 10^{-7}}{0.0375}(100\%) = 0.0013\%$$

$$\% \text{ extra HClO from hydrol} = \frac{4.8 \times 10^{-7}}{0.025}(100\%) = 0.0019\%$$

which validates Assumptions 2b and 6a. Thus, the use of (6.17) was justified.

6.12 Find the pH of a solution that is $0.0750\ M$ dichloroacetic acid, HA ($K_a = 3.32 \times 10^{-2}$), and $0.0250\ M$ NaA. Justify any assumptions.

$C_a + C_{cb} = 0.100$, $\alpha = 0.250$; the coordinates α, K_a in Fig. 6-4 indicate that (6.16) is the appropriate simplified equation. Solving

$$[H^+]^2 + 0.0582[H^+] - 0.00249 = 0$$

by (1.4) yields $[H^+] = 0.0287$, or pH = 1.543 (Assumption 1a is valid, and so $a = [H^+]$).

$$\% \text{ dissoc of HA} = \frac{0.0287}{0.0750}(100\%) = 38.3\%$$

(Assumption 2a is *not* valid) and

$$\% \text{ extra } A^- \text{ from dissoc of HA} = \frac{0.0287}{0.0250}(100\%) = 115\%$$

(Assumption 5a is *not* valid). Thus, only Assumption 1a is valid, and the use of (6.16) is justified.

6.13 When 90.0% of the way to the equivalence point in the titration of a solution of the weak base aniline, B, with HCl, the mixture has as the concentrations of the principal species: $[B] = 1.00 \times 10^{-3}$ and $[BH^+] = 9.00 \times 10^{-3}$. Given $K_b = 4.27 \times 10^{-10}$, find the pH. Justify any assumptions.

$C_b + C_{ca} = 0.0100$, $\alpha = 0.900$; the coordinates α, K_b in the "base" version of Fig. 6-4 indicate that the equation analogous to (6.18) is valid:

$$[H^+]^2 + (C_b + K_{ca})[H^+] - C_{ca}K_{ca} = 0 \quad \text{or} \quad [H^+]^2 + (1.02_3 \times 10^{-3})[H^+] - (2.11 \times 10^{-7}) = 0$$

Solving by (1.4), $[H^+] = 1.76 \times 10^{-4}$, or pH = 3.754 (Assumption 1a is valid). Since the solution is acidic, this indicates that the conjugate acid, BH^+, dissociates.

$$\% \text{ dissoc of } BH^+ = \frac{1.76 \times 10^{-4}}{9.00 \times 10^{-3}}(100\%) = 1.96\%$$

$$\% \text{ extra B from dissoc} = \frac{1.76 \times 10^{-4}}{1.00 \times 10^{-3}}(100\%) = 17.6\%$$

Thus, neither Assumption 2a nor Assumption 5b is valid; only Assumption 1a is valid, and so the "base" analog of (6.18) may be used.

6.14 What is the pH of a solution prepared by mixing 20.00 mL of $0.0250\ M$ sodium formate, NaA, with 30.00 mL of $0.100\ M$ HCl ($K_a = 1.77 \times 10^{-4}$)? Justify any assumptions.

The 3.00 mmol of HCl will react with 3.00 mmol of A^- to produce 3.00 mmol of HA, leaving $5.00 - 3.00 = 2.00$ mmol of A^-. Therefore, the concentrations of principal species are:

$$C_a = [HA] = \frac{3.00}{50.00} = 0.0600 \qquad C_{cb} = [A^-] = \frac{2.00}{50.00} = 0.0400$$

With $C_a + C_{cb} = 0.100$, the point $(\alpha, K_a) = (0.400, 1.77 \times 10^{-4})$ in Fig. 6-4 lies just within the area for (6.17). Thus,

$$[H^+] = \frac{(1.77 \times 10^{-4})(0.0600)}{0.0400} = 2.66 \times 10^{-4} \quad \text{or} \quad \text{pH} = 3.576$$

whence Assumption 1a is valid and $a = [H^+]$.

$$\% \text{ dissoc of HA} = \frac{2.66 \times 10^{-4}}{0.0600}(100\%) = 0.433\%$$

$$\% \text{ extra } A^- \text{ from dissoc} = \frac{2.66 \times 10^{-4}}{0.0400}(100\%) = 0.665\%$$

which validate Assumptions 2a and 5a, and thus the use of (6.17).

6.15 Find the percent errors in $[H^+]$ in Problems 6.12 and 6.13 if the simplest equation, (6.17), is used instead of the valid equation.

For Problem 6.12, (6.17) gives $[H^+] = 0.0996$, whereas the correct $[H^+]$ equals 0.0287. Hence the error is

$$\frac{0.0996 - 0.0287}{0.0287}(100\%) = 247\%$$

For Problem 6.13, the "base" (6.17) yields

$$[OH^-] = \frac{K_b C_b}{C_{ca}} = \frac{(4.27 \times 10^{-10})(0.00100)}{0.00900} = 4.74 \times 10^{-11}$$

so that $[H^+] = 2.11 \times 10^{-4}$, whereas the correct $[H^+]$ is 1.76×10^{-4}. Hence the error is

$$\frac{2.11 - 1.76}{1.76}(100\%) = 20\%$$

REGION I: pH GIVEN

6.16 A weak acid, HA, has dissociation constant 3.00×10^{-5}. Using (a) the appropriate simplified pH equation, (b) the exact equation, determine how many moles of HCl must be added to 0.200 mol of NaA and the mixture diluted to 1.00 L to produce a pH of 5.000.

We are given pH = 5.000, $K_a = 3.00 \times 10^{-5}$, and $C_a + C_{cb} = 0.200$. Since one mole of added HCl converts one mole of A^- to one mole of HA, and since the final volume is 1.00 L, the unknown amount is

$$\text{mols HCl} = \text{mols HA} = C_a$$

(a) Since pH < 6.00, Assumption 1a is valid; no other Assumption can be made at this stage. Thus, the relevant equation from Table 6-1 is (6.16):

$$[H^+]^2 + (C_{cb} + K_a)[H^+] - C_a K_a = 0$$

or

$$10^{-10} + \{(0.200 - C_a) + (3.00 \times 10^{-5})\}10^{-5} - (3.00 \times 10^{-5})C_a = 0$$

Solving, $C_a = 0.05001 = 0.0500$.

(b) Equation (6.4a) becomes

$$10^{-15} + \{(0.200 - C_a) + (3.00 \times 10^{-5})\}10^{-10} - \{(3.00 \times 10^{-5})C_a + 10^{-14}\}10^{-5} - (3.00 \times 10^{-19}) = 0$$

Solving, $C_a = 0.0500099995 = 0.0500$.

REGION I: BUFFER SOLUTIONS

6.17 What change in pH will occur when 0.050 mmol of HAc ($K_a = 1.78 \times 10^{-5}$) is added to 100.0 mL of a buffer solution that is 0.1000 M HAc and 0.1000 M NaAc? Assume no change in volume.

For the buffer solution, $C_a + C_{cb} = 0.1000 + 0.1000 = 0.2000$ and $\alpha = 0.5000$. Figure 6-4 shows that (6.17) will be valid, whence pH = $pK_a = 4.750$.

The new mixture has $C_a = 0.1005$ and $C_{cb} = 0.1000$. From (6.17),

$$[H^+] = \frac{(1.78 \times 10^{-5})(0.1005)}{0.1000} = 1.79 \times 10^{-5}$$

or pH = 4.747. Thus, ΔpH = -0.003.

6.18 What changes in pH would occur if 0.050 mmol of (a) NaAc, (b) HCl, (c) NaOH, was added to the original buffer solution of Problem 6.17?

(a) The new concentrations would now be $C_a = 0.1000$ and $C_{cb} = 0.1005$. Hence, (6.17) would still be valid, giving

$$[H^+] = \frac{(1.78 \times 10^{-5})(0.1000)}{0.1005} = 1.77 \times 10^{-5}$$

or pH = 4.752, for a pH change of +0.002.

(b) The added HCl would convert 0.050 mmol of Ac^- to 0.050 mmol of HAc, so that the concentrations would now be $C_a = 0.1005$ and $C_{cb} = 0.0995$. Since (6.17) would still be valid,

$$[H^+] = \frac{(1.78 \times 10^{-5})(0.1005)}{0.0995} = 1.80 \times 10^{-5}$$

or pH = 4.745, for a ΔpH of -0.005.

(c) The added NaOH converts 0.050 mmol of HAc to 0.050 mmol of NaAc, so that the concentrations are now $C_a = 0.0995$ and $C_{cb} = 0.1005$. Since (6.17) is still valid,

$$[H^+] = \frac{(1.78 \times 10^{-5})(0.0995)}{0.1005} = 1.76 \times 10^{-5}$$

or pH = 4.754, whence ΔpH = +0.004.

6.19   What change in pH would occur if the original buffer solution of Problem 6.17 was diluted by a factor of 10.0?

The dilution would cause the concentrations of the principal species all to decrease by a factor of 10.0, so that now $C_a + C_{cb} = 0.02000$; α would remain at 0.5000. Figure 6-4 would still indicate (6.17), so that pH = 4.750 as before, and ΔpH = 0.000.

REGION II: CALCULATING pH

6.20   What is the pH of a solution prepared by mixing 25.00 mL of 0.100 M HAc ($K_a = 1.78 \times 10^{-5}$) with 30.00 mL of 0.100 M NaOH? Justify any assumptions.

The situation is that of being 5.00 mL (20%) past the equivalence point in a titration of 25.00 mL of 0.100 M HAc with 0.100 M NaOH. The NaOH is present in stoichiometric excess, so that, following neutralization of HAc, concentrations are:

$$C_{cb} = [Ac^-] = \frac{2.50 \text{ mmol}}{55.00 \text{ mL}} = 0.0455 \qquad C_{sb} = \frac{0.50}{55.00} = 0.0091$$

We have $K_{cb} = K_w/K_a = 5.62 \times 10^{-10}$; in Fig. 6-6, the point $(20, 5.62 \times 10^{-10})$ indicates the use of (6.23). Thus,

$$[OH^-] = C_{sb} = 0.0091$$

or pOH = 2.04 (Assumption 1b is valid), whence pH = 11.96. Since K_{cb} is so small and since there is excess OH^-, the hydrolysis of Ac^- should be negligible. Taking $b \ll C_{cb}$ in (6.14), that equation gives

$$\frac{b}{C_{cb}} = \frac{K_{cb}}{[OH^-]} = \frac{K_{cb}}{C_{sb}} = \frac{5.62 \times 10^{-10}}{0.0091} = 6.18 \times 10^{-8}$$

showing that the hydrolysis is indeed negligible. Further, Assumption 7a is valid, because b is also small compared to C_{sb}.

6.21   Repeat Problem 6.20 for the case where 40.00 mL of NaOH was used.

Since this mixture is even further (60%) past the equivalence point, Fig. 6-6 indicates that (6.23) will still apply:

$$[OH^-] = C_{sb} = \frac{(4.00 - 2.50) \text{ mmol}}{65.00 \text{ mL}} = 0.0231$$

or pOH = 1.636, and pH = 12.364. Assumptions 1b and 7a are validated as in Problem 6.20.

6.22   Determine the pH of a solution made by mixing 20.00 mL of 0.0500 M HA ($K_a = 3.50 \times 10^{-12}$) with 22.00 mL of 0.0500 M NaOH. Justify any assumptions.

This amounts to being 2.00 mL (10%) past the equivalence point in a titration of 20.00 mL of 0.0500 M HA with 0.0500 M NaOH. Concentrations are:

$$C_{cb} = [A^-] = \frac{1.00 \text{ mmol}}{42.00 \text{ mL}} = 0.0238 \qquad C_{sb} = \frac{0.10}{42.00} = 0.0024$$

In Fig. 6-6, the coordinates $(10, K_{cb}) = (10, 2.86 \times 10^{-3})$ indicate (6.24) as the appropriate equation. Solving the quadratic by (1.4) results in $[OH^-] = 0.0240$, pOH = 1.619 (Assumption 1b is valid), and pH = 12.381.

6.23   What is the pH of a solution that has $C_{cb} = 0.0700$ and $C_{sb} = 0.0300$, if $K_a = 1.00 \times 10^{-16}$?

This solution is 100(0.0300)/0.0700 = 42.9% past the equivalence point in the titration of HA with a strong base. The point $(42.9, K_{cb}) = (42.9, 10^2)$ in Fig. 6-6 indicates (6.25), which gives

$$[OH^-] = C_{cb} + C_{sb} = 0.1000$$

pOH = 1.0000 (Assumption 1b is valid), and pH = 13.0000.

To check that the conjugate base is actually a strong base (Assumption 8a), use the analog of (5.14):

$$\% \text{ hydrol} = \frac{K_{cb}}{K_{cb} + [OH^-]}(100\%) = \frac{100}{100.1}(100\%) = 99.9\%$$

Assumption 8a is valid.

REGION II: pH GIVEN

6.24   Can you find the concentration, C_{cb}, of acetate in a solution formed by titrating a solution of HAc past the equivalence point with NaOH, given $C_{sb} = 0.0628$, pH = 12.798, and $K_a = 1.78 \times 10^{-5}$?

The pH corresponds to $[OH^-] = 0.0628$, so that the C_{sb} M excess hydroxide accounts completely for the measured pH. One cannot obtain a value for C_{cb}, for the pH is specified with insufficient precision to allow determination of the quantity b, which results from the hydrolysis of the acetate and contributes to the hydroxide concentration. Putting it another way, the general equation (6.7) yields a *negative* value for C_{cb} in the case $[OH^-] = C_{sb}$, which means that $[OH^-]$ must actually exceed C_{sb} (the excess may be immeasurably small).

pH COLOR INDICATORS

6.25   Solutions identical to those existing at the equivalence point in a weak acid or weak base titration are found in (a) Problem 5.16, (b) Problem 5.17, (c) Problem 5.21, (d) Problem 5.22. Which indicator from Table 6-3 would be appropriate for the determination of each end point?

Table 6-3

Indicator	pH Transition Range	Color	
		Acid Form	Base Form
Thymol blue	1.2– 2.8	red	yellow
Methyl orange	3.1– 4.4	red	yellow
Methyl red	4.2– 6.3	red	yellow
Bromthymol blue	6.0– 7.6	yellow	blue
Phenol red	6.4– 8.0	yellow	red
Phenolphthalein	8.0– 9.6	colorless	red
Thymolphthalein	9.3–10.5	colorless	blue
Alizarin yellow	10.1–12.0	colorless	violet

(a) pH = 5.267; so the appropriate indicator is methyl red, which changes color in a pH-range that is approximately centered on the equivalence-point pH.

(b) pH = 6.477; bromthymol blue.

(c) pH = 9.147; phenolphthalein.

(d) pH = 7.217; phenol red.

6.26 20.00 mL of 0.100 M HAc ($K_a = 1.78 \times 10^{-5}$) is titrated with 0.100 M NaOH. Since the pH at the equivalence point is 8.724, phenolphthalein is an appropriate indicator (see Table 6-3). What error would occur in the calculated concentration of HAc if alizarin yellow were used instead and the end-point pH identified (falsely) as 11.050?

Let V_b (mL) denote the volume of titrant added to produce pH 11.050. If (6.23) is valid,

$$C_{sb} = [OH^-] = 1.12 \times 10^{-3}$$

Since 0.100 V_b mmol of NaOH was added and (0.100)(20.00) = 2.00 mmol was used in neutralizing the HAc, the concentration of excess strong base is

$$C_{sb} = \frac{(0.100\ V_b - 2.00)\ \text{mmol}}{(20.00 + V_b)\ \text{mL}} = 1.12 \times 10^{-3}$$

Solving, $V_b = 20.4$ mL. Therefore, the calculated original concentration of weak acid is

$$\frac{(20.4\ \text{mL})(0.100\ M)}{20.00\ \text{mL}} = 0.102\ M$$

and the percent error is

$$\frac{0.102 - 0.100}{0.100}\,(100\%) = 2\%$$

Since $K_{cb} = K_w/K_a = 5.62 \times 10^{-10}$ and the solution is

$$\frac{0.4}{20.0}\,(100\%) = 2\%$$

past the equivalence point, the coordinates in Fig. 6-6 indicate that (6.23) is indeed valid.

Supplementary Problems

TITRATIONS: REGION I

6.27 At a particular stage in the titration of a benzoic acid ($K_a = 6.46 \times 10^{-5}$) solution, $C_a = [HA] = 0.0150$ and $C_{cb} = [A^-] = 0.0600$. Calculate α and estimate the pH of the solution.   *Ans.* $\alpha = 0.800$, pH 4.5–5.0

6.28 Calculate the pH of the solution of Problem 6.27.   *Ans.*   4.792

6.29 At a particular stage in the titration of dichloroacetic acid ($K_a = 3.32 \times 10^{-2}$), $C_a = [HA] = 0.0500$ and $C_{cb} = [A^-] = 0.0250$. Calculate α and estimate the pH of the solution.
Ans. $\alpha = 0.333$; if pH $\approx$ pK_a at $\alpha = 0.5$, then pH is about 1.0–1.3.

6.30 Calculate the pH of the solution of Problem 6.29.
Ans.   1.678 (the approximation pH $\approx$ pK_a at $\alpha = 0.5$ is not valid, since (6.17) does not apply)

6.31   At a particular stage in the titration of a solution of lactic acid ($K_a = 1.59 \times 10^{-4}$), $C_a = [HA] = 0.0350$ and $C_{cb} = [A^-] = 0.0650$. Calculate α and estimate the pH of the solution.   *Ans.* $\alpha = 0.650$, pH 4.0–4.2

6.32   Calculate the pH of the solution of Problem 6.31.   *Ans.*   4.067

6.33   At a particular stage in the titration of NH_4OH ($K_b = 1.80 \times 10^{-5}$), $C_b = [NH_4OH] = 0.0663$ and $C_{ca} = [NH_4^+] = 0.0224$. Calculate α and estimate the pH of the solution.   *Ans.* $\alpha = 0.253$, pH 9.6–9.9

6.34   Calculate the pH of the solution of Problem 6.33.   *Ans.*   9.727

6.35   At a particular stage in the titration of the weak base piperidine ($K_b = 1.60 \times 10^{-3}$) $C_b = [B] = 0.0275$ and $C_{ca} = [BH^+] = 0.0366$. Calculate α and estimate the pH of the solution.
Ans. $\alpha = 0.571$; if $pOH \approx pK_b$ when $\alpha = 0.5$, then pH is about 10.9–11.1.

6.36   Calculate the pH of the solution of Problem 6.35.   *Ans.*   11.049

pH GIVEN

6.37   An unknown amount of weak acid, HA, was dissolved in 50.00 mL of water and titrated with 0.0800 M NaOH. Data from a pH-meter showed an equivalence point upon addition of 32.80 mL of base. When only 26.00 mL was added, the pH was 5.300. Find K_a.   *Ans.* 1.9×10^{-5}

6.38   How many grams of NH_4Cl (MW = 53.49) must be added to 1.00 L of 0.02000 M NH_4OH ($K_b = 1.80 \times 10^{-5}$) to produce a solution with a pH of 10.300?   *Ans.*   0.0846 g

TITRATIONS: REGION II

6.39   25.00 mL of 0.100 M HAc ($K_a = 1.78 \times 10^{-5}$) is titrated with 0.100 M NaOH. What is the pH of the solution when the volume of added NaOH is 3.00 mL in excess of that needed to reach the equivalence point?   *Ans.*   11.753

6.40   35.00 mL of 0.124 M HF ($K_a = 6.76 \times 10^{-4}$) is titrated with 0.200 M NaOH. What is the pH of the solution when the volume of added NaOH is 15.00 mL in excess of that needed to reach the equivalence point?   *Ans.*   12.622

Chapter 7

Polyprotic Weak Acids

7.1 GENERAL PROPERTIES

Acids, such as H_2CO_3 and H_3PO_4, that contain more than one replaceable hydrogen ion are termed *polyprotic*. The hydrogen ion that is most easily removed gives rise to the largest K_a-value; succeeding K_a-values are progressively smaller. Separate equivalence points, each showing a large change in pH, will be observed in the titration of a polyprotic acid if the ratios of successive K-values are at least 10^4.

EXAMPLE 7.1 (a) Sulfurous acid, H_2SO_3, has $K_1 = 1.74 \times 10^{-2}$ and $K_2 = 6.17 \times 10^{-8}$. Since

$$\frac{K_1}{K_2} = 2.82 \times 10^5$$

sharp changes in pH will occur at both the first and second equivalence points in the titration of a (moderately concentrated, $C_a \approx 0.1\ M$) solution of H_2SO_3. (b) Figure 7-1 shows the titration of 10.00 mL of 0.100 M citric acid with 0.100 M NaOH. Here, $K_1 = 1.15 \times 10^{-3}$, $K_2 = 3.63 \times 10^{-5}$, and $K_3 = 1.51 \times 10^{-6}$. Since

$$\frac{K_1}{K_2} = 31.7 \qquad \frac{K_2}{K_3} = 24.0$$

only a single equivalence point, corresponding to the neutralization of all three hydrogen ions, is visible. (c) Figure 7-2 shows the titration of 10.00 mL of 0.100 M H_3PO_4 with 0.100 M NaOH. Here, $K_1 = 7.50 \times 10^{-3}$, $K_2 = 6.20 \times 10^{-8}$, and $K_3 = 4.80 \times 10^{-13}$, so that

$$K_1/K_2 = 1.21 \times 10^5 \qquad K_2/K_3 = 1.29 \times 10^5$$

and three visible equivalence points would be expected in Fig. 7-2. However, after the second equivalence point has been passed, there is insufficient difference in strength between the weak acid being titrated and the weak acid H_2O for there to be a visible third equivalence point. (This will be the case whenever $K_a < 10^{-9}$.)

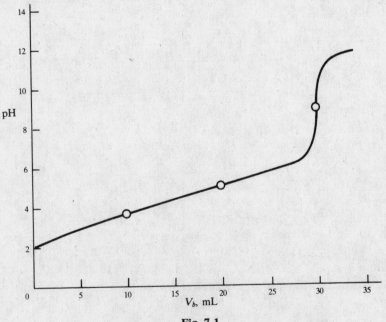

Fig. 7-1

104

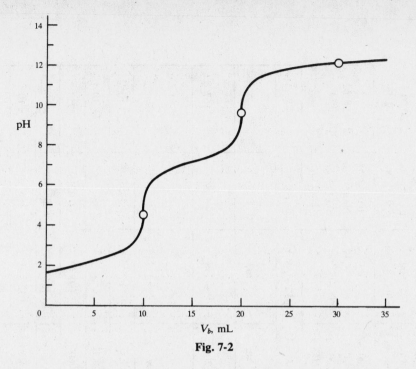

Fig. 7-2

If the *intermediate* equivalence points show sharp changes in pH, as with H_3PO_4 (Fig. 7-2), then the pH titration curve can be imagined as that of a mixture of separate weak acids, each of the same concentration and thus requiring the same amount of strong-base titrant. This is illustrated in Fig. 7-3, where the principal species at point *a* is the weak acid H_3X. In Region I-1 the principal species are H_3X and the conjugate base H_2X^-; H_2X^- is the single principal species at point E_1, the first equivalence point. In Region I-2 the principal species are H_2X^- and HX^{2-}, whereas HX^{2-} is the sole principal species at point E_2, the second equivalence point. Species HX^{2-} and X^{3-} are the principal species in Region I-3, whereas X^{3-} is the principal species at the third equivalence point, *c*. When excess NaOH is added, in Region II, the principal species will be X^{3-} and OH^-. This is also illustrated for the case of H_3PO_4 in Table 7-1. On the other hand, sharp changes in pH do not occur at the

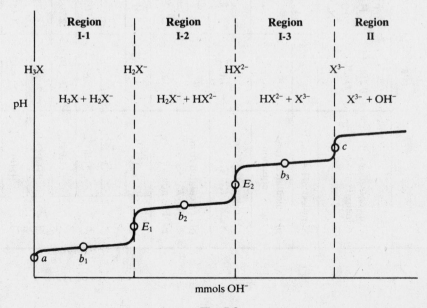

Fig. 7-3

Table 7-1. Titration of 10.00 mL of 0.100 M H_3PO_4 with 0.100 M NaOH

V_b, mL	0.00 (start)	5.00 (1/2 way to 1st eq. pt.)	10.00 (1st eq. pt.)	15.00 (1/2 way to 2nd eq. pt.)	20.00 (2nd eq. pt.)	25.00 (1/2 way to 3rd eq. pt.)	30.00 (3rd eq. pt.)
Concentrations of principal species	0.100 M H_3PO_4	0.0333 M H_3PO_4, 0.0333 M NaH_2PO_4	0.0500 M NaH_2PO_4	0.0200 M NaH_2PO_4, 0.0200 M Na_2HPO_4	0.0333 M Na_2HPO_4	0.0143 M Na_2HPO_4, 0.0143 M Na_3PO_4	0.0250 M Na_3PO_4
Simplest pH-equation	—	pK_1	$(pK_1 + pK_2)/2$	pK_2	$(pK_2 + pK_3)/2$	pK_3	—
pH calculated from simplest equation	—	2.125	4.666	7.208	9.763	12.319	
pH calculated from K_1, K_2, K_3, K_w	1.622	2.267	4.697	7.208	9.658	11.849	12.167
pH calculated from K_1, K_2, K_3, K_w, and EDHE non-ideality	1.624	2.227	4.533	6.894	9.265	11.465	11.869
Concentrations based on K_1, K_2, K_3, K_w — $[H^+]$	2.39×10^{-2}	5.41×10^{-3}	2.01×10^{-5}	6.20×10^{-8}	2.20×10^{-10}	1.42×10^{-12}	6.80×10^{-13}
$[OH^-]$	4.18×10^{-13}	1.85×10^{-12}	4.98×10^{-10}	1.61×10^{-7}	4.55×10^{-5}	7.06×10^{-3}	1.47×10^{-2}
$[Na^+]$	zero	0.0333	0.0500	0.0600	0.0667	0.0715	0.0750
$[H_3PO_4]$	7.61×10^{-2}	2.79×10^{-2}	1.34×10^{-4}	1.65×10^{-7}	3.46×10^{-12}	9.28×10^{-17}	1.46×10^{-17}
$[H_2PO_4^-]$	2.39×10^{-2}	3.87×10^{-2}	4.97×10^{-2}	2.00×10^{-2}	1.18×10^{-4}	4.90×10^{-7}	1.61×10^{-7}
$[HPO_4^{2-}]$	6.20×10^{-8}	4.44×10^{-7}	1.54×10^{-4}	2.00×10^{-2}	3.31×10^{-2}	2.14×10^{-2}	1.47×10^{-2}
$[PO_4^{3-}]$	1.25×10^{-18}	3.94×10^{-17}	3.68×10^{-12}	1.55×10^{-7}	7.27×10^{-5}	7.21×10^{-3}	1.03×10^{-2}

intermediate equivalence points in the case of citric acid, Fig. 7-1, and three different citrate species are present in most of the titration region.

A theoretical basis for the above observations will be provided in Example 7.8.

7.2 pH APPROXIMATIONS

As in the case of a simple weak acid (see Problem 6.3(*b*)), the pH halfway to each equivalence point in the titration of a polyprotic acid can be approximated by the pK, where K is the dissociation constant for that step of the titration. As is shown in Table 7-1, such approximations are quite good for the first two steps in the titration of H_3PO_4. Even for the third step, which involves the exceedingly weak acid HPO_4^{2-}, the approximation is still useful. And, as shown in Table 7-2, the acid being titrated does not even need to have K_1/K_2 and K_2/K_3 exceed 10^4; the approximations apply reasonably well to citric acid (see Example 7.1(*b*)).

Table 7-2. Titration of 10.00 mL of 0.100 M Citric Acid with 0.100 M NaOH

V_b, mL	0.00	5.00	10.00	15.00	20.00	25.00	30.00	35.00
Stage of titration, per Fig. 7-3	start	b_1	E_1	b_2	E_2	b_3	c	II
Approximate pH		2.939	3.690	4.440	5.131	5.821		
Calculated pH	1.991	2.930	3.693	4.432	5.134	5.867	9.109	12.046

The pH at the intermediate equivalence points can be approximated as follows:

At the first intermediate equivalence point, E_1, of Fig. 7-3,

$$pH \approx \tfrac{1}{2}(pK_1 + pK_2) \qquad (7.1)$$

At the second intermediate equivalence point, E_2, of Fig. 7-3,

$$pH \approx \tfrac{1}{2}(pK_2 + pK_3) \qquad (7.2)$$

The third equivalence point, c of Fig. 7-3, is the end of the titration and the pH is calculated in terms of the hydrolysis of the conjugate base X^{3-}, as discussed in Chapter 5.

EXAMPLE 7.2 The derivation of (7.1) is based on the shifts-to-equilibrium method of Chapters 5 and 6. At E_1, H_2X^- will be the principal weak-acid species; hence, $C_a = [H_2X^-]$ prior to any shifts. Small amounts of HX^{2-} and H_3X will exist in the equilibrated solution, the former resulting from the dissociation of H_2X^- and the latter from the hydrolysis of some H_2X^-. There will also be some X^{3-}, but in most cases its concentration will be so small that we may neglect X^{3-} and the associated reaction based on K_3. Therefore, the two important reactions are: (1) the dissociation of H_2X^-, and (2) the hydrolysis of H_2X^-. We will imagine that we have a solution that is C_a M H_2X^- in dissociated water, so that at the start, $[H_2X^-] = C_a$ and $[H^+] = [OH^-] = 10^{-7}$. Since the concentration of HX^{2-} is zero, a moles of H_2X^- per liter will dissociate to produce a moles each of H^+ and HX^{2-} per liter:

	initial	C_a	10^{-7}	0
		H_2X^- $\longrightarrow$	H^+	+ HX^{2-}
	equilibrium	$C_a - a - b$	$10^{-7} + a - w$	a

The constant for this reaction is K_2. The hydrolysis reaction will also proceed, since the starting concentration of H_3X is zero and b moles of H_2X^- per liter will hydrolyze to produce b moles each of H_3X and OH^- per liter:

$$
\begin{array}{ccccccc}
\textit{initial} & C_a & & & 0 & & 10^{-7} \\
& H_2X^- & + & H_2O & \longrightarrow & H_3O^+ & + & OH^- \\
\textit{equilibrium} & C_a - a - b & & & b & & 10^{-7} + b - w
\end{array}
$$

The constant for this reaction is $K_{cb1} = K_w/K_1$.

Since H^+ and OH^- are both produced, the water dissociation reaction is no longer at equilibrium, and w moles of H^+ per liter reacts with w moles of OH^- per liter to restore the equilibrium:

$$
\begin{array}{cccc}
\textit{initial} & 10^{-7} & & 10^{-7} \\
H_2O & \longrightarrow & H^+ & + & OH^- \\
\textit{equilibrium} & 10^{-7} + a - w & & 10^{-7} + b - w
\end{array}
$$

From these equations, we have:

$$
\begin{aligned}
[H^+] - \frac{K_w}{[H^+]} &= [H^+] - [OH^-] \\
&= (10^{-7} + a - w) - (10^{-7} + b - w) = a - b \\
&= [HX^{2-}] - [H_3X] \\
&= \frac{K_2[H_2X^-]}{[H^+]} - \frac{[H^+][H_2X^-]}{K_1}
\end{aligned}
$$

which may be rearranged as

$$
[H^+]^2 = K_1K_2 \frac{1 + (K_w/K_2[H_2X^-])}{1 + (K_1/[H_2X^-])} \tag{7.3}
$$

If $[H_2X^-] \gg K_w/K_2$ and $[H_2X^-] \gg K_1$, then the two ratios on the right of (7.3) are negligible compared to 1, and

$$
[H^+]^2 \approx K_1K_2
$$

which is equivalent to (7.1).

The same argument, in terms of the ions and K-values appropriate to E_2, establishes (7.2).

EXAMPLE 7.3 Assuming ideality, verify the calculations for the first equivalence point (column 3) in Table 7-1. How much in error is the value of $[H^+]$ provided by (7.1)?

At E_1, $C_a = [NaH_2PO_4] = 0.0500$. Assume that $[PO_4^{3-}]$ can be neglected relative to the other phosphate concentrations; assume, in addition, that $[H_2PO_4^-] \approx 0.0500$. Then, since

$$
\frac{K_w}{K_2} = \frac{10^{-14}}{6.20 \times 10^{-8}} = 1.61 \times 10^{-7} \ll 0.0500
$$

(7.3) reduces to

$$
[H^+]^2 = \frac{K_1K_2}{1 + (K_1/[H_2PO_4^-])} \tag{1}
$$

(Note that $K_1/[H_2PO_4^-] \approx 7.50 \times 10^{-3}/0.0500 = 0.15 \approx 1$, so that (1) may not be further simplified.) From (1), $[H^+] \approx 2.01 \times 10^{-5}$. Therefore,

$$
a = \frac{K_2[H_2PO_4^-]}{[H^+]} = \frac{(6.20 \times 10^{-8})(0.0500)}{2.01 \times 10^{-5}} = 1.54 \times 10^{-4}
$$

$$
b = \frac{[H^+][H_2PO_4^-]}{K_1} = \frac{(2.01 \times 10^{-5})(0.0500)}{7.50 \times 10^{-3}} = 1.34 \times 10^{-4}
$$

and a better, second, approximation for $[H_2PO_4^-]$ is $C_a - a - b = 0.0497$. Using this value in (1) results in $[H^+] = 2.01 \times 10^{-5}$ as before, so that further successive approximations are not needed. The pH is 4.697. Since (7.1) gives $[H^+] = 2.16 \times 10^{-5}$, the error involved in its use is

$$
\frac{2.16 - 2.01}{2.10}(100\%) = 7.5\%
$$

The concentrations of the other species in solution are: $[Na^+] = 0.0500$, $[OH^-] = K_w/[H^+] = 4.98 \times 10^{-10}$, and

$$[PO_4^{3-}] = \frac{K_3[HPO_4^{2-}]}{[H^+]} = \frac{(4.80 \times 10^{-13})(1.54 \times 10^{-4})}{2.01 \times 10^{-5}} = 3.68 \times 10^{-12}$$

(neglecting $[PO_4^{3-}]$ was justified). In addition, $a = [HPO_4^{2-}] = 1.54 \times 10^{-4}$ and $[H_3PO_4] = b = 1.34 \times 10^{-4}$.

7.3 GENERAL pH EQUATION

A completely general equation can also be derived for the titration of V_a volume units of C_a° M H_3X using V_b volume units of C_{sb}° M NaOH. There will be seven species in solution: H^+, OH^-, Na^+, H_3X, H_2X^-, HX^{2-}, and X^{3-}. The charge-balance equation is

$$[H^+] + [Na^+] = [OH^-] + [H_2X^-] + 2[HX^{2-}] + 3[X^{3-}] \qquad (7.4)$$

and the remaining six equations are: the mass balance for Na^+; the mass balance for total X; and the definitions of K_w, K_1, K_2, and K_3. Elimination of all unknowns but $[H^+]$ results in

$$V_b\left\{C_{sb}^\circ + [H^+] - \frac{K_w}{[H^+]}\right\} = V_a\left\{C_a^\circ \frac{K_1[H^+]^2 + 2K_1K_2[H^+] + 3K_1K_2K_3}{[H^+]^3 + K_1[H^+]^2 + K_1K_2[H^+] + K_1K_2K_3} - [H^+] + \frac{K_w}{[H^+]}\right\} \qquad (7.5)$$

The various polyprotic titration curves shown in this chapter were obtained by inverse use of (7.5) (cf. Example 6.1). With $K_3 = 0$, the equation applies to a diprotic acid; with $K_2 = K_3 = 0$, it reduces to (6.1). As (7.5) is fifth-degree in $[H^+]$, pH calculations for polyprotic acids are always more easily handled by the approximation methods already described.

7.4 IDENTIFICATION OF PRINCIPAL SPECIES

If a solution of a triprotic acid contains two of H_3X, H_2X^-, HX^{2-}, and X^{3-}—or if a solution of a diprotic acid contains two of H_2X, HX^-, and X^{2-}—the *lever principle* can be used to identify the two species and to approximate their concentrations. An accurate calculation of the concentrations may then be carried out with the *conservation-of-mass method*; this method also allows identifying the principal species.

Alternatively, a solution may be titrated and the amounts of titrant required to reach various equivalence points determined. From these data, it is possible to calculate the concentrations of the principal species in the original solution mixture.

EXAMPLE 7.4 25.00 mL of a phosphate solution was titrated with 0.100 M NaOH. It required 10.00 mL of NaOH to reach equivalence point E_1, at pH $\approx$ 4.7, and an *additional* 25.00 mL to reach E_2, at pH $\approx$ 9.7. What were the principal phosphate species and their original concentrations?

Since the amount of NaOH required to reach E_1 was less than the additional amount required to reach E_2, the original mixture must fall in Region I-1 of Fig. 7-3. Therefore, the principal species were H_3PO_4 and $H_2PO_4^-$. The first step in the titration involved the reaction

$$H_3PO_4 + OH^- \rightarrow H_2PO_4^- + H_2O$$

and required $(10.00)(0.100) = 1.00$ mmol of NaOH. Therefore, originally,

$$[H_3PO_4] = \frac{1.00 \text{ mmol}}{25.00 \text{ mL}} = 0.0400 \ M$$

The second step in the titration involved the reaction

$$H_2PO_4^- + OH^- \rightarrow HPO_4^{2-} + H_2O$$

and required $(25.00)(0.100) = 2.50$ mmol of NaOH. However, 1.00 mmol of this total was used for the second step in the titration of the original H_3PO_4. Therefore, only 1.50 mmol was used in reacting with the original $H_2PO_4^{2-}$, and so

$$[H_2PO_4^-] = \frac{1.50 \text{ mmol}}{25.00 \text{ mL}} = 0.0600 \ M$$

originally.

The mixture being analyzed need not have as the two principal species only those involving a weak acid. One might have H^+ and H_3X (in a mixture of a strong acid and a polyprotic acid) or, at the opposite extreme, X^{3-} and OH^- (in a mixture of the conjugate base of a polyprotic acid and a strong base; i.e., in Region II of Fig. 7-3).

The Lever Principle

If a bar bears a weight on each end, it is possible to balance that bar on a fulcrum if the lever arms of the bar are in the inverse ratio of the weights. A similar principle will apply to solutions that are mixtures of two species arising from simple weak acids, polyprotic acids, or the equivalent weak bases. In the chemical analogy, the two weights correspond to the amounts (mmols) of the two species, and the fulcrum represents the position on a titration curve that corresponds to the overall composition of the mixture.

EXAMPLE 7.5 Consider the case of a simple weak acid, HA, with the titration curve shown in Fig. 7-4. If 100 mL of a solution is 0.100 M HA and 0.300 M NaA, this corresponds to 10.0 mmol of HA and 30.0 mmol of A^-. The bar joining these two points must be divided so that the lever arm for HA is 30.0/10.0 = 3 times the lever arm for A^-. This puts the fulcrum 3/4 of the way between HA, the start of the titration curve, and A^-, the equivalence point in the titration curve. The solution mixture, therefore, is the same as would exist if 75% of a sample of the weak acid was titrated with NaOH, as shown by the solid circle in Fig. 7-4.

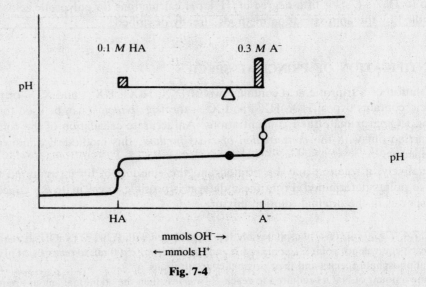

Fig. 7-4

EXAMPLE 7.6 Figure 7-5 applies to a mixture of two species of a polyprotic acid; a solution that is 0.5 M H_3X and 0.3 M Na_2HX is depicted. If the volume is 100 mL, then there is 50 mmol of H_3X and 30 mmol of HX^{2-}. The lever arm for HX^{2-} must be longer than the lever arm for H_3X; in fact, they must be in the proportion of 5 : 3. Therefore, the fulcrum must be 3/(5 + 3) = 3/8 of the way between points a and E_2. Since the distance between a and E_1 must be equal to the distance between E_1 and E_2 (each step in the titration requires the same number of mmols of NaOH), the fulcrum is (3/8)2 = 3/4 of the way between a and E_1. The lever arms of H_3X and H_2X^- are thus in the ratio 3 : 1; i.e.,

$$[H_2X^-] = 3[H_3X] \qquad (1)$$

In addition, the total amount of X in all forms must, by conservation of mass, be equal to the starting amount, 50 + 30 = 80 mmol; hence,

$$[H_2X^-] + [H_3X] = 0.8 \qquad (2)$$

Solving (1) and (2) simultaneously results in $[H_2X^-]$ = 0.6, $[H_3X]$ = 0.2. This corresponds to the titration solution given by the solid circle in Fig. 7-5.

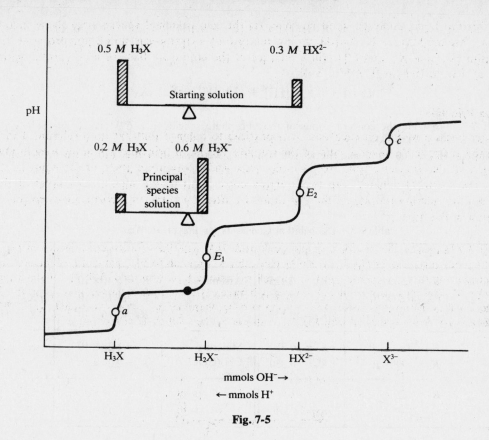

Fig. 7-5

Conservation-of-Mass Method

We illustrate this simpler alternative to the lever principle by reworking Example 7.6:

1. The original species are first "dissociated" into the simplest possible ions. Therefore, the $(0.5\ M\ H_3X) + (0.3\ M\ HX^{2-})$ is dissociated into H^+ and X^{3-} ions. This results in

$$[H^+] = 3(0.5) + 1(0.3) = 1.8 \qquad [X^{3-}] = 0.5 + 0.3 = 0.8$$

2. We next imagine that the $0.8\ M\ X^{3-}$ is "titrated" with the $1.8\ M\ H^+$ until the H^+ is used up. Therefore, $0.8\ M$ of the H^+ is used to convert $0.8\ M\ X^{3-}$ to $0.8\ M\ HX^{2-}$. Another $0.8\ M$ of the H^+ is used to convert the $0.8\ M\ HX^{2-}$ to $0.8\ M\ H_2X^-$. This two-step "titration" leaves

$$1.8 - 2(0.8) = 0.2\ M$$

of H^+, which is sufficient to convert only $0.2\ M$ of the $0.8\ M\ H_2X^-$ to $0.2\ M\ H_3X$; there remains $0.6\ M$ of H_2X^-. Thus, the concentrations of the principal species are: $[H_3X] = 0.2$, $[H_2X^-] = 0.6$.

7.5 SOLUTIONS OF KNOWN pH

In cases where the pH of the solution is known, it is possible to calculate the concentration of each weak-acid species as a fraction of C, the total concentration of all weak-acid species. For a triprotic acid, define

$$F_0 \equiv \frac{[H_3X]}{C} \qquad F_1 \equiv \frac{[H_2X^-]}{C} \qquad F_2 \equiv \frac{[HX^{2-}]}{C} \qquad F_3 \equiv \frac{[X^{3-}]}{C}$$

Then it can be shown that

$$F_0 = \frac{[H^+]^3}{D} \qquad F_1 = \frac{K_1[H^+]^2}{D} \qquad F_2 = \frac{K_1K_2[H^+]}{D} \qquad F_3 = \frac{K_1K_2K_3}{D} \qquad (7.6)$$

where

$$D \equiv [H^+]^3 + K_1[H^+]^2 + K_1K_2[H^+] + K_1K_2K_3 \qquad (7.7)$$

EXAMPLE 7.7 What are the concentrations of the phosphate species in a $C\,M$ phosphate solution of pH 5.000? Refer to Table 7-3 for K-values.

Substituting $[H^+] = 1.00 \times 10^{-5}$ and the K-values in (7.7), we obtain $D = 7.56 \times 10^{-13}$. Then (7.6) gives:

$$F_0C = [H_3PO_4] = 1.32 \times 10^{-3}C \qquad F_1C = [H_2PO_4^-] = 0.992C$$
$$F_2C = [HPO_4^{2-}] = 6.15 \times 10^{-3}C \qquad F_3C = [PO_4^{3-}] = 2.95 \times 10^{-10}C$$

Table 7-3. Dissociation Constants for Polyprotic Acids

Acid	K	pK	Acid	K	pK
Phosphoric, H_3PO_4	$K_1 = 7.50 \times 10^{-3}$ $K_2 = 6.20 \times 10^{-8}$ $K_3 = 4.80 \times 10^{-13}$	pK_1 = 2.125 pK_2 = 7.208 pK_3 = 12.319	Arsenic, H_3AsO_4	$K_1 = 6.46 \times 10^{-3}$ $K_2 = 1.15 \times 10^{-7}$ $K_3 = 3.16 \times 10^{-12}$	pK_1 = 2.190 pK_2 = 6.939 pK_3 = 11.500
Carbonic, H_2CO_3	$K_1 = 4.47 \times 10^{-7}$ $K_2 = 4.68 \times 10^{-11}$	pK_1 = 6.350 pK_2 = 10.330	Malonic, $CH_2(COOH)_2$	$K_1 = 1.38 \times 10^{-3}$ $K_2 = 2.00 \times 10^{-6}$	pK_1 = 2.860 pK_2 = 5.699
Oxalic, $H_2C_2O_4$	$K_1 = 6.46 \times 10^{-2}$ $K_2 = 6.17 \times 10^{-5}$	pK_1 = 1.190 pK_2 = 4.210	Sulfurous, H_2SO_3	$K_1 = 1.74 \times 10^{-2}$ $K_2 = 6.17 \times 10^{-8}$	pK_1 = 1.760 pK_2 = 7.210

EXAMPLE 7.8 A plot of the equations (7.6) for the fractional composition of a citric acid solution is given in Fig. 7-6. Since the K-values for this acid are close together, there exist three principal species at the two intermediate equivalence points, E_1 and E_2. Between these two points there is a small range in which all four citrate species qualify as principal species. The start of the titration, point a, involves H_3Cit and H_2Cit^- as principal species, but only Cit^{3-} is present as a principal species at the final equivalence point, c.

It is otherwise in the titration of phosphoric acid, whose F-values are plotted in Fig. 7-7. Since the ratios of successive K-values for H_3PO_4 exceed 10^4, there is only a single principal species at each of the intermediate

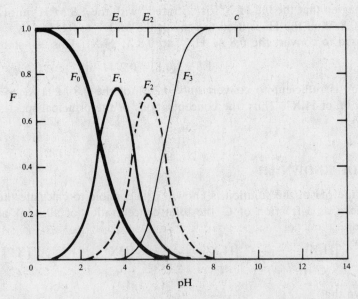

Fig. 7-6

equivalence points, E_1 and E_2. There are two principal species present in the equilibrium mixture at the start of the titration, point a, since K_1 is relatively large and there is more than 1% dissociation of H_3PO_4. The final equivalence point in the titration, point c, also involves two phosphate species, since the hydrolysis constant for the final product, PO_4^{3-}, is quite large ($K_{cb} = K_w/K_3 = 2.08 \times 10^{-2}$) and appreciable hydrolysis therefore occurs. Intermediate stages in the titration involve at most two phosphate species in appreciable concentration.

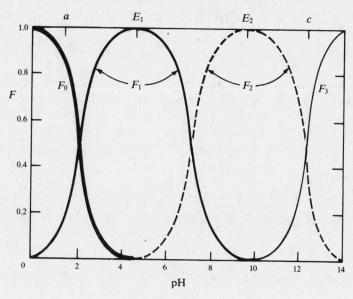

Fig. 7-7

Solved Problems

CALCULATION OF pH

7.1   Calculate the pH and the concentrations of all species at the start of the titration of 0.100 M H_3PO_4 and 0.100 M NaOH, and show that they agree with the values in column 1 of Table 7-1.

The principal reaction will involve K_1; we can neglect K_2, K_3, $[HPO_4^{2-}]$, and $[PO_4^{3-}]$ in the initial calculations. The point $(C_a, K_1) = (0.100, 7.50 \times 10^{-3})$ lies in Region 1 of Fig. 5-1. The solution to the quadratic is $[H^+] = 0.0239$, or pH = 1.622, and Assumption 1a and the use of the quadratic are validated.

$$[OH^-] = \frac{K_w}{[H^+]} = 4.18 \times 10^{-13} \qquad [H_3PO_4] = 0.100 - [H^+] = 0.076 \qquad [H_2PO_4^-] = [H^+] = 0.0239$$

$$[HPO_4^{2-}] = \frac{K_2[H_2PO_4^-]}{[H^+]} = K_2 = 6.20 \times 10^{-8} \qquad [PO_4^{3-}] = \frac{K_3[HPO_4^{2-}]}{[H^+]} = 1.25 \times 10^{-18}$$

The assumption that $[HPO_4^{2-}]$ and $[PO_4^{3-}]$ may be neglected in the initial calculations is justified.

7.2   Calculate the pH and the concentrations of all species in the solution of Problem 7.1 when corrections for non-ideality are introduced.

The concentrations calculated in Problem 7.1 can serve as a first approximation in the calculation of the ionic strength. From (3.8), $I = 0.0239$, and (3.7) gives

$$f_{H^+} = 0.883 \qquad f_{OH^-} = 0.855 \qquad f_{H_2PO_4^-} = 0.860 \qquad f_{HPO_4^{2-}} = 0.548 \qquad f_{PO_4^{3-}} = 0.258$$

and $f_{H_3PO_4} = 1.000$. Assumption 1a should also be valid here; the non-ideal equation valid in Region 1 of Fig. 5-1 is

$$[H^+]^2 + (K_1/K_{1f})[H^+] - C_a(K_1/K_{1f}) = 0 \quad \text{where} \quad K_{1f} = f_{H^+} f_{H_2PO_4^-}/f_{H_3PO_4} = 0.758$$

Solving the quadratic: $[H^+] = 0.0269$,

$$[OH^-] = \frac{K_w}{K_{wf}[H^+]} = 4.92 \times 10^{-13} \qquad [H_3PO_4] = 0.100 - [H^+] = 0.073 \qquad [H_2PO_4^-] = [H^+] = 0.0269$$

$$[HPO_4^{2-}] = \frac{K_2[H_2PO_4^-]}{K_{2f}[H^+]} = \frac{K_2}{K_{2f}} = 1.10 \times 10^{-7} \qquad [PO_4^{3-}] = \frac{K_3[HPO_4^{2-}]}{K_{3f}[H^+]} = 4.72 \times 10^{-18}$$

The recalculated ionic strength is $I = 0.0269$, and the recalculated coefficients for H^+, OH^-, and $H_2PO_4^-$ will be within 1% of those calculated before. There will be a slightly larger difference in the activity coefficients for HPO_4^{2-} and PO_4^{3-}, but only the calculated concentrations of those two ions will be affected thereby. Thus,

$$a_{H^+} = (0.0269)(0.883) = 0.0238 \qquad \text{and} \qquad pH = -\log a_{H^+} = 1.624$$

and this pH-value justifies Assumption 1a.

7.3 Calculate the pH and the concentrations of all species halfway to the first equivalence point in the titration of $0.100\ M\ H_3PO_4$ with $0.100\ M\ NaOH$, and show that they agree with the values in column 2 of Table 7-1.

Since the ratios of successive K-values for H_3PO_4 exceed 10^4, only two principal phosphate species will exist in each titration step, and each step can be treated as if it involved a single weak acid. At the halfway point b_1, we can neglect HPO_4^{2-} and PO_4^{3-} and consider the mixture to be the result of the weak-acid dissociation with constant $K_1 = 7.50 \times 10^{-3}$. The concentrations of principal species are:

$$[H_3PO_4] = C_a = [H_2PO_4^-] = C_{cb} = 0.0333$$

Since $\alpha = 0.5$ and $C_a + C_{cb} = 0.0666 \approx 0.1$, the point (α, K_1) falls in the region of Fig. 6-4 for (6.16); the solution to the quadratic is $[H^+] = 5.41 \times 10^{-3}$, or pH = 2.267. Hence, Assumption 1a is valid ($[H^+] = a$), and (6.9a) gives

$$[H_3PO_4] = C_a - a = 2.79 \times 10^{-2} \qquad [H_2PO_4^-] = C_{cb} + a = 3.87 \times 10^{-2}$$

Furthermore,

$$[HPO_4^{2-}] = \frac{K_2[H_2PO_4^-]}{[H^+]} = 4.44 \times 10^{-7} \qquad [PO_4^{3-}] = \frac{K_3[HPO_4^{2-}]}{[H^+]} = 3.94 \times 10^{-17}$$

As the only source of Na^+ is the original $0.0333\ M\ NaH_2PO_4$, $[Na^+] = 0.0333$.

7.4 Find the pH and concentrations of all species in the solution of Problem 7.3 when corrections are introduced for non-ideality.

The concentrations calculated in Problem 7.3 can serve as a first approximation in the calculation of the ionic strength. From (3.8),

$$I = \tfrac{1}{2}\{[H^+]1^2 + [OH^-](-1)^2 + [Na^+]1^2 + [H_2PO_4^-](-1)^2 + [HPO_4^{2-}](-2)^2 + [PO_4^{3-}](-3)^2\} = 0.0387$$

and (3.7) gives

$$f_{H^+} = 0.864 \qquad f_{Na^+} = f_{H_2PO_4^-} = 0.833 \qquad f_{OH^-} = 0.825 \qquad f_{HPO_4^{2-}} = 0.481 \qquad f_{PO_4^{3-}} = 0.193$$

and $f_{H_3PO_4} = 1.000$. The assumptions that allowed the use of (6.16) in Problem 7.3 should also be valid here. The non-ideal form of (6.16) is

$$[H^+]^2 + \{C_{cb} + (K_1/K_{1f})\}[H^+] - C_a(K_1/K_{1f}) = 0 \quad \text{where} \quad K_{1f} = f_{H^+} f_{H_2PO_4^-}/f_{H_3PO_4} = 0.720$$

The solution of the quadratic is $[H^+] = 0.00686$, so that

$$[H_3PO_4] = 0.0333 - [H^+] = 0.0264 \qquad [H_2PO_4^-] = 0.0333 + [H^+] = 0.0402$$
$$[HPO_4^{2-}] = 3.63 \times 10^{-7} \qquad\qquad [PO_4^{3-}] = 2.54 \times 10^{-17}$$

and $[Na^+] = 0.0333$.

The second approximation for the ionic strength is $I = 0.0402$; however, this new value of I results in activity coefficients that differ by less than 1% from those used, so that it is unnecessary to recalculate $[H^+]$ and the other ionic concentrations. Therefore,

$$pH = -\log a_{H^+} = -\log [H^+] - \log f_{H^+} = 2.227$$

7.5 Correct Example 7.3 for non-ideality.

The concentrations calculated in Example 7.3 can serve as a first approximation in the calculation of the ionic strength. From (3.8), $I = 0.0502$; from (3.7):

$$f_{H^+} = 0.854 \qquad f_{Na^+} = f_{H_2PO_4^-} = 0.817 \qquad f_{OH^-} = 0.807 \qquad f_{HPO_4^{2-}} = 0.445 \qquad f_{PO_4^{3-}} = 0.161$$

and $f_{H_3PO_4} = 1.000$. The assumptions that yielded (1) of Example 7.3 should also be valid here; thus,

$$[H^+]^2 = \frac{(K_1/K_{1f})(K_2/K_{2f})}{1 + \{(K_1/K_{1f})/[H_2PO_4^-]\}}$$

where
$$K_{1f} = f_{H^+} f_{H_2PO_4^-}/f_{H_3PO_4} = 0.698 \qquad K_{2f} = f_{H^+} f_{HPO_4^{2-}}/f_{H_2PO_4^-} = 0.465$$

Approximating $[H_2PO_4^{2-}]$ as $C_a = 0.0500$ results in $[H^+] = 3.43 \times 10^{-5}$, so that

$$a = [HPO_4^{2-}] = \frac{K_2[H_2PO_4^-]}{K_{2f}[H^+]} = 1.94 \times 10^{-4} \qquad b = [H_3PO_4] = \frac{K_{1f}[H^+][H_2PO_4^-]}{K_1} = 1.60 \times 10^{-4}$$

The second approximation for $[H_2PO_4^-]$ is $C_a - a - b = 0.0497$, which leads to $I = 0.0502$ (as before). The calculated activity coefficients will also be the same as before, so that $[H^+]$ remains at 3.43×10^{-5}. Therefore,

$$[OH^-] = \frac{K_w}{K_{wf}[H^+]} = 4.23 \times 10^{-10} \qquad [H_3PO_4] = 1.60 \times 10^{-4} \qquad [H_2PO_4^-] = 0.0497$$
$$[HPO_4^{2-}] = 1.94 \times 10^{-4} \qquad [PO_4^{3-}] = 8.89 \times 10^{-12}$$

and $[Na^+] = 0.0500$. It is easy to verify that the condition

$$\frac{K_w/K_{wf}}{(K_2/K_{2f})[H_2PO_4^-]} \ll 1$$

holds, so that the non-ideal $[H^+]$-equation used was valid. We then have:

$$a_{H^+} = [H^+]f_{H^+} = 2.93 \times 10^{-5} \qquad \text{and} \qquad pH = 4.533$$

7.6 Confirm column 4 of Table 7-1, which pertains to stage b_2, halfway to the second equivalence point.

We can neglect H_3PO_4 and PO_4^{3-} in the initial calculations and consider the mixture at b_2 to be the result solely of the dissociation with constant $K_2 = 6.20 \times 10^{-8}$. Concentrations of principal species are

$$C_a = [H_2PO_4^-] = C_{cb} = [HPO_4^{2-}] = 0.0200$$

so that $C_a + C_{ab} = 0.0400$. In Fig. 6-4, the point $(\alpha, K_2) = (0.500, 6.20 \times 10^{-8})$ indicates the use of (6.17), which gives

$$pH = pK_2 = 7.208$$

Then, from (6.11), $b = [OH^-] - [H^+] = (K_w/K_2) - K_2 = 9.93 \times 10^{-8}$, whence

$$[H_2PO_4^-] = C_a + b = 0.0200 \qquad [HPO_4^{2-}] = C_{cb} - b = 0.0200$$

Further,

$$[H_3PO_4] = \frac{[H^+][H_2PO_4^-]}{K_1} = 1.65 \times 10^{-7} \qquad [PO_4^{3-}] = \frac{K_3[HPO_4^{2-}]}{[H^+]} = 1.55 \times 10^{-7}$$

$$[OH^-] = \frac{K_w}{[H^+]} = 1.61 \times 10^{-7}$$

and $[Na^+] = 0.0600$.

7.7   Rework Problem 7.6 if the solution is non-ideal.

The concentration calculated in Problem 7.6 can serve as a first approximation in the calculation of the ionic strength. From (3.8), $I = 0.0800$, and (3.7) gives:

$$f_{H^+} = 0.835 \qquad f_{OH^-} = 0.772 \qquad f_{H_2PO_4^-} = 0.785 \qquad f_{HPO_4^{2-}} = 0.381 \qquad f_{PO_4^{3-}} = 0.114$$

and $f_{H_3PO_4} = 1.000$. Equation (6.17) should also be valid here, so that

$$[H^+] = \frac{K_2}{K_{2f}} = \frac{6.20 \times 10^{-8}}{0.405} = 1.53 \times 10^{-7}$$

and $$pH = -\log[H^+] - \log f_{H^+} = 6.894$$

Now the solution is slightly acidic, with $[OH^-] = K_w/K_{wf}[H^+] = 1.01 \times 10^{-7}$. From (6.9),

$$a = [H^+] - [OH^-] = 0.52 \times 10^{-7}$$

is so small that

$$[H_2PO_4^-] = 0.0200 - a = 0.0200 \qquad \text{and} \qquad [HPO_4^{2-}] = 0.0200 + a = 0.0200$$

Moreover,

$$[H_3PO_4] = \frac{[H^+][H_2PO_4^-]K_{1f}}{K_1} = 2.67 \times 10^{-7} \qquad [PO_4^{3-}] = \frac{K_3[HPO_4^{2-}]}{K_{3f}[H^+]} = 2.51 \times 10^{-7}$$

and $[Na^+] = 0.0600$.

In the second approximation, I remains at 0.0800; hence, the calculated concentrations are accepted as correct.

7.8   Confirm column 5 of Table 7-1, which pertains to the second equivalence point, E_2.

The concentrations of the principal species are $C_{cb} = 0.0333$ and $[Na^+] = 0.0667$. If there is negligible shift of equilibrium, we have, approximately,

$$[HPO_4^{2-}] = C_{cb} = 0.0333$$

Looking at the E_2-analog of (7.3), we see that

$$\frac{K_w}{K_3 C_{cb}} = \frac{10^{-14}}{(4.80 \times 10^{-13})(0.0333)} = 0.625 \approx 1 \qquad \frac{K_2}{C_{cb}} = 1.86 \times 10^{-6} \ll 1$$

so that the equation becomes

$$[H^+]^2 = K_2 K_3 \{1 + (K_w/K_3 C_{cb})\} \qquad\qquad (1)$$

whence $[H^+] = 2.20 \times 10^{-10}$, or $pH = 9.658$. Then:

$$[H_2PO_4^-] = \frac{[H^+][HPO_4^{2-}]}{K_2} = 1.18 \times 10^{-4} \qquad [H_3PO_4] = \frac{[H^+][H_2PO_4^-]}{K_1} = 3.46 \times 10^{-12}$$

$$[PO_4^{3-}] = \frac{K_3[HPO_4^{2-}]}{[H^+]} = 7.27 \times 10^{-5} \qquad [OH^-] = \frac{K_w}{[H^+]} = 4.55 \times 10^{-5}$$

The second approximation for $[HPO_4^{2-}]$ is

$$0.0333 - [H_2PO_4^-] - [PO_4^{3-}] = 0.0331$$

This would have a very small effect on the calculated concentrations.

7.9   Correct Problem 7.8 for non-ideality.

The concentrations calculated in Problem 7.8 can serve as a first approximation in the calculation of the ionic strength. From (3.8), $I = 0.100$; and from (3.7),

$$f_{H^+} = 0.826 \qquad f_{OH^-} = 0.754 \qquad f_{H_2PO_4^-} = 0.770 \qquad f_{HPO_4^{2-}} = 0.351 \qquad f_{PO_4^{3-}} = 0.0949$$

and $f_{H_3PO_4} = 1.000$. As in Problem 7.8, there should be negligible equilibrium shift, so that we can take $[HPO_4^{2-}] \approx C_{cb} = 0.0333$. We can also retain (1) of Problem 7.8, provided each K be replaced by K/K_f; this gives $[H^+] = 6.58 \times 10^{-10}$. So:

$$pH = -\log [H^+] - \log f_{H^+} = 9.265 \qquad\qquad [OH^-] = \frac{K_w}{K_{wf}[H^+]} = 2.44 \times 10^{-5}$$

$$[H_2PO_4^-] = \frac{[H^+][HPO_4^{2-}]K_{2f}}{K_2} = 1.33 \times 10^{-4} \qquad [H_3PO_4] = \frac{[H^+][H_2PO_4^-]K_{1f}}{K_1} = 7.42 \times 10^{-12}$$

$$[PO_4^{3-}] = \frac{K_3[HPO_4^{2-}]}{K_{3f}[H^+]} = 1.09 \times 10^{-4}$$

The second approximation for $[HPO_4^{2-}]$,

$$0.0333 - [H_2PO_4^-] - [PO_4^{3-}] = 0.0331$$

would have little effect on the calculated values.

7.10   Confirm column 6 of Table 7-1, which pertains to stage b_3, halfway to the third equivalence point.

We can neglect H_3PO_4 and $H_2PO_4^-$ in the initial calculations and consider the mixture to involve a single weak acid having dissociation constant $K_3 = 4.80 \times 10^{-13}$. Concentrations are

$$C_a = [HPO_4^{2-}] = C_{cb} = [PO_4^{3-}] = 0.0143 \qquad \text{and} \qquad [Na^+] = 0.0715$$

With $C_a + C_{cb} = 0.0286$ and $\alpha = 0.500$, the point (α, K_3) falls in the (6.18) area of Fig. 6-4. The solution of that equation is $[OH^-] = 7.04 \times 10^{-3}$, or $pOH = 2.152$ (Assumption 1b is valid; $[OH^-] = b$), or $pH = 11.848$. From (6.11),

$$[PO_4^{3-}] = C_{cb} - b = 7.2 \times 10^{-3} \qquad\qquad [HPO_4^{2-}] = C_a + b = 2.14 \times 10^{-2}$$

$$[H_2PO_4^-] = \frac{[H^+][HPO_4^{2-}]}{K_2} = 4.90 \times 10^{-7} \qquad [H_3PO_4] = \frac{[H^+][H_2PO_4^-]}{K_1} = 9.28 \times 10^{-17}$$

7.11   Correct Problem 7.10 for non-ideality.

The concentrations calculated in Problem 7.10 can serve as a first approximation in the calculation of the ionic strength. From (3.8), $I = 0.111$; from (3.7):

$$f_{H^+} = 0.821 \qquad f_{OH^-} = 0.745 \qquad f_{H_2PO_4^-} = 0.762 \qquad f_{HPO_4^{2-}} = 0.338 \qquad f_{PO_4^{3-}} = 0.0869$$

and $f_{H_3PO_4} = 1.000$. The non-ideal equivalent of (6.18) is

$$[OH^-]^2 + \{C_a + (K_w K_{3f}/K_{wf}K_3)\}[OH^-] - C_{cb}(K_w K_{3f}/K_{wf}K_3) = 0$$

Since $C_a = C_{cb} = 0.0143$, the solution to this quadratic is $[OH^-] = 4.03 \times 10^{-3}$. Therefore,

$$[H^+] = \frac{K_w}{K_{wf}[OH^-]} = 4.05 \times 10^{-12} \qquad \text{and} \qquad pH = -\log [H^+] - \log f_{H^+} = 11.478$$

As in Problem 7.10,

$$[PO_4^{3-}] = 0.0143 - [OH^-] = 0.0103 \qquad\qquad [HPO_4^{2-}] = 0.0143 + [OH^-] = 0.0183$$

$$[H_2PO_4^-] = \frac{[H^+][HPO_4^{2-}]K_{2f}}{K_2} = 4.35 \times 10^{-7} \qquad [H_3PO_4] = \frac{[H^+][H_2PO_4^-]K_{1f}}{K_1} = 1.47 \times 10^{-16}$$

The second approximation for the ionic strength, $I = 0.121$, is sufficiently different from the first approximation to require a repeat calculation. The results are:

$$f_{H^+} = 0.818 \qquad f_{OH^-} = 0.738 \qquad f_{H_2PO_4^-} = 0.756 \qquad f_{HPO_4^{2-}} = 0.327 \qquad f_{PO_4^{3-}} = 0.0807$$

and $f_{H_3PO_4} = 1.000$. The solution of the quadratic is $[OH^-] = 3.95 \times 10^{-3}$, so that $[H^+] = 4.19 \times 10^{-12}$ and pH = 11.465. We find:

$$[PO_4^{3-}] = 0.0104 \qquad [HPO_4^{2-}] = 0.0183 \qquad [H_2PO_4^-] = 4.38 \times 10^{-7} \qquad [H_3PO_4] = 1.51 \times 10^{-16}$$

These values result in a third approximation for I that agrees with the second approximation.

7.12 Confirm column 7 of Table 7-1, which pertains to stage c, the third equivalence point.

The principal reaction is the hydrolysis of the conjugate base, PO_4^{3-}, so that H_3PO_4 and $H_2PO_4^-$ can be neglected in the initial calculations.

$$[PO_4^{3-}] = C_{cb} = 0.0250 \qquad K_{cb3} = \frac{K_w}{K_3} = 2.08 \times 10^{-2}$$

In Fig. 5-1, the point (C_{cb}, K_{cb3}) indicates the "conjugate base" equation

$$[OH^-]^2 + K_{cb3}[OH^-] - C_{cb}K_{cb3} = 0 \qquad (1)$$

Substituting numerical values and solving,

$$[OH^-] = 1.47 \times 10^{-2} \qquad [H^+] = \frac{K_w}{[OH^-]} = 6.80 \times 10^{-13} \qquad pH = 12.167$$

Since Assumption 1b is valid, $b = [OH^-]$ represents the amount of PO_4^{3-} from hydrolysis. Therefore,

$$[PO_4^{3-}] = 0.0250 - [OH^-] = 0.0103 \qquad [HPO_4^{2-}] = [OH^-] = 0.0147$$

$$[H_2PO_4^-] = \frac{[H^+][HPO_4^-]}{K_2} = 1.61 \times 10^{-7} \qquad [H_3PO_4] = \frac{[H^+][H_2PO_4^-]}{K_1} = 1.46 \times 10^{-17}$$

and $[Na^+] = 0.0750$.

7.13 What are the concentrations and the pH if the solution of Problem 7.12 is non-ideal?

The concentrations calculated in Problem 7.12 can serve as a first approximation in the calculation of the ionic strength. From (3.8), $I = 0.121$, so that (3.7) results in

$$f_{H^+} = 0.818 \qquad f_{OH^-} = 0.738 \qquad f_{H_2PO_4^-} = 0.756 \qquad f_{HPO_4^{2-}} = 0.327 \qquad f_{PO_4^{3-}} = 0.0807$$

and $f_{H_3PO_4} = 1.000$. The non-ideal version of (1) of Problem 7.12,

$$[OH^-]^2 + (K_wK_{3f}/K_{wf}K_3)[OH^-] - C_{cb}(K_wK_{3f}/K_{wf}K_3) = 0$$

should be valid here. The solution is $[OH^-] = 1.02 \times 10^{-2}$, whence

$$[H^+] = \frac{K_w}{K_{wf}[OH^-]} = 1.62 \times 10^{-12} \qquad pH = -\log [H^+] - \log f_{H^+} = 11.878$$

As Assumption 1b is valid, $[OH^-] = b$; then,

$$[PO_4^{3-}] = 0.0250 - [OH^-] = 0.0148 \qquad [HPO_4^{2-}] = [OH^-] = 0.0102$$

$$[H_2PO_4^-] = \frac{[H^+][HPO_4^{2-}]K_{2f}}{K_2} = 9.43 \times 10^{-8} \qquad [H_3PO_4] = \frac{[H^+][H_2PO_4^-]K_{1f}}{K_1} = 1.26 \times 10^{-17}$$

and $[Na^+] = 0.0750$.

The second approximation of the ionic strength is $I = 0.130$, which is sufficiently different from the first approximation to require a repeat of the various calculations. The new values are:

$$f_{H^+} = 0.815 \qquad f_{OH^-} = 0.732 \qquad f_{H_2PO_4^-} = 0.751 \qquad f_{HPO_4^{2-}} = 0.318 \qquad f_{PO_4^{3-}} = 0.0758$$

and $f_{H_3PO_4} = 1.000$. Therefore, $[OH^-] = 1.01 \times 10^{-2}$ and

$$[H^+] = 1.66 \times 10^{-12} \qquad pH = -\log [H^+] - \log f_{H^+} = 11.869$$

Further,

$$[PO_4^{3-}] = 0.0149 \qquad [HPO_4^{2-}] = 0.0101 \qquad [H_2PO_4^-] = 9.33 \times 10^{-8} \qquad [H_3PO_4] = 1.26 \times 10^{-17}$$

and $[Na^+] = 0.0750$.

The third approximation for the ionic strength, $I = 0.130$, is identical to the second approximation, so that further calculations are unnecessary.

7.14 What are the errors in $[H^+]$ as calculated from the simplest approximation of the pH, for Problems 7.3, 7.6, 7.8, 7.10, and Example 7.3?

See Table 7-4.

Table 7-4

Titration Stage	[H⁺]		Error, %
	Approximate	Ideal-Solution	
b_1	7.50×10^{-3}	5.41×10^{-3}	38.6
E_1	2.16×10^{-5}	2.01×10^{-5}	7.5
b_2	6.20×10^{-8}	6.20×10^{-8}	zero
E_2	1.73×10^{-10}	2.20×10^{-10}	−21
b_3	4.80×10^{-13}	1.42×10^{-12}	−66

7.15 In Problems 7.1 through 7.13, what are the errors in $[H^+]$ as calculated for an assumed ideal solution, compared with the non-ideal a_{H^+}-values?

See Table 7-5.

Table 7-5

Titration Stage	[H⁺]	a_{H^+}	Error, %
a	2.39×10^{-2}	2.38×10^{-2}	0.4
b_1	5.41×10^{-3}	5.93×10^{-3}	−8.8
E_1	2.01×10^{-5}	2.93×10^{-5}	−31
b_2	6.20×10^{-8}	1.28×10^{-7}	−52
E_2	2.20×10^{-10}	5.44×10^{-10}	−59.6
b_3	1.42×10^{-12}	3.33×10^{-12}	−57.4
c	6.80×10^{-13}	1.35×10^{-12}	−50

7.16 Predict and then calculate the pH of a solution that is $0.0235\ M$ $NaHCO_3$ and $0.0148\ M$ Na_2CO_3. Justify assumptions.

This mixture corresponds to a solution between the first and second equivalence points in the titration of H_2CO_3. From (6.5),

$$\alpha = \frac{1}{1 + (0.0235/0.0148)} = 0.386$$

Since $pH \approx pK_2 = 10.330$ at $\alpha = 0.500$, the pH of this solution should be slightly less, perhaps in the range 10.0–10.3. With $C_a + C_{cb} = 0.0383$ and $K_2 = 4.68 \times 10^{-11}$, the point (α, K_2) in Fig. 6-4 indicates that (6.17) is (just barely) valid. Thus, $[H^+] = 7.43 \times 10^{-11}$, or pH = 10.129, and Assumption 1b is valid. Since $b = [OH^-] = K_w/[H^+] = 1.35 \times 10^{-4}$ is less than 1% of 0.0235 and of 0.0148, Assumptions 2b and 6a are also valid.

7.17 Predict and then calculate the pH of a mixture of 10.00 mL of 0.304 M NaOH and 25.00 mL of 0.248 M H_3AsO_4.

This mixture would result if $(25.00)(0.248) = 6.20$ mmol of H_3AsO_4 was titrated with $(10.00)(0.304) = 3.04$ mmol of NaOH. The concentrations of the principal species (Region I-1 of Fig. 7-3) would be:

$$[Na^+] = \frac{3.04 \text{ mmol}}{35.00 \text{ mL}} = 0.0869 \qquad [H_2AsO_4^-] = \frac{3.04}{35.00} = 0.0869 \qquad [H_3AsO_4] = \frac{6.20 - 3.04}{35.00} = 0.0903$$

By (6.5),

$$\alpha = \frac{1}{1 + (0.0903/0.0869)} = 0.490$$

so that the mixture is just slightly before point b_1 in Fig. 7-3. Since pH $\approx$ p$K_1 = 2.190$ at point b_1, the pH of this mixture should be about 2.0–2.2. With $C_a + C_{cb} = 0.177$ and $K_1 = 6.46 \times 10^{-3}$, the point (α, K_1) in Fig. 6-4 indicates that (6.16) is valid. Using (1.4), the solution of the quadratic is $[H^+] = 5.87 \times 10^{-3}$, whence pH $= 2.231$. (This is slightly greater than pK_1, rather than slightly less as predicted; pH $\approx$ pK_1 is *not* valid at b_1 for this solution.)

7.18 Predict and then calculate the pH of a mixture of 35.00 mL of 0.304 M NaOH and 25.00 mL of 0.248 M H_3AsO_4.

This mixture would result if $(25.00)(0.248) = 6.20$ mmol of H_3AsO_4 was titrated with $(35.00)(0.304) = 10.64$ mmol of NaOH. The resulting solution will be $10.64 - 6.20 = 4.44$ mmol past the first equivalence point, and the concentrations of the principal species will be:

$$[Na^+] = \frac{10.64 \text{ mmol}}{60.00 \text{ mL}} = 0.177 \qquad [HAsO_4^{2-}] = \frac{4.44}{60.00} = 0.0740 \qquad [H_2AsO_4^-] = \frac{6.20 - 4.44}{60.00} = 0.0293$$

According to (6.5),

$$\alpha = \frac{1}{1 + (0.0293/0.0740)} = 0.716$$

so that the mixture will be between points b_2 and E_2 in Fig. 7-3. Since pH $\approx$ p$K_2 = 6.939$ at b_2, the pH of this mixture will be slightly greater, perhaps 7.1–7.3. With $C_a + C_{cb} = 0.103$ and $K_2 = 1.15 \times 10^{-7}$, the point (α, K_2) in Fig. 6-4 indicates that (6.17) holds. Thus, $[H^+] = 4.55 \times 10^{-8}$ and pH $= 7.342$.

7.19 Predict and then calculate the pH of a mixture of 25.00 mL of 0.128 M H_3AsO_4 and 20.00 mL of 0.286 M Na_2HAsO_4.

This is a solution prepared by mixing $(25.00)(0.128) = 3.20$ mmol of H_3AsO_4 and $(20.00)(0.286) = 5.72$ mmol of Na_2HAsO_4. According to the lever principle (Section 7.4), the solution lies in Region I-2 of Fig. 7-3, there being more $HAsO_4^{2-}$ than H_3AsO_4. Therefore, principal arsenate species will be $H_2AsO_4^-$ and $HAsO_4^{2-}$, resulting from the reaction

$$H_3AsO_4 + HAsO_4^{2-} \rightarrow 2H_2AsO_4^-$$

In this solution, H_3AsO_4 is the limiting species, so that

$$(3.20 \text{ mmol } H_3AsO_4) + (5.72 \text{ mmol } HAsO_4^{2-}) \rightarrow (6.40 \text{ mmol } H_2AsO_4^-) + (2.52 \text{ mmol } HAsO_4^{2-} \text{ excess})$$

The concentrations of principal species are:

$$[Na^+] = \frac{2(5.72 \text{ mmol})}{45.00 \text{ mL}} = 0.254 \qquad [H_2AsO_4^-] = \frac{6.40}{45.00} = 0.142 \qquad [HAsO_4^{2-}] = \frac{2.52}{45.00} = 0.0560$$

With $C_a + C_{cb} = 0.198$, $\alpha = 0.0560/0.198 = 0.282$, and $K_2 = 1.15 \times 10^{-7}$, Fig. 6-4 shows that (6.17) is valid, so that $[H^+] = 2.92 \times 10^{-7}$ and pH $= 6.535$.

7.20 A mixture has as principal species either (i) HCl and H_2X, (ii) H_2X and NaHX, (iii) NaHX and Na_2X, or (iv) Na_2X and NaOH, where H_2X is a weak acid with $K_1 = 1.00 \times 10^{-4}$ and $K_2 = 1.00 \times 10^{-8}$. It is known that the concentrations of the principal species are in the range 0.01–0.5 M. The pH of a 0.1 M H_2X solution is about 2.5, the pH of a 0.1 M NaHX solution is about 6.0, and the pH of a 0.1 M Na_2X solution is about 11.0. These pH-values correspond to the various equivalence points as shown in Fig. 7-8. Given that 50.00 mL of this mixture required 14.00 mL of 0.250 M NaOH to reach an equivalence point at pH 6.0 and *an additional* 33.00 mL to reach an equivalence point at pH 11.0, which pair of principal species is present in the mixture and what are the two concentrations?

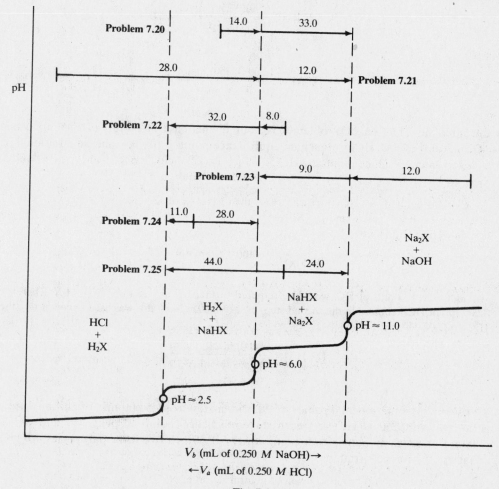

Fig. 7-8

Since it required less NaOH to reach pH 6.0 than to reach pH 11.0, the start of the titration must have been in the region of Fig. 7-8 where H_2X and NaHX are principal species. The 14.00 mL of NaOH converted the H_2X to HX^-, so that

$$[H_2X] = \frac{(14.00)(0.250) \text{ mmol}}{50.00 \text{ mL}} = 0.0700$$

Of the additional 33.00 mL, 14.00 mL was used in continuing the titration of the original H_2X. Therefore, $33.00 - 14.00 = 19.00$ mL was used in converting the original HX^- to X^{2-}, so that

$$[NaHX] = \frac{(19.00)(0.250) \text{ mmol}}{50.00 \text{ mL}} = 0.0950$$

7.21 Consider the situation of Problem 7.20, but in this case 50.00 mL of the mixture required 28.00 mL of 0.250 M NaOH to reach an equivalence point at pH 6.0, and an additional 12.00 mL to reach an equivalence point at pH 11.0. Which two principal species are present and what are their concentrations?

Since it required more NaOH to reach pH 6.0 than to reach pH 11.0, there must have been HCl present as a principal species, along with H_2X (see Fig. 7-8). Since it required 12.00 mL to convert HX^- (pH = 6.0) to X^{2-} (pH = 11.0), it must have also required 12.00 mL in the original conversion of H_2X (pH = 2.5) to HX^- (pH = 6.0). Therefore, $28.00 - 12.00 = 16.00$ mL of NaOH was required to neutralize the HCl, and

$$[HCl] = \frac{(16.00)(0.250) \text{ mmol}}{50.00 \text{ mL}} = 0.0800$$

whereas

$$[H_2X] = \frac{(12.00)(0.250) \text{ mmol}}{50.00 \text{ mL}} = 0.0600$$

7.22 Consider the situation of Problem 7.20, but in this case 50.00 mL of the mixture required 8.00 mL of 0.250 M HCl to reach an equivalence point at pH 6.0, and an additional 32.00 mL to reach an equivalence point at pH 2.5. Which two principal species are present and what are their concentrations?

Since it required less HCl to reach pH 6.0 than to reach pH 2.5, the original mixture must have had as principal species NaHX and Na_2X (see Fig. 7-8). The 8.00 mL of HCl was used in converting the original X^{2-} to HX^-, so that

$$[Na_2X] = \frac{(8.00)(0.250) \text{ mmol}}{50.00 \text{ mL}} = 0.0400$$

An additional 8.00 mL of HCl would be required further to convert the X^{2-} to H_2X. Therefore, of the 32.00 mL of HCl required in the second step, $32.00 - 8.00 = 24.00$ mL was used in converting the original HX^- to H_2X. Hence,

$$[NaHX] = \frac{(24.00)(0.250) \text{ mmol}}{50.00 \text{ mL}} = 0.120$$

7.23 Consider the situation of Problem 7.20, but in this case 50.00 mL of the mixture required 12.00 mL of 0.250 M HCl to reach an equivalence point at pH 11.0, and an additional 9.00 mL to reach an equivalence point at pH 6.0. Which two principal species are present and what are their concentrations?

The first step in the titration involved neutralization of NaOH, as shown in Fig. 7-8; therefore,

$$[NaOH] = \frac{(12.00)(0.250) \text{ mmol}}{50.00 \text{ mL}} = 0.0600$$

The second step involved converting X^{2-} to HX^-, so that

$$[Na_2X] = \frac{(9.00)(0.250) \text{ mmol}}{50.00 \text{ mL}} = 0.0450$$

7.24 Consider the situation of Problem 7.20, but in this case 50.00 mL of the mixture required 11.00 mL of 0.250 M HCl to reach an equivalence point at pH 2.5, and a separate 50.00 mL portion of the mixture required 28.00 mL of 0.250 M NaOH to reach an equivalence point at pH 6.0. Which two principal species are present and what are their concentrations?

Since the separate titrations are in opposite directions and end in the first two equivalence points, the mixture must be between these two equivalence points and the principal species must be H_2X and $NaHX$ (see Fig. 7-8). The HCl converts the HX^- to H_2X, so that

$$[NaHX] = \frac{(11.00)(0.250) \text{ mmol}}{50.00 \text{ mL}} = 0.0550$$

The NaOH converts the H_2X to HX^-, so that

$$[H_2X] = \frac{(28.00)(0.250) \text{ mmol}}{50.00 \text{ mL}} = 0.140$$

7.25 Consider the situation of Problem 7.20, but in this case 50.00 mL of the mixture required 24.00 mL of 0.250 M NaOH to reach an equivalence point at pH 11.0, and a separate 50.00 mL portion required 44.00 mL of 0.250 M HCl to reach an equivalence point at pH 2.5. Which two principal species are present and what are their concentrations?

The two equivalence points are separated by the equivalence point at pH 6.0. Since it required *fewer* mmols to reach the pH 11.0 equivalence point than to reach the pH 2.5 equivalence point, the original mixture must have been closer to the pH 11.0 point (see Fig. 7-8). The principal species are $NaHX$ and Na_2X. The NaOH converted the HX^- to X^{2-}, so that

$$[NaHX] = \frac{(24.00)(0.250) \text{ mmol}}{50.00 \text{ mL}} = 0.120$$

The HCl converted the $(24.00)(0.250) = 6.00$ mmol of HX^- to H_2X and also converted the original X^{2-} to H_2X. Therefore, the amount of HCl required for the latter conversion is $(44.00)(0.250) - 6.00 = 5.00$ mmol. Since two mmol of HCl was required for every mmol of X^{2-}, the original concentration of X^{2-} is

$$[Na_2X] = \frac{(5.00/2) \text{ mmol}}{50.00 \text{ mL}} = 0.0500$$

7.26 A solution was prepared by mixing 100.0 mL of 0.200 M H_3PO_4 with 100.0 mL of x M Na_2HPO_4. 50.00 mL of this mixture required 27.00 mL of 0.0800 M NaOH to reach an equivalence point at pH ≈ 4.7. Find the value of x.

From Table 7-1, pH ≈ 4.7 corresponds to the first equivalence point in the titration of H_3PO_4. Since NaOH was being added, the mixture must have been in Region I-1 of Fig. 7-3 and the principal species must have been H_3PO_4 and NaH_2PO_4. The NaOH converted the H_3PO_4 to $H_2PO_4^-$, so that there was $(27.00)(0.0800) = 2.16$ mmol of H_3PO_4 in the 50.00 mL sample. The total (200.0 mL) sample, therefore, contained $4(2.16) = 8.64$ mmol of H_3PO_4. Since $(100.0)(0.200) = 20.0$ mmol of H_3PO_4 was used to prepare the mixture, $20.0 - 8.64 = 11.3_6$ mmol of H_3PO_4 must have reacted with the $(100.0)(x) = 100.0\,x$ mmol of HPO_4^{2-}, according to

$$H_3PO_4 + HPO_4^{2-} \rightarrow 2H_2PO_4^-$$

Therefore, $100.0\,x = 11.3_6$, or $x = 0.114$.

7.27 Vitamin C (ascorbic acid) has the formula $C_6H_8O_6$, a molecular weight of 176, and two replaceable hydrogens for which $K_1 = 6.30 \times 10^{-5}$ and $K_2 = 2.50 \times 10^{-12}$. The molecular weights of the two sodium salts of ascorbic acid, NaHA and Na_2A, are 198 and 220, respectively. A vitamin-C supplement is a mixture of the two sodium salts plus soluble inert material. A 500 mg tablet of this supplement was dissolved in 50.00 mL of water and titrated with 0.0725 M HCl. It required 7.50 mL of the HCl to reach an equivalence point at pH ≈ 7.9, and an additional 30.00 mL to reach an equivalence point at pH ≈ 2.8 (the pH of a typical solution of ascorbic acid, H_2A). Find the wt % of NaHA and the wt % of Na_2A in the tablet.

Since $$\tfrac{1}{2}(pK_1 + pK_2) = \tfrac{1}{2}(4.201 + 11.602) = 7.902$$

the first equivalence point reached in the titration was the point E_1 for H_2A (see Fig. 7-3). This step in the titration converted A^{2-} to HA^-; hence, the tablet contained

$$(7.50)(0.0725) = 0.544 \text{ mmol}$$

of Na_2A. This corresponds to $(0.544)(220) = 120$ mg, so that the wt % of Na_2A was

$$\frac{120}{500}(100\%) = 24.0\%$$

The second step in the titration ended at a pH that corresponds to H_2A, point a in Fig. 7-3, and involves the conversion of HA^- to H_2A. A total of

$$(30.00)(0.0725) = 2.17_5 \text{ mmol}$$

of HA^- was converted. Since 0.544 mmol of this total involved the continuation of the titration of the original A^{2-}, the amount of HA^- in the sample was

$$2.17_5 - 0.544 = 1.63_1 \text{ mmol}$$

This corresponds to $(1.63_1)(198) = 323$ mg of Na_2A, so that the wt % of Na_2A was

$$\frac{323}{500}(100\%) = 64.6\%$$

pH KNOWN

7.28 What volume of 0.500 M NaOH should be added to 250.0 mL of 0.400 M H_3PO_4 to prepare 1.00 L of a pH 7.000 buffer?

From (7.6), $F_0 = 8.23 \times 10^{-6}$, $F_1 = 0.617$, $F_2 = 0.383$, $F_3 = 1.84 \times 10^{-6}$. The 1.00 L of buffer contains a total phosphate concentration

$$C = \frac{(250.0)(0.400)}{1000} = 0.100 \ M$$

so that

$$[H_3PO_4] = F_0C = 8.23 \times 10^{-7} \qquad [H_2PO_4^-] = F_1C = 0.0617$$

$$[HPO_4^{2-}] = F_2C = 0.0383 \qquad [PO_4^{3-}] = F_3C = 1.84 \times 10^{-7}$$

and $[H^+] = [OH^-] = 1.00 \times 10^{-7}$. The charge-balance expression is:

$$[H^+] + [Na^+] = [OH^-] + [H_2PO_4^-] + 2[HPO_4^{2-}] + 3[PO_4^{3-}]$$

All terms are known except $[Na^+]$, which, upon solving, is 0.1383. Therefore, the 1.00 L contained 138.3 mmol of Na^+, and

$$\frac{138.3 \text{ mmol}}{0.500 \ M} = 276.6 \text{ mL}$$

of NaOH was required.

7.29 A phosphate buffer solution is prepared by adding NaOH to a H_3PO_4 solution until the pH = 7.000. It is known that the total phosphate concentration in this solution is $C = 0.0200 \ M$; and, because of added NaCl, the ionic strength is $I = 0.1000$. For this solution,

$$f_{H^+} = 0.830 \qquad f_{OH^-} = 0.755 \qquad f_{Na^+} = 0.770 \qquad f_{Cl^-} = 0.755$$

$$f_{H_2PO_4^-} = 0.770 \qquad f_{HPO_4^{2-}} = 0.355 \qquad f_{PO_4^{3-}} = 0.0949$$

and $f_{H_3PO_4} = 1.000$. What are the concentrations of all species in solution?

A pH of 7.000 is just to the left of point b_2 in Fig. 7-3, so the principal species should be $H_2PO_4^-$ and HPO_4^{2-}. We shall therefore neglect H_3PO_4 and PO_4^{3-} in the initial calculations. From the two equations

$$K_2 = 6.20 \times 10^{-8} = \frac{a_{H^+}[HPO_4^{2-}]f_{HPO_4^{2-}}}{[H_2PO_4^-]f_{H_2PO_4^-}}$$

$$[HPO_4^{2-}] + [H_2PO_4^-] = 0.0200$$

where $a_{H^+} = 1.00 \times 10^{-7}$, we find:

$$[HPO_4^{2-}] = 0.0115 \qquad [H_2PO_4^-] = 8.55 \times 10^{-3}$$

In addition,

$$[H^+] = a_{H^+}/f_{H^+} = 1.20 \times 10^{-7} \qquad [OH^-] = a_{OH^-}/f_{OH^-} = \frac{1.00 \times 10^{-7}}{0.755} = 1.32 \times 10^{-7}$$

$$[H_3PO_4] = \frac{[H^+][H_2PO_4^-]K_{1f}}{K_1} = 8.74 \times 10^{-8} \qquad [PO_4^{3-}] = \frac{[HPO_4^{2-}]K_3}{K_{3f}[H^+]} = 2.07 \times 10^{-7}$$

(so that neglecting H_3PO_4 and PO_4^{3-} in the initial calculations was justified). Since $[H^+]$ is so very small, the concentration of the added NaOH equaled

$$[H_2PO_4^-] + 2[HPO_4^{2-}] + 3[PO_4^{3-}] = 0.0316$$

so that $[Na^+]$ from NaOH equaled 0.0316. Letting the additional $[Na^+] = [Cl^-] = x$, we have

$$I = 0.100 = \tfrac{1}{2}\{[H^+]1^2 + [OH^-]1^2 + [Na^+]1^2 + [Cl^-]1^2 + [H_2PO_4^-]1^2 + [HPO_4^{2-}](-2)^2 + [PO_4^{3-}](-3)^2\}$$

$$= \tfrac{1}{2}\{(1.20 \times 10^{-7}) + (1.32 \times 10^{-7}) + (0.0316 + x) + x + (8.55 \times 10^{-3}) + 0.0460 + (1.86 \times 10^{-6})\}$$

Therefore, $x = [Cl^-] = 0.0569$ and $[Na^+] = 0.0569 + 0.0316 = 0.0885$.

7.30   Titration of 664 mg of a pure organic carboxylic acid, H_2A, showed a rapid change in pH at 40.00 mL and at 80.00 mL of 0.100 M NaOH titrant. When 40.00 mL of NaOH was added, the pH was 5.85, and at 60.00 mL of NaOH the pH was 8.08. What is the molecular weight of H_2A and what are the approximate values of K_1 and K_2?

Since $(40.00)(0.100) = 4.00$ mmol of NaOH was required to reach the first equivalence point, the molecular weight was

$$\frac{664 \text{ mg}}{4.00 \text{ mmol}} = 166$$

60.00 mL of NaOH is represented by a point midway to the second equivalence point, at which $pH \approx pK_2$. Therefore, $pK_2 \approx 8.08$ and $K_2 \approx 8.3 \times 10^{-9}$. At the first equivalence point,

$$pH \approx \tfrac{1}{2}(pK_1 + pK_2) = 5.85$$

Since $pK_2 \approx 8.08$, $pK_1 \approx 3.62$ and $K_1 \approx 2.4 \times 10^{-4}$.

Supplementary Problems

7.31 If 20.00 mL of 0.0500 M H_2CO_3 is titrated with 0.0500 M NaOH, what is the initial pH of the carbonic acid?   *Ans.*   3.825

7.32 What was the pH in Problem 7.31 when (*a*) 10.00 mL, (*b*) 20.00 mL, (*c*) 30.00 mL, (*d*) 40.00 mL, (*e*) 50.00 mL, of NaOH was added?   *Ans.*   (*a*) 6.350; (*b*) 8.338; (*c*) 10.313; (*d*) 11.252; (*e*) 11.877

7.33 A solution has as principal species either (i) HCl and H_2SO_3, (ii) H_2SO_3 and $NaHSO_3$, (iii) $NaHSO_3$ and Na_2SO_3, or (iv) Na_2SO_3 and NaOH. It is known that the concentrations of the principal species are in the range 0.01–0.5 M. At the first equivalence point, E_1, in the H_2SO_3 titration,

$$pH \approx (pK_1 + pK_2)/2 = 4.5$$

The pH at the second equivalence point, c, in the H_2SO_3 titration is about 10.1. Titration of $NaHSO_3$ with HCl does not show a sharp change in pH at the H_2SO_3 equivalence point, since $K_1 > 10^{-3}$. If 25.00 mL of this mixture required 15.00 mL of 0.100 M NaOH to reach an equivalence point at pH 4.5, and an additional 20.00 mL to reach an equivalence point at pH 10.1, which principal species are present and what are their concentrations?   *Ans.* $[H_2SO_3] = 0.0600$, $[NaHSO_3] = 0.0200$

7.34 Consider the same situation as in Problem 7.33, but in this case 25.00 mL of the mixture required 15.00 mL of 0.100 M NaOH to reach an equivalence point at pH 4.5, and an additional 10.00 mL to reach an equivalence point at pH 10.1. Which principal species are present and what are their concentrations?   *Ans.* $[HCl] = 0.0200$, $[H_2SO_3] = 0.0400$

7.35 Consider the same situation as in Problem 7.33, but in this case 25.00 mL of the mixture required 10.00 mL of 0.100 M NaOH to reach an equivalence point at pH 10.1. A separate 25.00 mL portion of the mixture required 15.00 mL of 0.100 M HCl to reach an equivalence point at pH 4.5. Which principal species are present and what are their concentrations?
Ans. $[NaHCO_3] = 0.0400$, $[Na_2CO_3] = 0.0600$

7.36 Consider the same situation as in Problem 7.33, but in this case 25.00 mL of the mixture required 18.00 mL of 0.100 M HCl to reach an equivalence point at pH 10.1, and an additional 14.00 mL to reach an equivalence point at pH 4.5. Which principal species are present and what are their concentrations?
Ans. $[Na_2CO_3] = 0.0560$, $[NaOH] = 0.0720$

7.37 Solid NaOH is added to 1.00 L of 0.200 M citric acid, H_3Cit, until the pH is 4.000. Assuming that there was no change in volume, find the concentrations of all species in solution.
Ans. $[H_3Cit] = 0.0120$, $[H_2Cit^-] = 0.137$, $[HCit^{2-}] = 0.0498$, $[Cit^{3-}] = 7.52 \times 10^{-4}$, $[H^+] = 1.00 \times 10^{-4}$, $[OH^-] = 1.00 \times 10^{-10}$, $[Na^+] = 0.239$

Chapter 8

Precipitates and Solubilities

8.1 SOLUBILITY PRODUCT

Thermodynamic equilibrium constants also apply to the solubilities of compounds. In this chapter, we shall consider only the solubility of salts that dissociate completely in solution into ions. We shall neglect the formation of complexes such as $AgCl_2^-$; that is the subject of Chapter 9.

The simplest salt has the general formula AB, and the equation governing the dissolving of the precipitate is one of

$$AB(s) \rightarrow A^+ + B^- \tag{8.1a}$$

$$AB(s) \rightarrow A^{2+} + B^{2-} \tag{8.1b}$$

$$AB(s) \rightarrow A^{3+} + B^{3-} \tag{8.1c}$$

The first reaction is the most common; e.g., $AgBr(s) \rightarrow Ag^+ + Br^-$. The thermodynamic equilibrium constant for (8.1a) is

$$K_s = \frac{a_{A^+} a_{B^-}}{a_{AB}(s)} \tag{8.2}$$

But, if the pressure on the solution and the precipitate is at most a few atmospheres, then, as mentioned in Section 3.2, the activity of the solid can be approximated as 1.00. Therefore,

$$K_s \approx a_{A^+} a_{B^-} \equiv K_{sp} \tag{8.3}$$

where K_{sp} is called the (*thermodynamic*) *solubility product*. It should be stressed that this and other equations for the solubility product are valid only if some precipitate—be it only a tiny crystal—exists in equilibrium with the ions in solution. With no precipitate, there is no activity term for the solid, and the activities of the ions may have any values such that their product is less than K_{sp}.

In terms of relative concentrations, (8.3) becomes

$$\frac{K_{sp}}{K_{spf}} = [A^+][B^-] \tag{8.4}$$

in which

$$K_{spf} \equiv f_{A^+} f_{B^-} \tag{8.5}$$

Similar expressions can be obtained for salts with different stoichiometry. For instance, if the salt dissociates according to

$$A_2B(s) \rightarrow 2A^+ + B^{2-} \tag{8.6}$$

then

$$\frac{K_{sp}}{K_{spf}} = [A^+]^2[B^{2-}] \tag{8.7}$$

where

$$K_{spf} = (f_{A^+})^2 f_{B^{2-}} \tag{8.8}$$

8.2 SOLUBILITY OF SALTS

If, at equilibrium, S moles per liter of a precipitate have dissolved, then S is defined to be the *solubility* of the precipitate. Supposing that the solution is ideal ($K_{spf} = 1.00$), that the ions formed do not undergo further reaction—through hydrolysis or with other species in solution—and that the solution does not contain ions in common with those arising from the precipitate (this rules out hydroxide precipitates), then the solubility can be directly related to the solubility product. Thus, if S

mol/L of $AB(s)$ dissolves to give S mol/L of A^+ and S mol/L of B^-, then, by (8.4), $K_{sp} = [A^+][B^-] = S^2$. If S mol/L of $A_2B(s)$ dissolves to give $2S$ mol/L of A^+ and S mol/L of B^{2-}, then, by (8.7), $K_{sp} = [A^+]^2[B^{2-}] = 4S^3$. A number of similar results are collected in Table 8-1.

Table 8-1

Formula of precipitate	Expression for K_{sp}	
	In terms of concentrations	In terms of solubility
AB	$[A^+][B^-]$ $[A^{2+}][B^{2-}]$ $[A^{3+}][B^{3-}]$	S^2 S^2 S^2
A_2B	$[A^+]^2[B^{2-}]$	$4S^3$
AB_2	$[A^{2+}][B^-]^2$	$4S^3$
A_3B	$[A^+]^3[B^{3-}]$	$27S^4$
AB_3	$[A^{3+}][B^-]^3$	$27S^4$
A_2B_3	$[A^{3+}]^2[B^{2-}]^3$	$108S^5$
A_3B_2	$[A^{2+}]^3[B^{3-}]^2$	$108S^5$
ABC	$[A^{2+}][B^+][C^{3-}]$	S^3

EXAMPLE 8.1 Find the solubility of $PbBr_2(s)$, (a) in an ideal solution; (b) in a non-ideal solution with an amount of added $NaNO_3$ such that, at equilibrium, the ionic strength is 0.100. The solubility product of the solid is $K_{sp} = 3.90 \times 10^{-5}$.

(a) From Table 8-1, $K_{sp} = 3.90 \times 10^{-5} = 4S^3$ for a molecule with the formula AB_2. Solving, $S = 0.0214$.

(b) According to Table 3-3, $f_{Pb^{2+}} = 0.377$ and $f_{Br^-} = 0.754$. Therefore,

$$K_{spf} = (0.377)(0.754)^2 = 0.214 \qquad \text{and} \qquad \frac{K_{sp}}{K_{spf}} = 1.82 \times 10^{-4} = 4S^3$$

so that $S = 0.0357$.

8.3 PRECIPITATION TITRATIONS

The concentrations of ions that form precipitates can be determined in a way similar to the pH titration analysis of a weak acid. Although the types of electrochemical cells that are appropriate for such analyses will not be discussed until Chapter 11, we can still consider here the general principles of precipitation titrations. What is done is to devise an electrochemical cell the output voltage of which is proportional to $-\log[X] \equiv pX$, where X is either the ion whose concentration is being determined or the titrant ion.

EXAMPLE 8.2 Plotted in Fig. 8-1 is pAg, corresponding to the titrant ion in the precipitation titration of 20.00 mL of 0.0825 M Cl^- solution with 0.100 M $AgNO_3$. 16.50 mL of $AgNO_3$ is required to reach the equivalence point, which is marked by a sharp decrease in pAg. Since AgCl is very insoluble ($K_{sp} = 1.78 \times 10^{-10}$), almost all Ag^+, prior to the equivalence point, is converted to $AgCl(s)$. In this titration region, the amount of Ag^+ at equilibrium remains quite low; hence, pAg remains large and shows only a small decline. As the equivalence point is reached, [Cl⁻] begins to decrease markedly and the corresponding allowable equilibrium [Ag⁺] increases markedly; pAg shows a sharp decrease in this region. Once slightly past the equivalence point, [Cl⁻] remains very small and, therefore, [Ag⁺] remains large. As a result, pAg is low and shows little further change, for added Ag⁺ produces only a small additional effect beyond that of the large amount of Ag⁺ already in solution.

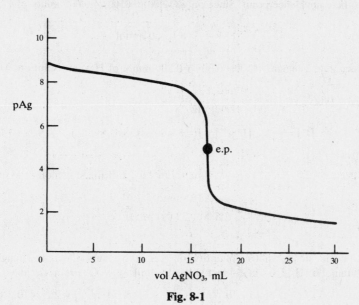

Fig. 8-1

EXAMPLE 8.3 Calculate pAg (a) at the point where 10.00 mL of AgNO₃ was added, and (b) at the equivalence point, in the titration of Fig. 8-1.

(a) There is $(0.0825)(20.00) = 1.65$ mmol of NaCl and $(0.100)(10.00) = 1.00$ mmol of AgNO₃. Therefore, almost all of the Ag⁺ will be precipitated as AgCl(s),

$$[\text{Cl}^-] = \frac{(1.65 - 1.00)\ \text{mmol}}{30.00\ \text{mL}} = 2.2 \times 10^{-2} \qquad [\text{Ag}^+] = \frac{K_{sp}}{[\text{Cl}^-]} = \frac{1.78 \times 10^{-10}}{2.2 \times 10^{-2}} = 8.1 \times 10^{-9}$$

and pAg = 8.09.

(b) At the equivalence point, any remaining Ag⁺ and Cl⁻ in solution must be in a stoichiometric ratio, so that $[\text{Ag}^+] = [\text{Cl}^-]$. Therefore,

$$K_{sp} = 1.78 \times 10^{-10} = [\text{Ag}^+][\text{Cl}^-] = [\text{Ag}^+]^2 \qquad \text{or} \qquad [\text{Ag}^+] = 1.33 \times 10^{-5}$$

and pAg = 4.875.

Precipitation titrations are practical only if the reaction product is very insoluble, in which case it can usually be assumed that almost all of the ion present in the limiting stoichiometric amount has been precipitated. Thus, prior to the equivalence point, very little titrant ion is present in solution. Following the equivalence point, at which stage excess titrant is present, very little analyte ion will be present in solution. Knowing how much of the limiting ion is available allows use of stoichiometry to calculate the concentration of the other ion. From this and from the K_{sp}-value, the concentration of the limiting ion can be found. Only in the immediate vicinity of the equivalence point (but not at the equivalence point) is it necessary to consider the solubility equilibrium in detail.

EXAMPLE 8.4 Mercurous iodide is very insoluble ($K_{sp} = 4.5 \times 10^{-29}$) and undergoes the reaction

$$\text{Hg}_2\text{I}_2(s) \rightarrow \text{Hg}_2^{2+} + 2\text{I}^-$$

since the mercurous ion is a dimer. 25.00 mL of 0.0200 M Hg₂²⁺ was titrated with 0.0500 M NaI. Calculate pI after (a) 12.00 mL, (b) 25.00 mL, (c) 20.00 mL, of NaI has been added.

(a)   The $(25.00)(0.0200) = 0.500$ mmol of Hg_2^{2+} requires $2(0.500) = 1.000$ mmol of I^- and, therefore, a volume of

$$\frac{1.000}{0.0500} = 20.00 \text{ mL}$$

of NaI, to reach the equivalence point. Since only $(12.00)(0.0500) = 0.600$ mmol of I^- was added,

$$\frac{0.600}{2} = 0.300 \text{ mmol}$$

of Hg_2^{2+} was precipitated, leaving $0.500 - 0.300 = 0.200$ mmol of Hg_2^{2+} in solution. Therefore,

$$[Hg_2^{2+}] = \frac{0.200 \text{ mmol}}{(25.00 + 12.00) \text{ mL}} = 5.41 \times 10^{-3}$$

$$[I^-] = \sqrt{K_{sp}/[Hg_2^{2+}]} = [(4.5 \times 10^{-29})/(5.41 \times 10^{-3})]^{1/2} = 9.1_2 \times 10^{-14}$$

so that pI = 13.04.

(b)   Since $(25.00)(0.0500) = 1.250$ mmol I^- is 0.250 mmol in excess of the stoichiometric requirement,

$$[I^-] = \frac{0.250 \text{ mmol}}{(25.00 + 25.00) \text{ mL}} = 5.00 \times 10^{-3}$$

and pI = 2.301.

(c)   Upon addition of the stoichiometric quantity of NaI, the solution can be imagined as one initially containing 0.500 mmol of $Hg_2I_2(s)$ in contact with 45.00 mL of H_2O. From Table 8-1,

$$K_{sp} = 4S^3 = 4.5 \times 10^{-29} \qquad \text{or} \qquad S = [Hg_2^{2+}] = 2.2_4 \times 10^{-10}$$

$[I^-] = 2S = 4.4_8 \times 10^{-10}$ and pI = 9.35.

Analysis of Mixtures

Certain mixtures of anions or cations can be analyzed in a precipitation titration. If the ions each form a precipitate with the titrant, and if there are sufficient differences in solubilities (if, say, the solubility ratios exceed 10^2), then separate equivalence points will be observed.

EXAMPLE 8.5   Shown in Fig. 8-2 is the precipitation titration response when 0.0500 M $AgNO_3$ is added to 25.00 mL of a solution that is 0.0150 M NaCl, 0.0100 M NaBr, and 0.0250 M NaI. The order of precipitation can be determined by calculating the concentrations of Ag^+ required to begin precipitation of each of the silver halides.

To begin precipitating $AgCl(s)$,

$$[Ag^+] = \frac{K_{sp}}{[Cl^-]} = \frac{1.78 \times 10^{-10}}{0.0150} = 1.19 \times 10^{-8}$$

To begin precipitating $AgBr(s)$,

$$[Ag^+] = \frac{K_{sp}}{[Br^-]} = \frac{5.25 \times 10^{-13}}{0.0100} = 5.25 \times 10^{-11}$$

To begin precipitating $AgI(s)$,

$$[Ag^+] = \frac{K_{sp}}{[I^-]} = \frac{8.31 \times 10^{-17}}{0.0250} = 3.32 \times 10^{-15}$$

Since $AgI(s)$ requires the least Ag^+, it will precipitate first, with equivalence point a of Fig. 8-2 reached after

$$\frac{(0.0250)(25.00)}{0.0500} = 12.5 \text{ mL}$$

of $AgNO_3$ has been added. $AgBr(s)$ will precipitate next, with *an extra*

$$\frac{(0.0100)(25.00)}{0.0500} = 5.00 \text{ mL}$$

of $AgNO_3$ required to reach point b. Finally, $AgCl(s)$ will precipitate, with a further

$$\frac{(0.0150)(25.00)}{0.0500} = 7.50 \text{ mL}$$

of $AgNO_3$ required to reach point c.

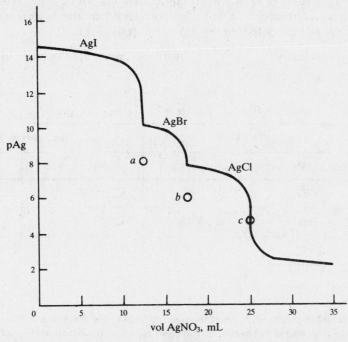

Fig. 8-2

The equivalence points a, b, and c correspond to the condition $[Ag^+] = \sqrt{K_{sp}}$. At a and b, the pAg-value is slightly less than the "corner" value of pAg for which the next halide begins to precipitate. However, since the titration curve is nearly vertical going into the corner, very little error results when the abscissa of the corner is taken as the equivalence volume of $AgNO_3$. As for point c, it is identified as though the NaCl alone were being titrated.

8.4 CLASSIFICATION OF PRECIPITATES

There are three general classes of precipitates. The first consists of *salts of strong acids* whose ions do not undergo hydrolysis and are not hydroxides. These include $AgCl$, PbI_2, Hg_2I_2, $TlBr$; their solubilities may be calculated from Table 8-1.

The second class are *hydroxide precipitates*; their solubilities depend on the amount of OH^- arising from the precipitate, as compared with the OH^- from the water equilibrium.

The third class, which is often the most difficult to treat, consists of *salts of weak acids*. For these, it is necessary to consider the K_{sp} equilibrium, the water equilibrium, and the acid dissociation reaction(s) for the weak acid.

Hydroxide Precipitates

Hydroxide precipitates may be divided into three subclasses, depending on the concentration of hydroxide ion produced when the precipitate dissolves in water:

(1) If the concentration of hydroxide produced exceeds $1.00 \times 10^{-6} M$, then the water equilibrium can be neglected (Assumption 1b, Chapters 5 and 6). Almost all $M(OH)_2(s)$ precipitates belong to this subclass, as indicated in Table 8-3.

(2) If the concentration of hydroxide produced is less than $1.99 \times 10^{-9} M$, then the hydroxide that results from the precipitate can be neglected and the pH will be 7.000 (Assumption 3b, Chapters 5 and 6). Almost all $M(OH)_3(s)$ precipitates belong to this subclass, as shown in Table 8-3.

(3) If neither (1) nor (2) is valid, the water equilibrium cannot be neglected and the equilibrium shift in the water dissociation must be considered in the calculation. Of the common hydroxide precipitates, only $Be(OH)_2(s)$ and $Cr(OH)_3(s)$ fall in this subclass.

The value of K_{sp} will determine whether a hydroxide precipitate belongs to subclass (1), (2), or (3); see Table 8-2.

Table 8-2

Compound	In Subclass (1) if:		In Subclass (2) if:	
	Solubility Product	Equation for S	Solubility Product	Equation for S
$MOH(s)$	$K_{sp} > 1.00 \times 10^{-12}$	$K_{sp} = S^2$	$K_{sp} < 1.99 \times 10^{-16}$	$K_{sp} = 1.00 \times 10^{-7} S$
$M(OH)_2(s)$	$K_{sp} > 5.00 \times 10^{-19}$	$K_{sp} = 4S^3$	$K_{sp} < 9.95 \times 10^{-24}$	$K_{sp} = 1.00 \times 10^{-14} S$
$M(OH)_3(s)$	$K_{sp} > 3.33 \times 10^{-25}$	$K_{sp} = 27S^4$	$K_{sp} < 6.63 \times 10^{-31}$	$K_{sp} = 1.00 \times 10^{-21} S$

EXAMPLE 8.6 (a) For what values of K_{sp} is it possible to neglect the water equilibrium when calculating the solubility, S, of $M(OH)_2(s)$ in water? (b) For what values of K_{sp} is it possible to assume that $[OH^-] = 1.00 \times 10^{-7}$ at equilibrium when calculating the solubility, S, of $M(OH)_3(s)$ in water?

(a) If S moles per liter dissolve, then $[M^{2+}] = S$, $[OH^-] = 2S$, and $K_{sp} = 4S^3$. If the water equilibrium can be neglected [subclass (1)],

$$[OH^-] = 2S > 1.00 \times 10^{-6} \qquad \text{or} \qquad S > 5.00 \times 10^{-7}$$

Therefore, $K_{sp} > 4(5.00 \times 10^{-7})^3 = 5.00 \times 10^{-19}$, as given in Table 8-2.

(b) If S moles per liter of $M(OH)_3(s)$ dissolve, there are produced S moles per liter of M^{3+} and $3S$ moles per liter of OH^-. If $[OH^-] = 1.00 \times 10^{-7}$ [subclass (2)], the OH^- arising from the precipitate must be such that

$$3S < 1.99 \times 10^{-9} \qquad \text{or} \qquad S < 6.63 \times 10^{-10}$$

Since $K_{sp} = S[OH^-]^3$,

$$K_{sp} < (6.63 \times 10^{-10})(1.00 \times 10^{-7})^3 = 6.63 \times 10^{-31}$$

as given in Table 8-2.

Hydroxide precipitates could also be added to solutions that contain acids, to buffer solutions, or to solutions that contain the precipitate cation as a common ion. Table 8-2 does *not* apply to these solutions, since it assumes that the precipitates are added to water.

EXAMPLE 8.7 How many moles of $Ca(OH)_2(s)$ will dissolve in 1.00 L of 1.00 M HCl? $K_{sp} = 5.50 \times 10^{-6}$ and $CaCl_2$ is soluble.

We can imagine the process as taking place in two steps. First, the acid will dissolve 0.500 mol of $Ca(OH)_2(s)$, resulting temporarily in a solution that has $[Ca^{2+}] = 0.500$ and $[Cl^-] = 1.00$. Then, an additional x mol of $Ca(OH)_2(s)$ will dissolve according to the reaction

Table 8-3. Data for Solubility in Water

Compound	Type	K_{sp}	No hydrolysis occurs. (Salts of Strong Acids)	Water equilibrium negligible; $[OH^-] > 10^{-6}$. Subclass (1). (Hydroxides)	Negligible anion hydrolysis; $h < 1.99 \times 10^{-9}$. Subclass (2). (Hydroxides)	Hydrolysis of anion less than 1%. Subclass (1). (Salts of Weak Acids)	Hydrolysis exceeds 1%, but $h < 1.99 \times 10^{-9}$. Subclass (2).	Water equilibrium negligible; $[OH^-] > 10^{-6}$. Subclass (3).	No simplifying assumptions are valid to 1%. Subclass (4).
		Table for criterion	—	8-2	8-2	—	—	—	—
		Table for equation	8-1	8-1	8-2	8-1	—	—	—
AgBr	AB	5.25×10^{-13}	×						
AgCl	AB	1.78×10^{-10}	×						
Ag$_2$CrO$_4$	A$_2$B	2.45×10^{-12}					×		
AgI	AB	8.31×10^{-17}	×						
AgIO$_3$	AB	3.02×10^{-8}				×			
Al(OH)$_3$	AB$_3$	2×10^{-32}			×				
BaCO$_3$	AB	8.1×10^{-9}						×	
BaCrO$_4$	AB	2.4×10^{-10}							×
BaSO$_4$	AB	1.08×10^{-10}				×			
CaCO$_3$	AB	4.8×10^{-9}						×	
CaF$_2$	AB$_2$	4.0×10^{-11}				×			
Ca(OH)$_2$	AB$_2$	5.5×10^{-6}		×					
CuS	AB	9×10^{-36}					×		
Fe(OH)$_2$	AB$_2$	8×10^{-16}		×					
Fe(OH)$_3$	AB$_3$	4×10^{-38}			×				
Hg$_2$Br$_2$*	AB$_2$	5.8×10^{-23}	×						
Hg$_2$Cl$_2$*	AB$_2$	1.3×10^{-18}	×						
Hg$_2$I$_2$*	AB$_2$	4.5×10^{-29}	×						
HgS	AB	4×10^{-53}					×		
Mg(OH)$_2$	AB$_2$	1.2×10^{-11}		×					
PbCrO$_4$	AB	1.8×10^{-14}							×
PbF$_2$	AB$_2$	3.7×10^{-8}				×			
PbI$_2$	AB$_2$	7.1×10^{-9}	×						
PbSO$_4$	AB	1.6×10^{-8}				×			
SrF$_2$	AB$_2$	2.8×10^{-9}				×			
TlBr	AB	3.4×10^{-6}	×						
TlCl	AB	1.7×10^{-4}	×						
TlI	AB	6.5×10^{-8}	×						

* Hg(I) ion is a dimer in solution: Hg_2^{2+}.

$$
\begin{array}{c}
\textit{start:} \qquad \textit{some} \qquad\quad 0.500 \qquad\quad 10^{-7} \\
\text{Ca(OH)}_2(s) \longrightarrow \text{Ca}^{2+} \ + \ 2\text{OH}^- \\
\textit{equilibrium:} \qquad \textit{some} \qquad 0.500 + x \qquad 10^{-7} + 2x - w
\end{array}
$$

which requires that the water equilibrium shift by w mol/L:

$$
\begin{array}{c}
\textit{start:} \qquad\quad 10^{-7} \qquad\quad 10^{-7} \\
\text{H}_2\text{O} \longrightarrow \text{H}^+ \ + \ \text{OH}^- \\
\textit{equilibrium:} \qquad 10^{-7} - w \qquad 10^{-7} + 2x - w
\end{array}
$$

Let us assume (i) that pH > 8.00, so that we can neglect the water equilibrium and write $[\text{OH}^-] \approx 2x$, and (ii) that

$$
\frac{x}{0.500} < 0.01
$$

so that $[\text{Ca}^{2+}] \approx 0.500$. Then,

$$
K_{sp} = (0.500)(2x)^2 \qquad \text{or} \qquad x = 1.66 \times 10^{-3}
$$

which validates assumption (ii). Since $[\text{OH}^-] = 2x = 3.32 \times 10^{-3}$, pOH $= 2.479$, and pH $= 11.521$; assumption (i) is also valid. The total amount of dissolved $\text{Ca(OH)}_2(s)$ is then

Weak-Acid Precipitates

Many salts of weak acids are relatively insoluble; common examples are BaCO_3, PbF_2, and Ag_3PO_4. In all cases the anion, which is a conjugate base, will undergo one or more hydrolysis steps, each with its $K_h = K_w/K_a$, and so the calculation of the solubility can be exceedingly complicated. Fortunately, various approximations will apply to most calculations.

The weak-acid precipitates can be divided into four subclasses depending on the degree of hydrolysis of, and the amount of hydroxide produced by, the conjugate-base anion when the salt is added to water.

(1) If there is less than 1% hydrolysis of the anion (conjugate base), then the algebraic expression for the solubility, S, will be given by Table 8-1. Consider a salt of the general type $A_x B_y(s)$ which results in the ions A^{m+} and B^{n-}. The hydrolysis reaction of the conjugate base, B^{n-}, is

$$
B^{n-} + H_2O \rightarrow HB^{(n-1)-} + OH^-
$$

Now, using Table 5-1 to convert Fig. 5-1 from "acid" to "conjugate base" form, we can see that C_s/K_{cb} exceeds 10^4 at every point of Region 1-2, the less-than-1%-hydrolysis region. In other words, $[HB^{(n-1)-}]$ will be less than 1% of $[B^{n-}]$ only if $C_s/K_{cb} > 10^4$. But, since

$$
[B^{n-}] \approx C_s \qquad \text{and} \qquad K_{cb} = \frac{K_w}{K_a} = \frac{10^{-14}}{K_a}
$$

the requirement becomes

$$
[B^{n-}] > \frac{10^{-10}}{K_a} \tag{8.9}
$$

This translates into a condition on K_{sp} that involves K_a and therefore depends on the specific weak-acid anion. In general, subclass (1) contains salts of the stronger weak acids, such as HF ($pK_a = 3.17$), HIO_3 ($pK_a = 0.79$), or HSO_4^- ($pK_a = 1.92$), since these acids have very weak conjugate bases and, therefore, undergo very little hydrolysis.

EXAMPLE 8.8 For what values of K_{sp} is it possible to neglect the hydrolysis of the anion in the aqueous reaction

$$
A_3B_2(s) \rightarrow 3A^{2+} + 2B^{3-}
$$

where B^{3-} is the anion of a weak triprotic acid such as H_3PO_4?

If S mol/L dissolves and if the hydrolysis can be neglected, then the concentrations will be $[A^{2+}] = 3S$ and $[B^{3-}] = 2S$. Hence, $K_{sp} = 108S^5$, as shown in Table 8-1. In order to be able to neglect the hydrolysis,

$$[B^{3-}] = 2S > \frac{10^{-10}}{K_a}$$

(where $K_a = K_3$ for the triprotic acid). Therefore,

$$S > \frac{10^{-10}}{2K_a} \quad \text{or} \quad K_{sp} > 108\left(\frac{10^{-10}}{2K_a}\right)^5 = \frac{3.38 \times 10^{-50}}{K_a^5}$$

EXAMPLE 8.9 May the hydrolysis of PO_4^{3-} be neglected in calculating the solubility of $Ca_3(PO_4)_2(s)$, which has $K_{sp} = 2.00 \times 10^{-29}$? $K_a = K_3 = 4.78 \times 10^{-13}$.

No; from Example 8.8, a necessary condition for neglecting the hydrolysis is

$$K_{sp} > \frac{3.38 \times 10^{-50}}{(4.78 \times 10^{-13})^5} = 1.35 \times 10^{12}$$

(2) Again consider a salt of the general type A_xB_y which results in the ions A^{m+} and B^{n-}. If h mol/L of B^{n-} undergoes hydrolysis according to

$$B^{n-} + H_2O \rightarrow HB^{(n-1)} + OH^-$$

there will be produced h mol/L of OH^- in addition to that available from the water dissociation reaction, and also h mol/L of $HB^{(n-1)-}$. Subclass (2) salts show more than 1% hydrolysis of the anion, but have $h < 1.99 \times 10^{-9}$, so that $[OH^-] \approx 1.00 \times 10^{-7}$. The hydrolysis constant is thus

$$K_h = \frac{K_w}{K_a} = \frac{[HB^{(n-1)-}][OH^-]}{[B^{n-}]} = \frac{h(1.00 \times 10^{-7})}{[B^{n-}]}$$

which implies, since $h < 1.99 \times 10^{-9}$,

$$[B^{n-}] < 1.99 \times 10^{-2}K_a \qquad (8.10)$$

EXAMPLE 8.10 For what values of K_{sp} is it possible to assume that $[OH^-] = 1.00 \times 10^{-7}$ when the salt of a weak acid, $AB_2(s)$, is added to water?

When S mol/L of the salt dissolves according to

$$AB_2(s) \rightarrow A^{2+} + 2B^-$$

then, $[A^{2+}] = S$ and $[B^-] = 2S - h$, where h mol/L of B^- undergoes hydrolysis to produce h mol/L of HB and h mol/L of OH^-. The two conditions

$$[B^-] = 2S - h < 1.99 \times 10^{-2}K_a \qquad \text{and} \qquad h < 1.99 \times 10^{-9}$$

imply

$$S < \frac{1.99}{2}(10^{-2}K_a + 10^{-9})$$

Consequently,

$$K_{sp} = S[B^-]^2 < \frac{1.99}{2}(10^{-2}K_a + 10^{-9})(1.99 \times 10^2K_a)^2$$

EXAMPLE 8.11 If a salt of a weak acid, AB_2, has a K_{sp} which satisfies the criterion found in Example 8.10 (therefore, $[OH^-] = 1.00 \times 10^{-7}$), what is the expression for K_{sp} as a function of the solubility, S?

In the notation of Example 8.10,

$$K_{sp} = S(2S - h)^2 \qquad (1)$$

But,
$$\frac{K_w}{K_a} = \frac{h(1.00 \times 10^{-7})}{[B^-]} = \frac{h(1.00 \times 10^{-7})}{2S - h} \qquad (2)$$

Elimination of h between (1) and (2) leads to

$$K_{sp} = 4S^3 \left(\frac{10^{-7}K_a}{10^{-7}K_a + K_w} \right)^2$$

(3) Here the amount of hydrolysis exceeds 1% and, in addition, the pH exceeds 8.00. Consider again a salt of the general type A_xB_y which results in the ions A^{m+} and B^{n-}. If h mol/L of B^{n-} undergoes hydrolysis according to

$$B^{n-} + H_2O \rightarrow HB^{(n-1)-} + OH^-$$

and if the amount of OH^- produced by hydrolysis results in $[OH^-] > 10^{-6}$, then the water equilibrium can be neglected (pOH < 6). For this set of conditions,

$$[HB^{(n-1)-}] = [OH^-] = h > 1.00 \times 10^{-6}$$

Since $K_w/K_a = h^2/[B^{n-}]$,

$$[B^{n-}] = \frac{h^2 K_a}{K_w} > \frac{10^{-12}K_a}{K_w} = 10^2 K_a \qquad (8.11)$$

EXAMPLE 8.12 For what values of K_{sp} is it possible to assume that $[OH^-] > 1.00 \times 10^{-6}$ when the salt of a weak acid, A_2B_3, is added to water?

When S mol/L of $A_2B_3(s)$ dissolves, the equilibrium concentrations will be $[A^{3+}] = 2S$ and $[B^{2-}] = 3S - h$, where h mol/L of B^{2-} hydrolyzes. The two conditions

$$h > 1.00 \times 10^{-6} \qquad \text{and} \qquad [B^{2-}] = 3S - h > 10^2 K_a$$

imply

$$S > \tfrac{1}{3}(10^2 K_a + 10^{-6}) \qquad \text{and} \qquad K_{sp} = (2S)^2[B^{2-}]^3 > \tfrac{4}{9}(10^2 K_a + 10^{-6})^2(10^2 K_a)^3$$

Unfortunately, there is no simple way to express K_{sp} in terms of S for this subclass of weak-acid salts. The most direct approach is to convert the expression for K_{sp} into an equation with h as the only variable, to solve for h (by successive approximations, for example), and then to solve for S in the equation for the hydrolysis constant. For such a calculation, see Problem 8.48.

(4) This subclass includes all weak-acid salts that do not fall into subclasses (1) through (3). Since no approximations are valid here, at least three equilibria are involved simultaneously (K_w, K_{sp}, K_a), and the calculation of the solubility in water is usually extremely complicated and requires methods such as successive approximations.

Other Solutions of Weak-Acid Salts

Salts of weak acids might be added to solutions that contain a common ion, that are buffered, or that contain an acid or a base. In some cases this would simplify considerably the calculation of the solubility; for instance, if the pH was fixed, as in a buffered solution, then the fraction F of each acid species in solution could be calculated by the methods described in Chapter 7.

Solved Problems

SALTS OF STRONG ACIDS

8.1 . What is the solubility of $TlBr(s)$?

From Tables 8-1 and 8-3, $K_{sp} = 3.4 \times 10^{-6} = S^2$, whence $S = 1.8 \times 10^{-3}$.

8.2 Determine the solubility of $TlBr(s)$ in a solution of $NaNO_3$ of equilibrium ionic strength 0.080. Refer to Problem 8.1.

From Tables 3-2 and 3-3, $f_{Tl^+} = f_{Br^-} = 0.772$. Therefore,

$$\frac{K_{sp}}{K_{spf}} = \frac{3.4 \times 10^{-6}}{(0.772)^2} = 5.7 \times 10^{-6} = S^2 \qquad \text{or} \qquad S = 2.4 \times 10^{-3}$$

8.3 What is the solubility of $PbI_2(s)$?

From Tables 8-1 and 8-3, $K_{sp} = 7.1 \times 10^{-9} = 4S^3$, whence $S = 1.2 \times 10^{-3}$.

8.4 Determine the solubility of $PbI_2(s)$ in a solution of $NaNO_3$ of equilibrium ionic strength 0.100. Refer to Problem 8.3.

From Tables 3-2 and 3-3, $f_{Pb^{2+}} = 0.377$ and $f_{I^-} = 0.754$. Therefore,

$$\frac{K_{sp}}{K_{spf}} = \frac{7.1 \times 10^{-9}}{(0.377)(0.754)^2} = 3.3 \times 10^{-8} = 4S^3 \qquad \text{or} \qquad S = 2.0 \times 10^{-3}$$

8.5 If $K_{sp} = 8.1 \times 10^{-19}$ for $BiI_3(s) \rightarrow Bi^{3+} + 3I^-$ and if hydrolysis of Bi^{3+} is neglected, what is the solubility of $BiI_3(s)$?

From Table 8-1, $K_{sp} = 27S^4$. Solving, $S = 1.3 \times 10^{-5}$.

8.6 What is the solubility of $BiI_3(s)$ in a solution such that, at equilibrium, $f_{Bi^{3+}} = 0.110$ and $f_{I^-} = 0.760$? Refer to Problem 8.5.

$$\frac{K_{sp}}{K_{spf}} = \frac{8.1 \times 10^{-19}}{(0.110)(0.760)^3} = 1.6_8 \times 10^{-17} = 27S^4$$

so that $S = 2.8 \times 10^{-5}$

8.7 The solubility of cuprous bromide, CuBr, is $S = 7.21 \times 10^{-5}$; find K_{sp}.

From Table 8-1, $K_{sp} = S^2 = (7.21 \times 10^{-5})^2 = 5.20 \times 10^{-9}$.

8.8 The solubility of mercurous iodide, Hg_2I_2, is $2.2_4 \times 10^{-10}$. Since the mercurous ion is a dimer, the solubility equation is

$$Hg_2I_2(s) \rightarrow Hg_2^{2+} + 2I^-$$

Find K_{sp}.

From Table 8-1, $K_{sp} = 4S^3 = 4(2.2_4 \times 10^{-10})^3 = 4.5 \times 10^{-29}$.

8.9 What is the solubility S of $AgCl(s)$ in a 0.0250 M NaCl solution?

When S mol/L of AgCl dissolves, the concentrations will be $[Ag^+] = S$ and $[Cl^-] = 0.0250 + S$. Since

$$S = \sqrt{K_{sp}} = 1.33 \times 10^{-5}$$

when Cl^- is not present, the solubility in NaCl will be much less and we should be able to neglect S as compared with 0.0250. Thus,

$$K_{sp} = 1.78 \times 10^{-10} = (S)(0.0250) \qquad \text{or} \qquad S = 7.12 \times 10^{-9}$$

(neglecting S was justified).

8.10 What is the solubility S of $PbI_2(s)$ in a 0.0426 M $Pb(NO_3)_2$ solution?

When S mol/L of $PbI_2(s)$ dissolves, the concentrations will be $[Pb^{2+}] = 0.0426 + S$ and $[I^-] = 2S$. Since

$$S = \left(\frac{K_{sp}}{4}\right)^{1/3} = 1.2 \times 10^{-3}$$

when Pb^{2+} is not present as a common ion, the solubility should be much smaller in the presence of 0.0426 M Pb^{2+}. Therefore, we should be able to neglect S as compared with 0.0426, so that

$$K_{sp} = 7.1 \times 10^{-9} = (0.0426)(2S)^2 \qquad \text{or} \qquad S = 2.0 \times 10^{-4}$$

and the assumption is justified.

8.11 What is the solubility S of $TlCl(s)$ in a 0.0168 M NaCl solution?

When S mol/L dissolves, the concentrations will be $[Tl^+] = S$ and $[Cl^-] = 0.0168 + S$. Since

$$S = \sqrt{K_{sp}} = 0.0130$$

when extra Cl^- is not present, the solubility will be less in the presence of 0.0168 M Cl^-; however, we probably will not be able to neglect S in the $[Cl^-]$ expression, since the normal solubility is about the same size as the added $[Cl^-]$. Therefore,

$$K_{sp} = 1.7 \times 10^{-4} = (S)(0.0168 + S) \qquad \text{or} \qquad S^2 + 0.0168\,S - (1.7 \times 10^{-4}) = 0$$

Solving by (1.4), $S = 0.0071$.

8.12 Excess $PbBr_2(s)$ was added to a 0.0252 M NaBr solution. Analysis of the equilibrium solution showed that $[Pb^{2+}] = 1.39 \times 10^{-2}$. What is K_{sp} for $PbBr_2(s)$.

If $[Pb^{2+}] = 1.39 \times 10^{-2}$, total $[Br^-]$ must be $0.0252 + 2(1.39 \times 10^{-2}) = 5.30 \times 10^{-2}$, so that

$$K_{sp} = [Pb^{2+}][Br^-]^2 = (1.39 \times 10^{-2})(5.30 \times 10^{-2})^2 = 3.90 \times 10^{-5}$$

8.13 If 25.00 mL of 0.0640 M $AgNO_3$ is added to 20.00 mL of 0.0250 M NaBr, determine the equilibrium concentrations of Ag^+ and Br^-.

Since the $(0.0640)(25.00) = 1.60$ mmol of Ag^+ is in excess of the $(0.0250)(20.00) = 0.500$ mmol of Br^-, we can assume initially that all of the Br^- will be converted to $AgBr(s)$. Just prior to equilibrium,

$$[Ag^+] = \frac{(1.60 - 0.500) \text{ mmol}}{45.00 \text{ mL}} = 2.44 \times 10^{-2}$$

Only a small amount, x, of $AgBr(s)$ will dissolve, so that, at equilibrium,

$$[Ag^+] = (2.44 \times 10^{-2}) + x \qquad \text{and} \qquad [Br^-] = x$$

On the assumption that x may be neglected as compared with 2.44×10^{-2},

$$K_{sp} = 5.25 \times 10^{-13} = (2.44 \times 10^{-2})(x) \qquad \text{or} \qquad x = 2.15 \times 10^{-11} = [Br^-]$$

The assumption is justified, and $[Ag^+] = 2.44 \times 10^{-2}$.

8.14 When 0.100 mol of AgCl(s) is added to 1.00 L of water, it is found that

$$[Cl^-] = \sqrt{K_{sp}} = 1.33 \times 10^{-5}$$

Find $[Cl^-]$ if 0.100 mol of $AgNO_3$ were added to the solution.

    The situation is as if the 0.100 mol of AgCl(s) were added to 1.00 L of 0.100 M $AgNO_3$. S mol of AgCl(s) would dissolve, giving $[Ag^+] = 0.100 + S$ and $[Cl^-] = S$. We should be able to neglect S as compared with 0.100, so that

$$K_{sp} = (0.100)(S) = 1.78 \times 10^{-10} \qquad \text{or} \qquad S = [Cl^-] = 1.78 \times 10^{-9}$$

8.15 Determine $[Cl^-]$ if 0.150 mol of NaI is added to the original solution of Problem 8.14.

    Since AgI(s) is much more insoluble than AgCl(s), we can imagine the initial conditions to be a 0.100 M NaCl solution with 0.100 mol AgI(s) and 0.050 M NaI. If S mol of AgI(s) dissolves, $[Ag^+] = S$ and $[I^-] = 0.050 + S$. Since $S = \sqrt{K_{sp}} = 9.12 \times 10^{-9}$ when no additional I^- is present, we will be able to neglect S as compared with 0.050. Hence,

$$K_{sp} = 8.31 \times 10^{-17} = (S)(0.050) \qquad \text{or} \qquad S = 1.7 \times 10^{-15}$$

and neglecting S was justified. The assumption that AgCl(s) was not present because it is much more soluble is also justified, since

$$[Ag^+][Cl^-] = (1.7 \times 10^{-15})(0.100) = 1.7 \times 10^{-16} < K_{sp} = 1.78 \times 10^{-10}$$

We conclude that $[Cl^-] = 0.100$.

8.16 0.100 mol of AgCl(s) is added to 1.00 L of H_2O. Next, crystals of NaBr are added until 75.00% of the AgCl(s) is converted to AgBr(s), the less soluble silver halide. What is $[Br^-]$ at this point?

    If 75.00% of the AgCl(s) is converted to AgBr(s), $[Cl^-]$ must equal 0.075. Since some AgCl(s) remains, its equilibrium requires that

$$[Ag^+] = \frac{K_{sp}(AgCl)}{[Cl^-]} = \frac{1.78 \times 10^{-10}}{0.075} = 2.4 \times 10^{-9}$$

Since some AgBr(s) is present, its equilibrium requires that

$$[Br^-] = \frac{K_{sp}(AgBr)}{[Ag^+]} = \frac{5.25 \times 10^{-13}}{2.4 \times 10^{-9}} = 2.2 \times 10^{-4}$$

8.17 NaI crystals are slowly added to a solution that is 0.100 M $Pb(NO_3)_2$ and 0.100 M $AgNO_3$. Which will precipitate first, AgI or PbI_2?

    In order that AgI(s) begin to precipitate,

$$[I^-] = \frac{K_{sp}(AgI)}{[Ag^+]} = \frac{8.31 \times 10^{-17}}{0.100} = 8.31 \times 10^{-16}$$

In order that PbI_2(s) begin to precipitate,

$$[I^-] = \left(\frac{K_{sp}(PbI_2)}{[Pb^{2+}]}\right)^{1/2} = \left(\frac{7.1 \times 10^{-9}}{0.100}\right)^{1/2} = 2.7 \times 10^{-4}$$

Since a lower $[I^-]$ is required for AgI precipitation, it will precipitate first.

8.18 For the solution of Problem 8.17, will any $PbI_2(s)$ have precipitated when 99.90% of the Ag^+ is precipitated?

If 99.90% of the Ag^+ is precipitated,

$$[Ag^+] = (0.100)(1.0 \times 10^{-3}) = 1.0 \times 10^{-4} \qquad [I^-] = \frac{K_{sp}(AgI)}{[Ag^+]} = \frac{8.31 \times 10^{-17}}{1.0 \times 10^{-4}} = 8.3 \times 10^{-13}$$

Then,

$$[Pb^{2+}][I^-]^2 = (0.100)(8.3 \times 10^{-13})^2 = 6.9 \times 10^{-26} < K_{sp}(PbI_2) = 7.1 \times 10^{-9}$$

and so no PbI_2 will have precipitated.

8.19 Evaluate $[Ag^+]$ in the solution of Problem 8.17 when the first crystals of $PbI_2(s)$ are formed.

This occurs when

$$[I^-] = \left(\frac{K_{sp}(PbI_2)}{[Pb^{2+}]}\right)^{1/2} = \left(\frac{7.1 \times 10^{-9}}{0.100}\right)^{1/2} = 2.7 \times 10^{-4}$$

Then

$$[Ag^+] = \frac{K_{sp}(AgI)}{[I^-]} = \frac{8.31 \times 10^{-17}}{2.7 \times 10^{-4}} = 3.1 \times 10^{-13}$$

8.20 If $0.200\ M$ NaI is slowly added to $20.00\ mL$ of a solution that is $0.0500\ M\ Pb^{2+}$, $0.0400\ M\ Hg_2^{2+}$, $0.0600\ M\ Ag^+$, and $0.0800\ M\ Tl^+$, in what order do the iodides precipitate?

The iodide concentrations needed to begin precipitation are, in increasing order:

For AgI

$$[I^-] = \frac{K_{sp}}{[Ag^+]} = \frac{8.31 \times 10^{-17}}{0.0600} = 1.39 \times 10^{-15}$$

For Hg_2I_2

$$[I^-] = \left(\frac{K_{sp}}{[Hg_2^{2+}]}\right)^{1/2} = \left(\frac{4.5 \times 10^{-29}}{0.0400}\right)^{1/2} = 3.4 \times 10^{-14}$$

For TlI

$$[I^-] = \frac{K_{sp}}{[Tl^+]} = \frac{6.5 \times 10^{-8}}{0.0800} = 8.1 \times 10^{-7}$$

For PbI_2

$$[I^-] = \left(\frac{K_{sp}}{[Pb^{2+}]}\right)^{1/2} = \left(\frac{7.1 \times 10^{-9}}{0.0500}\right)^{1/2} = 3.8 \times 10^{-4}$$

The precipitates form in this same order.

8.21 What precipitates are present, and what are the ionic concentrations, when $4.00\ mL$ of $0.200\ M$ NaI is added to the solution of Problem 8.20?

The stoichiometric amounts of $0.200\ M$ NaI required for the cations are:

$$\frac{(0.0600)(20.00)}{0.200} = 6.00\ mL,\ for\ Ag^+$$

$$\frac{2(0.0400)(20.00)}{0.200} = 8.00\ mL,\ for\ Hg_2^{2+}$$

$$\frac{(0.0800)(20.00)}{0.200} = 8.00\ mL,\ for\ Tl^+$$

$$\frac{2(0.0500)(20.00)}{0.200} = 10.00\ mL,\ for\ Pb^{2+}$$

Since only $4.00\ mL$ is added, let us assume that only $AgI(s)$ is precipitated. (This may not be the case, as the $[I^-]$ required for $Hg_2I_2(s)$ precipitation is, by Problem 8.20, comparable to that required for

AgI(s) precipitation.) Since $(0.200)(4.00) = 0.800$ mmol of I^- is added, we can imagine that initially there will be 0.800 mmol of AgI(s), so that

$$[Ag^+] = \frac{[(0.0600)(20.00) - 0.800] \text{ mmol}}{24.00 \text{ mL}} = 1.7 \times 10^{-2} \qquad [Hg_2^{2+}] = \frac{(0.0400)(20.00)}{24.00} = 3.33 \times 10^{-2}$$

$$[Tl^+] = \frac{(0.0800)(20.00)}{24.00} = 6.67 \times 10^{-2} \qquad\qquad [Pb^{2+}] = \frac{(0.0500)(20.00)}{24.00} = 4.17 \times 10^{-2}$$

Since x mol/L of AgI will dissolve, $[Ag^+] = (1.7 \times 10^{-2}) + x$ and $[I^-] = x$. But x should be negligible compared to 1.7×10^{-2}, so that

$$K_{sp} = [Ag^+][I^-] = 8.31 \times 10^{-17} = (1.7 \times 10^{-2})(x) \qquad \text{or} \qquad x = [I^-] = 4.9 \times 10^{-15}$$

(neglecting x was justified).

We now justify our assumption that only AgI(s) was present, as follows:

$$[Hg_2^{2+}][I^-]^2 = (3.33 \times 10^{-2})(4.9 \times 10^{-15})^2 = 8.0 \times 10^{-31} < K_{sp} = 4.5 \times 10^{-29}$$
$$[Tl^+][I^-] = (6.67 \times 10^{-2})(4.9 \times 10^{-15}) = 3.3 \times 10^{-16} < K_{sp} = 6.5 \times 10^{-8}$$
$$[Pb^{2+}][I^-]^2 = (4.17 \times 10^{-2})(4.9 \times 10^{-15})^2 = 1.0 \times 10^{-30} < K_{sp} = 7.1 \times 10^{-9}$$

8.22 What precipitates are present, and what are the concentrations of ions, when 20.00 mL of 0.200 M NaI is added to the solution of Problem 8.20?

According to the calculations of Problem 8.21, $6.00 + 8.00 = 14.00$ mL of NaI would precipitate all of the Ag^+ and Hg_2^{2+}, and the remaining 6.00 mL would precipitate $6.00/8.00 = 75\%$ of the TlI. Therefore, let us assume that none of the Pb^{2+} is precipitated, resulting in an initial solution with $(0.0600)(20.00) = 1.20$ mmol of AgI(s), $(0.0400)(20.00) = 0.800$ mmol of $Hg_2I_2(s)$, and $(0.750)(0.0800)(20.00) = 1.20$ mmol of TlI(s). The remaining Tl^+ and all of the Pb^{2+} will be in solution, so that

$$[Tl^+] = \frac{(0.250)(0.0800)(20.00) \text{ mmol}}{40.00 \text{ mL}} = 1.00 \times 10^{-2} \quad \text{and} \quad [Pb^{2+}] = \frac{(0.0500)(20.00) \text{ mmol}}{40.00 \text{ mL}} = 2.50 \times 10^{-2}$$

When x mol/L of TlI dissolves, $[Tl^+] = (1.00 \times 10^{-2}) + x$ and $[I^-] = x$. Let us assume that $x \ll 1.00 \times 10^{-2}$; then,

$$K_{sp} = 6.5 \times 10^{-8} = [Tl^+][I^-] = (1.00 \times 10^{-2})(x) \qquad \text{or} \qquad x = 6.5 \times 10^{-6}$$

(neglecting x was justified). Since

$$[Pb^{2+}][I^-]^2 = (2.50 \times 10^{-2})(6.5 \times 10^{-6})^2 = 1.1 \times 10^{-12} < K_{sp} = 7.1 \times 10^{-9}$$

no PbI$_2(s)$ is present, as was assumed.

$$[Ag^+] = \frac{K_{sp}}{[I^-]} = \frac{8.31 \times 10^{-17}}{6.5 \times 10^{-6}} = 1.3 \times 10^{-11}$$

$$[Hg_2^{2+}] = \frac{K_{sp}}{[I^-]^2} = \frac{4.5 \times 10^{-29}}{(6.5 \times 10^{-6})^2} = 1.1 \times 10^{-18}$$

HYDROXIDE PRECIPITATES

8.23 What is the solubility of Ca(OH)$_2(s)$?

From Table 8-2, since $K_{sp} = 5.5 \times 10^{-6} > 5.00 \times 10^{-19}$, the water equilibrium can be neglected, and $K_{sp} = 4S^3$. Solving, $S = 1.1 \times 10^{-2}$.

8.24 What is the solubility of Fe(OH)$_3(s)$?

From Table 8-2, since $K_{sp} = 4 \times 10^{-38} < 6.66 \times 10^{-31}$, the precipitate contributes negligible OH^- to the $[OH^-]$. Thus, we can approximate $[OH^-]$ as 1.00×10^{-7}, so that

$$K_{sp} = S(1.00 \times 10^{-7})^3 = 4 \times 10^{-38} \qquad \text{or} \qquad S = 4 \times 10^{-17}$$

8.25   Find the solubility of $Ca(OH)_2(s)$ in a solution of $NaNO_3$ of equilibrium ionic strength 0.050.

From Tables 3-2 and 3-3, $f_{Ca^{2+}} = 0.483$ and $f_{OH^-} = 0.807$, so that

$$\frac{K_{sp}}{K_{spf}} = \frac{5.5 \times 10^{-6}}{(0.483)(0.807)^2} = 1.7 \times 10^{-5} = 4S^3 \qquad \text{or} \qquad S = 1.6 \times 10^{-2}$$

8.26   Find the solubility of $Fe(OH)_3(s)$ in a solution of $NaNO_3$ of equilibrium ionic strength 0.020.

From Tables 3-2 and 3-3, $f_{Fe^{3+}} = 0.350$, $f_{OH^-} = 0.865$, and $f_{H^+} = 0.890$. Since the amount of OH^- resulting from the $Fe(OH)_3$ is negligible compared with that from water, the principal source of H^+ and OH^- is the water dissociation: $[H^+] = [OH^-]$. Therefore,

$$K_w = a_{H^+} \, a_{OH^-} = [H^+]f_{H^+} [OH^-]f_{OH^-} = [OH^-]^2(0.890)(0.865) = 1.00 \times 10^{-14}$$

whence $[OH^-] = 1.14 \times 10^{-7}$. Then,

$$\frac{K_{sp}}{K_{spf}} = \frac{4 \times 10^{-38}}{(0.350)(0.865)^3} = 1.8 \times 10^{-37} = [Fe^{3+}][OH^-]^3 = S(1.14 \times 10^{-7})^3$$

or $S = 1.2 \times 10^{-16}$.

8.27   What is the solubility of $Ca(OH)_2(s)$ in 0.0500 M NaOH?

If S mol/L of $Ca(OH)_2$ dissolves, there result $[Ca^{2+}] = S$, $[OH^-] = 2S + 0.0500$. Since the solubility in water is, from Problem 8.23, 1.1×10^{-2}, we shall probably not be able to neglect $2S$ as compared with 0.0500. Therefore, $K_{sp} = 5.5 \times 10^{-6} = S(2S + 0.0500)^2$, or

$$4S^3 + (0.200)S^2 + (2.50 \times 10^{-3})S - (5.5 \times 10^{-6}) = 0$$

Solving this cubic either by successive approximations or by the graphical intersection method (Section 1.4) results in $S = 1.9 \times 10^{-3}$.

8.28   What is the solubility of $Ca(OH)_2(s)$ in a solution buffered to pH 12.000?

Since $[OH^-] = 1.00 \times 10^{-2}$,

$$K_{sp} = 5.5 \times 10^{-6} = S(1.00 \times 10^{-2})^2 \qquad \text{or} \qquad S = 5.5 \times 10^{-2}$$

8.29   What is the solubility of $Fe(OH)_3(s)$ in a solution buffered to pH 5.000?

Since $[OH^-] = 1.00 \times 10^{-9}$,

$$K_{sp} = 4 \times 10^{-38} = S(1.00 \times 10^{-9})^3 \qquad \text{or} \qquad S = 4 \times 10^{-11}$$

8.30   If 0.250 M NaOH is slowly added to 20.00 mL of a solution that is 0.0500 M $Mg(NO_3)_2$, 0.100 M $Ca(NO_3)_2$, and 0.0300 M $Fe(NO_3)_2$, in what order do the hydroxides precipitate?

The concentrations of $[OH^-]$ needed to form a precipitate increase in the following sequence:

For $Fe(OH)_2(s)$ $[OH^-] = \left(\dfrac{K_{sp}}{0.0300}\right)^{1/2} = 2 \times 10^{-7}$

For $Mg(OH)_2(s)$ $[OH^-] = \left(\dfrac{K_{sp}}{0.0500}\right)^{1/2} = 1.5 \times 10^{-5}$

For $Ca(OH)_2(s)$ $[OH^-] = \left(\dfrac{K_{sp}}{0.100}\right)^{1/2} = 7.41 \times 10^{-3}$

The precipitates form in the same sequence.

8.31 What precipitates are present, and what are the ionic concentrations, when 3.00 mL of 0.250 M NaOH is added to the solution of Problem 8.30?

The stoichiometric amounts of 0.250 M NaOH required for the cations are:

$$\frac{2(0.0300)(20.00)}{0.250} = 4.80 \text{ mL, for } Fe(OH)_2$$

$$\frac{2(0.0500)(20.00)}{0.250} = 8.00 \text{ mL, for } Mg(OH)_2$$

$$\frac{2(0.100)(20.00)}{0.250} = 16.0 \text{ mL, for } Ca(OH)_2$$

Since only 3.00 mL of NaOH is added, we may assume that only a portion of the Fe^{2+} is precipitated. Since $(0.250)(3.00) = 0.750$ mmol of OH^- was added, $0.750/2 = 0.375$ mmol of $Fe(OH)_2$ was precipitated, leaving in solution

$$[Fe^{2+}] = \frac{[(0.0300)(20.00) - 0.375] \text{ mmol}}{23.00 \text{ mL}} = 9.78 \times 10^{-3}$$

If $Fe(OH)_2(s)$ was added to water, then, according to Table 8-3, the water equilibrium could be neglected. (From $K_{sp} = 4S^3$ and $[OH^-] = 2S$, we have $[OH^-] \approx 1.17 \times 10^{-5} > 10^{-6}$.) In the present case, the existence of Fe^{2+} in solution will cause less $Fe(OH)_2$ to dissolve; the $[OH^-]$ will probably be smaller than 10^{-6}, requiring that the water equilibrium be considered. If x mol/L of $Fe(OH)_2(s)$ dissolves and if w mol/L each of H^+ and OH^- react with each other to restore the water equilibrium, then the concentration shifts are as follows:

start:	*some*	9.78×10^{-3}	10^{-7}
	$Fe(OH)_2(s) \longrightarrow$	Fe^{2+} +	OH^-
equilibrium:	*some*	$(9.78 \times 10^{-3}) + x$	$10^{-7} + 2x - w$
start:		10^{-7}	10^{-7}
	$H_2O \longrightarrow$	H^+ +	OH^-
equilibrium:		$10^{-7} - w$	$10^{-7} + 2x - w$

Since $2x$ will probably be less than 10^{-6}, we should be able to neglect x as compared with 9.78×10^{-3}; therefore,

$$[OH^-] = \left(\frac{K_{sp}}{[Fe^{2+}]}\right)^{1/2} = \left(\frac{8 \times 10^{-16}}{9.78 \times 10^{-3}}\right)^{1/2} = 2._9 \times 10^{-7}$$

We can check our assumption by calculating x from the relation

$$2x = [OH^-] - [H^+] = [OH^-] - \frac{K_w}{[OH^-]}$$

We find: $x = 1.3 \times 10^{-7}$ (the assumption is valid).

The other ionic concentrations are:

$$[Mg^{2+}] = \frac{(0.0500)(20.00)}{23.00} = 4.35 \times 10^{-2} \qquad [Ca^{2+}] = \frac{(0.100)(20.00)}{23.00} = 8.70 \times 10^{-2}$$

8.32 What precipitates are present, and what are the ionic concentrations, when 24.32 mL of 0.250 M NaOH is added to the solution of Problem 8.30?

Refer to Problem 8.31. Since 4.80 mL of NaOH is required to precipitate $Fe(OH)_2$, and an additional 8.00 mL to precipitate $Mg(OH)_2$, only $24.32 - 12.80 = 11.52$ mL is available to precipitate $Ca(OH)_2$. Therefore, only $(0.0250)(11.52) = 2.88$ mmol of OH^- is used in precipitating $Ca(OH)_2$, so that $2.88/2 = 1.44$ mmol of Ca^{2+} is precipitated. There remains in solution

$$[Ca^{2+}] = \frac{[(0.100)(20.00) - 1.44] \text{ mmol}}{44.32 \text{ mL}} = 1.2_6 \times 10^{-2}$$

When x mol/L of $Ca(OH)_2(s)$ dissolves, $[Ca^{2+}] = (1.2_6 \times 10^{-2}) + x$ and $[OH^-]$ equals $2x$ plus any additional concentration from the water reaction. Let us assume that $2x > 10^{-6}$, so that the water equilibrium can be neglected. (This assumption is clearly valid if $Ca(OH)_2(s)$ is added to water, as was shown in Problem 8.23. Moreover, the additional $[Ca^{2+}]$ being of the same order of magnitude as the $[Ca^{2+}]$ produced when $Ca(OH)_2$ is added to water, it does not affect the solubility markedly.) However, we shall probably *not* be able to neglect x as compared with $1.2_6 \times 10^{-2}$. Thus,

$$K_{sp} = 5.5 \times 10^{-6} = [(1.2_6 \times 10^{-2}) + x](2x)^2 = (5.0_4 \times 10^{-2})x^2 + 4x^3$$

Solving this cubic by successive approximations yields $x = 7.9 \times 10^{-3}$, so that

$$[Ca^{2+}] = 2.07 \times 10^{-2} \qquad \text{and} \qquad [OH^-] = 2x = 1.6 \times 10^{-2}$$

Therefore,

$$[Mg^{2+}] = \frac{K_{sp}}{[OH^-]^2} = \frac{1.2 \times 10^{-11}}{(1.6 \times 10^{-2})^2} = 4.7 \times 10^{-8} \qquad [Fe^{2+}] = \frac{K_{sp}}{[OH^-]^2} = \frac{8 \times 10^{-16}}{(1.6 \times 10^{-2})^2} = 3 \times 10^{-12}$$

8.33 What are the ionic concentrations when 28.80 mL of 0.250 M NaOH is added to the solution of Problem 8.30?

This is the stoichiometric volume needed to precipitate all three hydroxides. Therefore, the solution can be imagined as arising from the addition of the three hydroxide precipitates to water. All three solubility equilibria must be satisfied, since all three precipitates will be present. However, the precipitate that results in the highest $[OH^-]$ will primarily determine the total $[OH^-]$ in solution. In this case, $Ca(OH)_2$ is the most soluble. From Problem 8.23, the $[OH^-]$ arising from $Ca(OH)_2(s)$ is $2S = 2.2 \times 10^{-2}$. We now determine if the next-most-soluble precipitate contributes any appreciable additional OH^-. That precipitate is $Mg(OH)_2$, and, using the $[OH^-]$ from $Ca(OH)_2$, we have

$$[Mg^{2+}] = \frac{K_{sp}}{[OH^-]^2} = \frac{1.2 \times 10^{-11}}{(2.2 \times 10^{-2})^2} = 2.5 \times 10^{-8}$$

Thus, the $Mg(OH)_2$ contributes only $2(2.5 \times 10^{-8}) = 5.0 \times 10^{-8}$ to the $[OH^-]$; so the equilibrium $[OH^-]$ is the value already calculated, 2.2×10^{-2}, and

$$[Fe^{2+}] = \frac{K_{sp}}{[OH^-]^2} = \frac{6 \times 10^{-16}}{(2.2 \times 10^{-2})^2} = 1 \times 10^{-12}$$

8.34 How large a concentration of (a) $Fe(NO_3)_2$, (b) $Fe(NO_3)_3$, and (c) $Mg(NO_3)_2$ can exist without hydroxide precipitation?

Since $[OH^-] = 1.00 \times 10^{-7}$, the maximum concentrations of the nitrate salts are governed by the hydroxide K_{sp}-values.

(a)
$$[Fe^{2+}] = \frac{K_{sp}}{[OH^-]^2} = \frac{8 \times 10^{-16}}{(1.00 \times 10^{-7})^2} = 8 \times 10^{-2}$$

(b)
$$[Fe^{3+}] = \frac{K_{sp}}{[OH^-]^3} = \frac{4 \times 10^{-38}}{(1.00 \times 10^{-7})^3} = 4 \times 10^{-17}$$

(Because this value is so low, it is necessary to acidify Fe^{3+} solutions to prevent $Fe(OH)_3$ precipitation.)

(c)
$$[Mg^{2+}] = \frac{K_{sp}}{[OH^-]^2} = \frac{1.2 \times 10^{-11}}{(1.00 \times 10^{-7})^2} = 1.2 \times 10^{+3}$$

($Mg(NO_3)_2$ solutions, therefore, can be prepared up to the limit of the solubility of $Mg(NO_3)_2$, which is about 2 M.)

8.35 If a solution is 0.0400 M $FeCl_2$, 0.0200 M $FeCl_3$, and 0.0100 M HCl, how large may be its pH without there being precipitation of either $Fe(OH)_2$ or $Fe(OH)_3$?

$Fe(OH)_2$ will begin to precipitate when

$$[OH^-] = \left(\frac{K_{sp}}{[Fe^{2+}]}\right)^{1/2} = \left(\frac{8 \times 10^{-16}}{4.00 \times 10^{-2}}\right)^{1/2} = 1._4 \times 10^{-7}$$

which is pH 7.1_5. $Fe(OH)_3$ will begin to precipitate when

$$[OH^-] = \left(\frac{K_{sp}}{[Fe^{3+}]}\right)^{1/3} = \left(\frac{4 \times 10^{-38}}{2.00 \times 10^{-2}}\right)^{1/3} = 1._3 \times 10^{-12}$$

which is pH 2.1_0. The maximum pH is 2.1_0.

8.36 For the solution of Problem 8.35, what pH-range will allow at least 99.990 % of the Fe^{3+} to be precipitated while Fe^{2+} remains in solution?

If $[Fe^{3+}] < (0.01\%)(2.00 \times 10^{-2}) = 2.0 \times 10^{-6}$, one must have

$$[OH^-] > \left(\frac{K_{sp}}{2.0 \times 10^{-6}}\right)^{1/3} = \left(\frac{4 \times 10^{-38}}{2.0 \times 10^{-6}}\right)^{1/3} = 2._7 \times 10^{-11}$$

which implies a pH greater than 3.4_3. From Problem 8.35, $Fe(OH)_2$ will begin to precipitate when pH = 7.1_5. Therefore, the pH-range is 3.4_3–7.1_5.

8.37 For the solution of Problem 8.35, what minimum pH will allow precipitation of at least 99.990 % of Fe^{3+} and also at least 99.990 % of Fe^{2+}?

According to Problem 8.36, more than 99.990 % of Fe^{3+} will have precipitated at pH 7.1_5, where $Fe(OH)_2$ just begins to precipitate. Therefore, we need to calculate the pH required to precipitate 99.990 % of Fe^{2+}. This would be accomplished when

$$[Fe^{2+}] = (0.01\%)(4.00 \times 10^{-2}) = 4.0 \times 10^{-6}$$

i.e., when

$$[OH^-] = \left(\frac{K_{sp}}{4.0 \times 10^{-6}}\right)^{1/2} = \left(\frac{8 \times 10^{-16}}{4.0 \times 10^{-6}}\right)^{1/2} = 1._4 \times 10^{-5}$$

which corresponds to pH 9.1_5.

8.38 Determine the concentrations of all ions in solution when 0.100 mol of $Fe(NO_3)_3$ is added to 1.00 L of water. Assume no change in volume.

Because $[Fe^{3+}][OH^-] = (0.100)(1.00 \times 10^{-7})^3 = 1.00 \times 10^{-22}$ is considerably greater than $K_{sp} = 4 \times 10^{-38}$, there should be extensive hydrolysis of Fe^{3+}:

$$Fe^{3+} + 3H_2O \rightarrow Fe(OH)_3(s) + 3H^+$$

for which $K_h = K_w^3/K_{sp} = 2._5 \times 10^{-5}$. Since K_h is moderately large, let us assume that the $[H^+]$ produced by the hydrolysis reaction is sufficient to allow us to neglect the water equilibrium. Therefore, when x mol/L of Fe^{3+} hydrolyzes, $[Fe^{3+}] = 0.100 - x$ and $[H^+] = 3x$. We have no reason to suppose x negligible compared with 0.100; so

$$K_h = 2._5 \times 10^{-5} = \frac{[H^+]^3}{[Fe^{3+}]} = \frac{(3x)^3}{0.100 - x}$$

Solving this cubic by successive approximations, $x = 4._5 \times 10^{-3}$, whence

$[Fe^{3+}] = 0.096$ $[H^+] = 1._4 \times 10^{-2}$ $[OH^-] = K_w/[H^+] = 7._4 \times 10^{-13}$ $[NO_3^-] = 0.300$

(Because of the high $[H^+]$ resulting from Fe^{3+} hydrolysis, soluble ferric salts (or aluminum salts, which also hydrolyze extensively) are often added to the soil for plants such as azaleas, rhododendrons, or blueberries, which require highly acidic conditions.)

SALTS OF WEAK ACIDS

8.39 What is the solubility of $BaSO_4(s)$ in water?

Since HSO_4^- is a relatively strong acid ($K_2 = 0.012$), we assume that there is less than 1% SO_4^{2-} hydrolysis (i.e., the salt belongs to subclass (1)). Then,

$$K_{sp} = 1.08 \times 10^{-10} = S^2 \qquad \text{or} \qquad S = 1.04 \times 10^{-5}$$

(From (8.9), $[SO_4^{2-}] = S > 10^{-10}/K_2 = 8.3 \times 10^{-9}$, and our assumption is valid.)

8.40 What is the solubility of $BaSO_4(s)$ in a $0.100\ M$ $NaNO_3$ solution when corrected for non-ideality?

The solubility will be comparable to that in Problem 8.39, and, therefore, can be neglected in calculating the ionic strength. Hence, $I = 0.100$. From Tables 3-2 and 3-3,

$$f_{Ba^{2+}} = 0.377 \qquad \text{and} \qquad f_{SO_4^{2-}} = 0.351$$

so that

$$\frac{K_{sp}}{K_{spf}} = \frac{1.08 \times 10^{-10}}{(0.377)(0.351)} = 8.16 \times 10^{-10} = S^2 \qquad \text{or} \qquad S = 2.86 \times 10^{-5}$$

8.41 What is the solubility of $BaSO_4(s)$ in a $1.00 \times 10^{-3}\ M$ solution of $Ba(NO_3)_2$?

Assume that the hydrolysis of SO_4^{2-} can be neglected, as in Problem 8.39. When S mol/L dissolves, there result

$$[Ba^{2+}] = (1.00 \times 10^{-3}) + S \qquad \text{and} \qquad [SO_4^{2-}] = S$$

Assume that S can be neglected compared with 1.00×10^{-3} (a logical assumption, since, without added Ba^{2+}, $S = 1.04 \times 10^{-5}$, according to Problem 8.39). Then,

$$K_{sp} = 1.08 \times 10^{-10} = (1.00 \times 10^{-3})S \qquad \text{or} \qquad S = 1.08 \times 10^{-7}$$

8.42 What is the solubility of $BaSO_4(s)$ in $0.250\ M$ H_2SO_4?

Assuming that (i) the hydrolysis of SO_4^{2-} can be neglected, and (ii) the SO_4^{2-} resulting from $BaSO_4$ can be neglected as compared with 0.250, we have:

$$K_{sp} = 1.08 \times 10^{-10} = S(0.250) \qquad \text{or} \qquad S = 4.32 \times 10^{-10}$$

(From (8.9), $[SO_4^{2-}] = 0.250 \gg 10^{-10}/K_2 = 8.3 \times 10^{-9}$, and Assumption (i) is valid. Assumption (ii) is valid, since $S \ll 0.250$.)

8.43 What is the solubility of $PbF_2(s)$ in water?

Since HF is a moderately weak acid ($K_a = 6.76 \times 10^{-4}$), there should be less than 1% hydrolysis, so that

$$K_{sp} = 3.7 \times 10^{-8} = 4S^3 \qquad \text{or} \qquad S = 2.1 \times 10^{-3}$$

(From (8.9), $[F^-] = 2S = 4.2 \times 10^{-3} > 10^{-10}/K_a = 1.48 \times 10^{-7}$, and the neglect of hydrolysis is justified.)

8.44 What is the solubility of $PbF_2(s)$ in a solution of $NaNO_3$ of equilibrium ionic strength 0.0800?

From Tables 3-2 and 3-3, $f_{Pb^{2+}} = 0.405$ and $f_{F^-} = 0.772$. Therefore,

$$\frac{K_{sp}}{K_{spf}} = \frac{3.7 \times 10^{-8}}{(0.405)(0.772)^2} = 1.5 \times 10^{-7} = 4S^3 \qquad \text{or} \qquad S = 3.4 \times 10^{-3}$$

8.45 What is the solubility of $PbF_2(s)$ in a $2.50 \times 10^{-3}\ M$ NaF solution?

Since the hydrolysis of F^- could be neglected in Problem 8.43, it can also be neglected here; for the total F^- in solution will be greater and percent hydrolysis will be less. When S mol/L of PbF_2 dissolves, there result $[Pb^{2+}] = S$ and $[F^-] = (2.50 \times 10^{-3}) + 2S$. Since $2S$ will probably be of the same order of magnitude as the added F^- ($2S = 4.2 \times 10^{-3}$ in water, according to Problem 8.43), we will not be able to neglect $2S$. Therefore,

$$K_{sp} = 3.7 \times 10^{-8} = (S)\{(2.50 \times 10^{-3}) + 2S\}^2$$

Solving this cubic by successive approximations gives $S = 1.3_6 \times 10^{-3}$.

8.46 What is the largest value of K_{sp} for a metal sulfide, $MS(s)$, that allows $[OH^-]$ to be taken as 1.00×10^{-7}?

We know from Assumption 3b of Section 5.2 that the condition for pH 7.000 is that the total $[OH^-]$ from hydrolysis be less than 1.99×10^{-9}. This converts to a condition on K_{sp} as follows.

The fractions of the total sulfur in solution present as H_2S, HS^-, and S^{2-} can be calculated from expressions analogous to (7.6) and (7.7). Thus, for a diprotic acid,

$$F_0 = \frac{[H^+]^2}{D} \qquad F_1 = \frac{K_1[H^+]}{D} \qquad F_2 = \frac{K_1 K_2}{D}$$

where

$$D = [H^+]^2 + K_1[H^+] + K_1 K_2$$

Substituting $[H^+] = 1.00 \times 10^{-7}$, $K_1 = 1.02 \times 10^{-7}$, $K_2 = 1.10 \times 10^{-15}$, we obtain:

$$F_0 = 0.495 \qquad F_1 = 0.505 \qquad F_2 = 5.55 \times 10^{-9}$$

Therefore, if C mol/L of $MS(s)$ dissolves and pH = 7.000, there will result $[M^{2+}] = C$ and $[S^{2-}] = (5.55 \times 10^{-9})C$. The total concentration of OH^- produced by hydrolysis of S^{2-} and the further hydrolysis of HS^- is

$$[OH^-] = [HS^-] + 2[H_2S] = (0.505)C + 2(0.495)C = (1.495)C$$

The pH will indeed be 7.000 if $(1.495)C < 1.99 \times 10^{-9}$, or $C < 1.33 \times 10^{-9}$, or

$$K_{sp} = (C)(5.55 \times 10^{-9}C) < (1.33 \times 10^{-9})(5.55 \times 10^{-9})(1.33 \times 10^{-9}) = 9.83 \times 10^{-27}$$

8.47 What is the solubility of $CuS(s)$ in water, and what are the concentrations of the ions in solution?

Here, $K_{sp} = 9 \times 10^{-36} < 9.83 \times 10^{-27}$, so that, by Problem 8.46, pH = 7.000 and $K_{sp} = (5.55 \times 10^{-9})C^2$. Solving,

$$C = 4._0 \times 10^{-14} = [Cu^{2+}] \qquad [H^+] = [OH^-] = 1.00 \times 10^{-7}$$
$$[H_2S] = F_0 C = 2 \times 10^{-14} \qquad [HS^-] = F_1 C = 2 \times 10^{-14} \qquad [S^{2-}] = F_2 C = 2._2 \times 10^{-22}$$

8.48 What is the solubility of $BaCO_3(s)$ in water?

If there was at most 1% hydrolysis of $BaCO_3$, then $K_{sp} = 8.1 \times 10^{-9} = S^2$, or $S = 9.0 \times 10^{-5}$. But this solubility would give

$$[CO_3^{2-}] = S < \frac{10^{-10}}{K_2}$$

because, from Table 7-3, $K_2 = 4.68 \times 10^{-11}$. Consequently, from (8.9), the hydrolysis must exceed 1%.

Since the hydrolysis constant is moderately large ($K_h = K_{cb2} = 2.14 \times 10^{-4}$) and since S will exceed $\sqrt{K_{sp}} = 9.0 \times 10^{-5}$, it is possible that the concentration of OH^- produced by hydrolysis is greater than 10^{-6} M, in which case the water equilibrium can be neglected. Let us assume this to be the case. Further, since K_{cb} is not extremely large, the amount of hydrolysis of CO_3^{2-} should still be relatively small; and the amount of hydrolysis in the second step,

$$HCO_3^- + H_2O \to H_2CO_3 + OH^- \qquad (K_h = K_{cb1} = 2.24 \times 10^{-8})$$

ought to be negligible. Thus we need only consider the concentration shifts

start:	*some*	0	0
	$BaCO_3(s) \longrightarrow$	Ba^{2+} +	CO_3^{2-}
equilibrium:	*some*	S	$S - h$

start:	0	0	0
	$CO_3^{2-} + H_2O \longrightarrow$	HCO_3^- +	OH^-
equilibrium:	$S - h$	h	h

where S mol/L of $BaCO_3(s)$ dissolves to produce S mol/L of Ba^{2+} and S mol/L of CO_3^{2-}, and h mol/L of CO_3^{2-} hydrolyzes to produce h mol/L each of HCO_3^- and OH^-. From the second reaction,

$$K_h = \frac{K_w}{K_2} = \frac{h^2}{S - h} \qquad \text{or} \qquad S = \frac{K_2 h^2 + K_w h}{K_w}$$

Then, from the first reaction,

$$K_{sp} = [Ba^{2+}][CO_3^{2-}] = (S)(S - h) = \frac{K_2^2 h^4 + K_2 K_w h^3}{K_w^2} = 8.1 \times 10^{-9}$$

By successive approximations, this biquadratic may be solved to give $h = 1.0_5 \times 10^{-4}$, from which

$$S = \frac{K_2 h^2 + K_w h}{K_w} = 1.5_7 \times 10^{-4}$$

Because $[OH^-] = h > 10^{-6}$, neglect of the water equilibrium is justified. Moreover, using Table 7-3, we have:

$$\frac{[H_2CO_3]}{[HCO_3^-]} = \frac{[H^+]}{K_1} = \frac{K_w/h}{K_1} \approx 2 \times 10^{-4}$$

and this justifies neglect of the second hydrolysis step.

8.49 If 4.00 mmol of $SrF_2(s)$ is added to 1.00 L of x M HNO_3, what is the smallest value of x that will allow all but the last tiny crystal of $SrF_2(s)$ to dissolve? For HF, $K_a = 6.76 \times 10^{-4}$.

Since essentially all of the salt dissolves, $[Sr^{2+}] = 4.00 \times 10^{-3}$ and, initially, $[F^-] = 8.00 \times 10^{-3}$. However, y mol/L of the F^- reacts with the x mol/L of H^+ available from the HNO_3, so that, at equilibrium,

$$[F^-] = (8.00 \times 10^{-3}) - y \qquad [HF] = y \qquad [H^+] = x - y$$

where the last formula reflects the assumption that, HNO_3 being a strong acid, the water equilibrium can be neglected. The equilibrium shifts are as follows:

start:	*some*	0	0
	$SrF_2(s) \longrightarrow$	Sr^{2+} +	$2F^-$
equilibrium:	*tiny crystal*	4.00×10^{-3}	$(8.00 \times 10^{-3}) - y$

start:	0	x	0
	$HF \longrightarrow$	H^+ +	F^-
equilibrium:	y	$x - y$	$(8.00 \times 10^{-3}) - y$

Solving $K_{sp} = [Sr^{2+}][F^-]^2 = (4.00 \times 10^{-3})\{(8.00 \times 10^{-3}) - y\}^2 = 2.8 \times 10^{-9}$, we obtain $y = 7.16 \times 10^{-3}$. Then solving

$$K_a = 6.76 \times 10^{-4} = \frac{[H^+][F^-]}{[HF]} = \frac{(x-y)\{(8.00 \times 10^{-3}) - y\}}{y}$$

we get $x = 1.29 \times 10^{-2}$.

8.50 How many moles of $SrF_2(s)$ will dissolve in 1.00 L of a pH 4.000 buffer solution? $K_a = 6.76 \times 10^{-4}$ for HF.

After S mol dissolves, $[Sr^{2+}] = S$ and $[F^-] = 2SF_1$, where

$$F_1 = \frac{[F^-]}{[F^-] + [HF]}$$

is the fraction of fluoride in the form F^-. But, $[HF] = [H^+][F^-]/K_a$, so that, upon rearrangement,

$$F_1 = \frac{K_a}{[H^+] + K_a} = \frac{6.76 \times 10^{-4}}{(1.00 \times 10^{-4}) + (6.76 \times 10^{-4})} = 0.871$$

Then, $K_{sp} = 2.8 \times 10^{-9} = (S)2^2(0.871\ S)^2$, whence $S = 9.7 \times 10^{-4}$ (i.e., 0.97 mmol).

8.51 How many moles of $BaCrO_4(s)$ will dissolve in 1.00 L of a pH 5.000 buffer? $K_1 = 9.55$ and $K_2 = 3.16 \times 10^{-7}$ for H_2CrO_4.

When S mol dissolves, there result $[Ba^{2+}] = S$ and $[CrO_4^{2-}] = F_2S$, where

$$F_2 = \frac{K_1K_2}{[H^+]^2 + K_1[H^+] + K_1K_2}$$

is the fraction of the total chromate in solution in the form CrO_4^{2-}. Substituting numerical values,

$$F_2 = 3.06 \times 10^{-2} \qquad \text{and} \qquad K_{sp} = 2.4 \times 10^{-10} = (S)(3.06 \times 10^{-2}S)$$

Solving, $S = 8.8_5 \times 10^{-5}$.

8.52 How many moles of $Ag_2CrO_4(s)$ will dissolve in 1.00 L of a pH 5.00 buffer solution?

When S mol dissolves, there are produced $[Ag^+] = 2S$ and $[CrO_4^{2-}] = F_2S$. From Problem 8.51, $F_2 = 3.06 \times 10^{-2}$ at pH 5.000. Therefore,

$$K_{sp} = 2.45 \times 10^{-12} = (2S)^2(3.06 \times 10^{-2}S) \qquad \text{or} \qquad S = 2.72 \times 10^{-4}$$

8.53 Determine the ratio $[Sr^{2+}]/[Ca^{2+}]$ when an excess of both $SrF_2(s)$ and $CaF_2(s)$ is added to 1.00 L of (a) pH 4.000 buffer and (b) pH 7.000 buffer.

Since

$$\frac{[Sr^{2+}]}{[Ca^{2+}]} = \frac{[Sr^{2+}][F^-]^2}{[Ca^{2+}][F^-]^2} = \frac{K_{sp}(SrF_2)}{K_{sp}(CaF_2)} = \text{constant}$$

the ratio of the concentrations of these two ions is independent of the pH. The absolute amounts of each, however, will vary with the pH (by Problem 8.50, F_1 depends on the pH).

PRECIPITATION TITRATIONS

8.54 25.00 mL of a solution is $x\ M$ NaCl, $y\ M$ NaBr, and $z\ M$ NaI. The solution is titrated with $0.100\ M$ $AgNO_3$ and the titration followed potentiometrically. A plot of pAg versus vol $AgNO_3$ shows breaks in the curve at 12.00 mL and 30.00 mL, and a further sharp decline in pAg at 38.00 mL. What are the concentrations x, y, and z?

Since comparable volumes of $AgNO_3$ are needed to titrate the three halides, their concentrations must be comparable and they must precipitate in the order of increasing solubility: AgI first, $AgBr$ second, and $AgCl$ last. Since 12.00 mL was required to precipitate AgI,

$$z = \frac{(0.100)(12.00)}{25.00} = 0.0480$$

Since $30.00 - 12.00 = 18.00$ mL of $AgNO_3$ was required to precipitate $AgBr$,

$$y = \frac{(0.100)(18.00)}{25.00} = 0.0720$$

Since $38.00 - 30.00 = 8.00$ mL of $AgNO_3$ was required to precipitate $AgCl$,

$$x = \frac{(0.100)(8.00)}{25.00} = 0.0320$$

8.55 20.00 mL of a solution contains Bi^{3+}, Ag^+, and Cu^+, each in a concentration of about 0.2 M. The solution is titrated with 0.200 M NaI and pI determined as a function of vol NaI. In what order do the iodides precipitate, if their K_{sp}-values are 8.1×10^{-19} for BiI_3, 8.31×10^{-17} for AgI, and 5.1×10^{-12} for CuI?

The values of $[I^-]$ needed to begin precipitation of the iodides are $(K_{sp}/0.2)^{1/3} = 1._6 \times 10^{-6}$ for BiI_3, $K_{sp}/0.2 = 4._2 \times 10^{-16}$ for AgI, and $2._6 \times 10^{-11}$ for CuI. Therefore, AgI precipitates first (it requires the lowest $[I^-]$), CuI precipitates next, and BiI_3 precipitates last.

8.56 Breaks occur in the titration curve of Problem 8.55 at 38.00 and 57.00 mL NaI, and a sharp decrease in pI also occurs at 93.00 mL NaI. What are the concentrations of Bi^{3+}, Ag^+, and Cu^+?

Since AgI precipitates first and 38.00 mL is required for this step,

$$[Ag^+] = \frac{(0.200)(38.00)}{20.00} = 0.380$$

Since $57.00 - 38.00 = 19.00$ mL is required to precipitate CuI,

$$[Cu^+] = \frac{(0.200)(19.00)}{20.00} = 0.190$$

Since $93.00 - 57.00 = 36.00$ mL is required to precipitate BiI_3,

$$[Bi^{3+}] = \frac{\frac{1}{3}(0.200)(36.00)}{20.00} = 0.120$$

8.57 What was $[Ag^+]$ at the stage where CuI just began to precipitate in the titration of Problem 8.55?

At this stage,

$$[Cu^+] = \frac{(0.190)(20.00) \text{ mmol}}{58.00 \text{ mL}} = 0.0655$$

Therefore

$$[I^-] = \frac{K_{sp}}{[Cu^+]} = \frac{5.1 \times 10^{-12}}{0.0655} = 7.8 \times 10^{-11}$$

$$[Ag^+] = \frac{K_{sp}}{[I^-]} = \frac{8.31 \times 10^{-17}}{7.8 \times 10^{-11}} = 1.1 \times 10^{-6}$$

8.58 Consider the potentiometric titration of 20.00 mL of 0.150 M NaBr with 0.100 M AgNO$_3$. What is pAg at the start of the titration, when the first tiny crystal of AgBr is formed?

Since very, very little Br$^-$ was lost by precipitation,

$$[Ag^+] = \frac{K_{sp}}{[Br^-]} = \frac{5.25 \times 10^{-13}}{0.150} = 3.50 \times 10^{-12} \qquad \text{or} \qquad pAg = 11.456$$

8.59 What is pAg when 10.00 mL of AgNO$_3$ is added in the titration of Problem 8.58?

The $(10.00)(0.100) = 1.00$ mmol of Ag$^+$ reacts with 1.00 mmol of Br$^-$, leaving

$$(20.00)(0.150) - 1.00 = 2.00 \text{ mmol}$$

of Br$^-$. Therefore,

$$[Br^-] = \frac{2.00 \text{ mmol}}{30.00 \text{ mL}} = 0.0667 \qquad [Ag^+] = \frac{K_{sp}}{[Br^-]} = \frac{5.25 \times 10^{-13}}{0.0667} = 7.87 \times 10^{-12}$$

and pAg = 11.104.

8.60 What is pAg at the equivalence point in the titration of Problem 8.58?

At this point, $[Ag^+] = [Br^-]$, so that $[Ag^+] = \sqrt{K_{sp}} = 7.25 \times 10^{-7}$, or pAg = 6.140.

8.61 What is pAg when 40.00 mL of AgNO$_3$ is added in the titration of Problem 8.58?

This is after the equivalence point and only 3.00 mmol of the 4.00 mmol of Ag$^+$ added is consumed by reaction with Br$^-$.

$$[Ag^+] = \frac{1.00 \text{ mmol}}{60.00 \text{ mL}} = 1.67 \times 10^{-2} \qquad \text{or} \qquad pAg = 1.778$$

8.62 Consider the potentiometric titration of 20.00 mL of 0.0600 M Hg$_2$(NO$_3$)$_2$ with 0.150 M NaCl to form the precipitate Hg$_2$Cl$_2$. What is pCl when the first tiny crystal of Hg$_2$Cl$_2(s)$ is formed? [Remember that mercurous ion is a dimer, Hg$_2^{2+}$.]

Since essentially no Hg$_2^{2+}$ is precipitated,

$$[Hg_2^{2+}] = 0.0600 \qquad [Cl^-] = \left(\frac{K_{sp}}{[Hg_2^{2+}]}\right)^{1/2} = \left(\frac{1.3 \times 10^{-18}}{0.0600}\right)^{1/2} = 4.7 \times 10^{-9}$$

and pCl = 8.33.

8.63 What is pCl when 8.00 mL of NaCl is added in the titration of Problem 8.62?

This is prior to the equivalence point, since the $(8.00)(0.150) = 1.20$ mmol of Cl$^-$ reacts with $1.20/2 = 0.600$ mmol of Hg$_2^{2+}$ out of an original $(20.00)(0.0600) = 1.20$ mmol. Therefore, there remains 0.60 mmol of Hg$_2^{2+}$,

$$[Hg_2^{2+}] = \frac{0.60 \text{ mmol}}{28.00 \text{ mL}} = 2.1 \times 10^{-2} \qquad [Cl^-] = \left(\frac{1.3 \times 10^{-18}}{2.1 \times 10^{-2}}\right)^{1/2} = 7.9 \times 10^{-9}$$

and pCl = 8.10.

8.64 What is pCl when 16.00 mL of NaCl is added in the titration of Problem 8.62?

This is the equivalence point in the titration, since $(16.00)(0.150) = 2.40$ mmol of Cl$^-$ reacts stoichiometrically with 1.20 mmol of Hg$_2^{2+}$. Since $K_{sp} = 4S^3 = 1.3 \times 10^{-18}$,

$$S = [Hg_2^{2+}] = 6.9 \times 10^{-7} \qquad 2S = [Cl^-] = 1.4 \times 10^{-6}$$

and pCl = 5.86.

8.65 What is pCl when 32.00 mL of NaCl is added in the titration of Problem 8.62?

The $(32.00)(0.150) = 4.80$ mmol of Cl^- is $4.80 - 2.40 = 2.40$ mmol in excess of that required for complete precipitation. Therefore,

$$[Cl^-] = \frac{2.40 \text{ mmol}}{52.00 \text{ mL}} = 4.62 \times 10^{-2} \qquad \text{or} \qquad pCl = 1.336$$

8.66 If the original solution of Problem 8.62 was also $0.0200\,M$ $AgNO_3$, would it have been possible to observe distinct end points for the Ag^+ and Hg_2^{2+} ions?

The $[Cl^-]$ required to begin precipitation of $AgCl(s)$ is

$$[Cl^-] = \frac{K_{sp}}{[Ag^+]} = \frac{1.78 \times 10^{-10}}{0.0200} = 8.90 \times 10^{-9}$$

which, according to Problem 8.62, is just about the same concentration as required to begin precipitation of $Hg_2Cl_2(s)$. Separate end points will not be seen.

Supplementary Problems

SALTS OF STRONG ACIDS

8.67 What is the solubility of $TlI(s)$? *Ans.* $2.5 \times 10^{-4}\,M$

8.68 What is the solubility of $TlI(s)$ in a solution of such ionic strength that $f_{Tl^+} = f_{I^-} = 0.807$?
Ans. $3.2 \times 10^{-4}\,M$

8.69 What is the solubility of mercurous bromide, $Hg_2Br_2(s)$? *Ans.* $2.4 \times 10^{-8}\,M$

8.70 What is the solubility of $Hg_2Br_2(s)$ in an ionic solution such that $f_{Hg^{2+}} = 0.381$ and $f_{Br^-} = 0.772$?
Ans. $4.0 \times 10^{-8}\,M$

8.71 Find the solubility of $TlI(s)$ in (a) $0.150\,M$ $TlNO_3$, (b) $1.50 \times 10^{-4}\,M$ $TlNO_3$, (c) $1.50 \times 10^{-4}\,M$ NaI, (d) a non-ideal $1.50 \times 10^{-4}\,M$ $TlNO_3$ solution such that $f_{Tl^+} = f_{I^-} = 0.772$.
Ans. (a) $4.3 \times 10^{-7}\,M$; (b) $1.9 \times 10^{-4}\,M$; (c) $1.9 \times 10^{-4}\,M$; (d) $2.6 \times 10^{-4}\,M$

8.72 What is the solubility of $Hg_2Br_2(s)$ in a $1.75 \times 10^{-3}\,M$ NaBr solution? *Ans.* $1.9 \times 10^{-17}\,M$

8.73 What is the solubility of $Hg_2Br_2(s)$ in a $3.28 \times 10^{-4}\,M$ $Hg_2(NO_3)_2$ solution? *Ans.* $2.1 \times 10^{-10}\,M$

8.74 When $CuBr(s)$ is added to water, $[Cu^+]$ is found to be 7.2×10^{-5}; find K_{sp}. *Ans.* 5.2×10^{-9}

8.75 When $CuCl(s)$ is added to a $2.60 \times 10^{-3}\,M$ NaCl solution, $[Cu^+]$ is found to be 4.0×10^{-4}; find K_{sp} for $CuCl(s)$. *Ans.* 1.2×10^{-6}

8.76 25.00 mL of a solution is $0.0600\,M$ $TlNO_3$, $0.0300\,M$ $AgNO_3$, $0.0400\,M$ $Hg_2(NO_3)_2$, and $0.0500\,M$ $Pb(NO_3)_2$. In what order do the iodides precipitate when $0.100\,M$ NaI is slowly added to the solution? *Ans.* $AgI(s)$ first, followed by $Hg_2I_2(s)$, $TlI(s)$, and $PbI_2(s)$.

8.77 What volumes of $0.100\,M$ NaI are required to precipitate the iodides in Problem 8.76?
Ans. 7.50 mL for $AgI(s)$, 20.0 mL for $Hg_2I_2(s)$, 15.0 mL for $TlI(s)$, 25.0 mL for $PbI_2(s)$

8.78 (a) What precipitates are present, and (b) what are the ionic concentrations, when 30.00 mL of 0.100 NaI is added to the solution of Problem 8.76?

Ans. (a) AgI(s), $Hg_2I_2(s)$, TlI(s); (b) $[Tl^+] = 2.3 \times 10^{-2}$, $[I^-] = 2.9 \times 10^{-6}$, $[Pb^{2+}] = 2.27 \times 10^{-2}$, $[Ag^+] = 2.9 \times 10^{-11}$, $[Hg_2^{2+}] = 5.5 \times 10^{-18}$

8.79 (a) What precipitates are present, and (b) what are the ionic concentrations, when 67.50 mL of 0.100 M NaI is added to the solution of Problem 8.76?

Ans. (a) This is the stoichiometric volume required to precipitate all four iodides. (b) $[I^-] = 2.4 \times 10^{-3}$, $[Pb^{2+}] = 1.2 \times 10^{-3}$, $[Tl^+] = 2.7 \times 10^{-5}$, $[Hg_2^{2+}] = 7.6 \times 10^{-24}$, $[Ag^+] = 3.4 \times 10^{-14}$.

8.80 What are the ionic concentrations when 75.00 mL of 0.100 M NaI is added to the solution of Problem 8.76? (This volume is 7.50 mL past the last stoichiometric equivalence point.)

Ans. $[I^-] = 7.50 \times 10^{-3}$, $[Ag^+] = 1.11 \times 10^{-14}$, $[Hg_2^{2+}] = 8.0 \times 10^{-25}$, $[Tl^+] = 8.7 \times 10^{-6}$, $[Pb^{2+}] = 1.3 \times 10^{-4}$

HYDROXIDE PRECIPITATES

8.81 What is the solubility of $Fe(OH)_2(s)$?   Ans. 6×10^{-6} M

8.82 What is the solubility of $Al(OH)_3(s)$?   Ans. 2×10^{-11} M

8.83 What is the solubility of $Fe(OH)_2(s)$ in a solution buffered to pH 9.000?   Ans. 8×10^{-6} M

8.84 What are the concentrations of the various ions when 0.250 mol of $Al(NO_3)_3$ is added to 1.00 L of water?

Ans. $[Al^{3+}] = 0.250$, $[NO_3^-] = 0.750$, $[H^+] = 2 \times 10^{-4}$, $[OH^-] = 4.3 \times 10^{-11}$

SALTS OF WEAK ACIDS

8.85 What is the solubility of $CaF_2(s)$?   Ans. 2.2×10^{-4} M

8.86 What is the solubility of $CaF_2(s)$ in a 0.0250 M $Ca(NO_3)_2$ solution?   Ans. 2.0×10^{-5} M

8.87 What is the solubility of $CaF_2(s)$ in a 0.0344 M NaF solution?   Ans. 3.4×10^{-8} M

8.88 What is the solubility of $CaF_2(s)$ in a solution buffered to pH 4.000, if $K_a = 6.76 \times 10^{-4}$ for HF?

Ans. 2.4×10^{-4} M

PRECIPITATION TITRATIONS

8.89 If 25.00 mL of a solution that is 0.0426 M $Pb(NO_3)_2$ and 0.0708 M $AgNO_3$ is titrated with 0.100 M NaI, in what order do the precipitates form?   Ans. AgI(s) first, $PbI_2(s)$ second.

8.90 What volumes of 0.100 M NaI are needed to reach the equivalence points in the titration of Problem 8.89?   Ans. 17.7 mL to precipitate AgI(s), and an additional 21.3 mL to precipitate $PbI_2(s)$.

8.91 What was pI when the first crystal of AgI(s) was formed in the titration of Problem 8.89?

Ans. 14.930

8.92 What was pI when half of the Ag^+ was precipitated in the titration of Problem 8.89?   Ans. 14.498

8.93 What was pI when PbI(s) began to precipitate in the titration of Problem 8.89?   Ans. 3.27

8.94 What was pI when half of the Pb^{2+} was precipitated in the titration of Problem 8.89?   Ans. 3.07

8.95   What was pI at the $PbI_2(s)$ equivalence point in the titration of Problem 8.89?   *Ans.*   2.62

8.96   What was pI when 50.00 mL of 0.100 M NaI was added in the titration of Problem 8.89?
Ans.   1.834

PRECIPITATE MIXTURES WITH A COMMON ION

8.97   If 0.400 mol of NaI is slowly added to 1.00 L of a solution that is 0.100 M $TlNO_3$, 0.100 M $AgNO_3$, and 0.200 M $Pb(NO_3)_2$, in what order do the precipitates form?
Ans. $AgI(s)$ first, then $TlI(s)$, and $PbI_2(s)$ last.

8.98   Which precipitates are present in the final solution of Problem 8.97?   *Ans.*   All three.

8.99   Which metal ion (besides Na^+) is present in significant concentration in the solution of Problem 8.97, and what is that concentration?   *Ans.* $[Pb^{2+}] = 0.100$

8.100   What are the concentrations of I^-, Ag^+, and Tl^+ in the solution of Problem 8.97?
Ans. $[I^-] = 2.7 \times 10^{-4}$, $[Ag^+] = 3.1 \times 10^{-13}$, $[Tl^+] = 2.4 \times 10^{-4}$

Chapter 9

Complex Ion Equilibria

9.1 INTRODUCTION

Many ions are able to form covalent bonds with certain chemical species called *ligands*, *complexing agents*, or, where more than one covalent bond is formed, *chelating agents*. For example, CN^-, NH_3, and EDTA (ethylenediaminetetraacetic acid) can form strongly bonded complexes with a variety of metal ions. Even halide ions can be considered ligands in certain cases; as is shown below, the solubility of precipitates such as $AgCl(s)$ is more complicated than the simple

$$AgCl(s) \rightarrow Ag^+ + Cl^-$$

would suggest.

9.2 COMMON-ION EFFECT

In any simplified discussion of solubilities of precipitates, it is usually noted that an increase in the concentration of one of the ions common to the precipitate will result in the precipitation of a greater fraction of the other ion. Thus, if you are analyzing a sample for Ag^+ by forming the precipitate $AgCl(s)$, you would infer from Le Chatelier's principle (Section 3.4) that an increase in $[Cl^-]$ beyond the stoichiometric concentration would result in a greater fraction of Ag^+ in the form of $AgCl(s)$ precipitate.

However, silver can exist in solution in forms other than Ag^+; the presence of $AgCl(aq)$, $AgCl_2^-$, $AgCl_3^{2-}, \ldots$, can result in a residual silver solubility greater than that predicted by the simple solubility expression. The various equilibria may be indicated as follows:

Reaction	Equilibrium Constant
$AgCl(s) \rightarrow AgCl(aq)$	S°
$AgCl(s) \rightarrow Ag^+ + Cl^-$	K_{sp}
$AgCl(aq) + Cl^- \rightarrow AgCl_2^-$	K_2
$AgCl_2^- + Cl^- \rightarrow AgCl_3^{2-}$	K_3

· ·

EXAMPLE 9.1 What is the total solubility, S, of $AgCl(s)$ as a function of the $[Cl^-]$ of the solution?

$$S = [AgCl(aq)] + [Ag^+] + [AgCl_2^-] + [AgCl_3^{2-}] + \cdots$$

But

$$[AgCl(aq)] = S^\circ$$
$$[Ag^+] = K_{sp}/[Cl^-]$$
$$[AgCl_2^-] = K_2[AgCl(aq)][Cl^-] = K_2 S^\circ[Cl^-]$$
$$[AgCl_3^{2-}] = K_3[AgCl_2^-][Cl^-] = K_2 K_3 S^\circ[Cl^-]^2$$

· ·

Hence

$$S = S^\circ + \frac{K_{sp}}{[Cl^-]} + K_2 S^\circ[Cl^-] + K_2 K_3 S^\circ[Cl^-]^2 + \cdots \qquad (1)$$

As is shown in Problem 9.5, the function (1) eventually becomes increasing; i.e., the amount of Ag precipitated eventually decreases with increasing $[Cl^-]$. (Le Chatelier's principle would predict this same result if applied to *all* steps, not just to the dissolution of $AgCl(s)$.)

9.3 COMPLEXATION

In some cases it may be useful to adjust the concentrations of a solution so that no precipitate forms or so that a particular precipitate is selectively dissolved. For instance, the adjustment of pH will often allow (or prevent) precipitation of salts of weak acids. Another method of preventing or specializing precipitation is the formation of strongly bonded complexes.

Consider the following general case where no excess common ion is present; where $S°$ is negligible; where the complexing ligand, L, may be a neutral molecule such as NH_3 or H_2O, or an ion such as CN^-, F^-, or $S_2O_3^{2-}$; and where M is a metal ion such as Ag^+ or Zn^{2+}.

Reaction	Stepwise Formation Constant
$M + L \rightarrow ML$	K_1
$ML + L \rightarrow ML_2$	K_2
$ML_2 + L \rightarrow ML_3$	K_3
. . . .	. . .
$ML_{n-1} + L \rightarrow ML_n$	K_n

Stepwise formation constants for several complexes are listed in Table 9-1.

Table 9-1. Stepwise Formation Constants

Ligand L	Metal Ion M	K_1	K_2	K_3	K_4
NH_3	Ag^+	2.34×10^3	6.92×10^3	—	—
NH_3	Cu^{2+}	2.04×10^4	4.68×10^3	1.10×10^3	2.00×10^2
NH_3	Zn^{2+}	1.51×10^2	1.78×10^2	2.04×10^2	9.12×10^1
NH_3	Cd^{2+}	4.47×10^2	1.26×10^2	2.75×10^1	8.51
CN^-	Ag^+	$K_{s2} = K_1 K_2 = 7.08 \times 10^{19}$		—	—
$S_2O_3^{2-}$	Ag^+	6.61×10^8	4.37×10^4	4.90	—

In terms of the stepwise formation constants, we may define *summed*, or *overall*, *formation constants*:

Reaction	Summed Formation Constant
$M + L \rightarrow ML$	$K_{s1} = K_1$
$M + 2L \rightarrow ML_2$	$K_{s2} = K_1 K_2$
$M + 3L \rightarrow ML_3$	$K_{s3} = K_1 K_2 K_3$
. . . .	. . . .
$M + nL \rightarrow ML_n$	$K_{sn} = K_1 K_2 K_3 \cdots K_n$

Another quantity which is useful in calculations of complex ion equilibria is the *formality* of M (see Table 1-4), which is the sum C_M of the concentrations of all forms of M in solution:

$$C_M = [M] + [ML] + [ML_2] + [ML_3] + \cdots + [ML_n] \qquad (9.1)$$

or, in terms of the summed formation constants,

$$C_M = [M] + K_{s1}[M][L] + K_{s2}[M][L]^2 + K_{s3}[M][L]^3 + \cdots + K_{sn}[M][L]^n \equiv [M]D_M \qquad (9.2)$$

The fractions of M present in the several forms can be expressed in a manner similar to that used for polyprotic acids:

$$F_{0M} \equiv \frac{[M]}{C_M} = \frac{1}{D_M}$$

$$F_{1M} \equiv \frac{[ML]}{C_M} = \frac{K_{s1}[L]}{D_M}$$

$$F_{2M} \equiv \frac{[ML_2]}{C_M} = \frac{K_{s2}[L]^2}{D_M} \tag{9.3}$$

$$\cdots\cdots\cdots\cdots\cdots\cdots$$

$$F_{nM} \equiv \frac{[ML_n]}{C_M} = \frac{K_{sn}[L]^n}{D_M}$$

EXAMPLE 9.2   Determine the solubility of $AgBr(s)$ in a solution of NH_4OH where, following equilibrium,

$$[NH_4OH] (=[NH_3]) = 0.800$$

$K_{sp} = 5.25 \times 10^{-13}$; stepwise formation constants are given in Table 9-1. Assume that S° is negligible and that there is negligible Br^- complexation of AgBr.

When S mol/L of $AgBr(s)$ dissolves, there will be produced S mol/L of Br^- and a total silver concentration of

$$C_{Ag^+} = [Ag^+] + [Ag(NH_3)^+] + [Ag(NH_3)_2^+] = S$$

From (9.3),

$$5.25 \times 10^{-13} = K_{sp} = [Ag^+][Br^-] = (C_{Ag^+} F_{0Ag^+})[Br^-] = (S/D_{Ag^+})S = S^2/D_{Ag^+}$$

and from (9.2),

$$D_{Ag^+} = 1 + (2.34 \times 10^3)(0.800) + (2.34 \times 10^3)(6.92 \times 10^3)(0.800)^2 = 1.04 \times 10^7$$

Therefore,

$$S = \sqrt{(1.04 \times 10^7)(5.25 \times 10^{-13})} = 2.34 \times 10^{-3}$$

EDTA Complexes

EDTA is a polyprotic acid, H_4Y, with dissociation constants

$$K_1 = 1.00 \times 10^{-2} \qquad K_2 = 2.16 \times 10^{-3} \qquad K_3 = 6.92 \times 10^{-7} \qquad K_4 = 5.50 \times 10^{-11}$$

The EDTA ion, Y (charge 4−), forms very strong complexes with many metal ions M, according to $M + Y \rightarrow MY$; hence, the *formation constant*, K_{form}, is given by

$$K_{form} = \frac{[MY]}{[M][Y]} \tag{9.4}$$

Some values are given in Table 9-2.

Table 9-2.   EDTA Formation Constants

Metal Ion M	K_{form}	Metal Ion M	K_{form}
Ag^+	2.09×10^7	Zn^{2+}	3.20×10^{16}
Cu^{2+}	6.17×10^{18}	Ca^{2+}	5.01×10^{10}
Fe^{2+}	2.14×10^{14}	Ni^{2+}	3.63×10^{18}
Mn^{2+}	3.80×10^{13}	Cd^{2+}	3.89×10^{16}

If C'_{EDTA} is the total EDTA in solution that is *not* in the form MY, then

$$C'_{\text{EDTA}} = [H_4Y] + [H_3Y^-] + [H_2Y^{2-}] + [HY^{3-}] + [Y^{4-}] \qquad (9.5)$$

and we have, analogous to (7.6) and (7.7),

$$F_4 \equiv \frac{[Y^{4-}]}{C'_{\text{EDTA}}} = \frac{K_{s4}}{D'_{\text{EDTA}}} = \frac{8.22 \times 10^{-22}}{D'_{\text{EDTA}}} \qquad (9.6)$$

where

$$D'_{\text{EDTA}} = [H^+]^4 + K_{s1}[H^+]^3 + K_{s2}[H^+]^2 + K_{s3}[H^+] + K_{s4} \qquad (9.7)$$

Conditional Formation Constants

For most metal ion–EDTA complexes, K_{form} is very large, as shown in Table 9-2. This means that the reaction to form the complex will proceed almost to completion, so that [MY] can be assumed equal to either the original [M] or the original [EDTA], whichever is added in the lesser amount.

EXAMPLE 9.3 If a solution is prepared with an initial Ni^{2+} concentration of $2.50 \times 10^{-2}\ M$ and an initial EDTA concentration of $1.36 \times 10^{-2}\ M$, then EDTA is the limiting reagent and we can assume that

$$[NiY] = 1.36 \times 10^{-2} \qquad C_{Ni^{2+}} = (2.50 - 1.36) \times 10^{-2} = 1.14 \times 10^{-2} \qquad C'_{\text{EDTA}} \approx 0$$

To handle calculations of this type, it is convenient to define a new constant based on (9.4). In place of [M] in the denominator of (9.4) there is substituted $F_{0M}C'_M$, where C'_M is the concentration of M in all forms but MY. Also, by (9.6), $F_4 C'_{\text{EDTA}}$ is substituted for [Y]. Then (9.4) becomes

$$K_{\text{form}} = \frac{[MY]}{(F_{0M}C'_M)(F_4 C'_{\text{EDTA}})} \equiv \frac{K'_{\text{form}}}{F_{0M}F_4}. \qquad (9.8)$$

defining the *conditional formation constant*, K'_{form}. For a solution of a metal ion M in a known ligand concentration [L], $F_{0M} = 1/D_M$ will have a fixed value according to (9.2); for a given pH, F_4 will have a fixed value according to (9.6) and (9.7).

EXAMPLE 9.4 Evaluate the conditional formation constant for the reaction $Zn^{2+} + Y^{4-} \rightarrow ZnY^{2-}$ in a buffer solution that has $[NH_4^+] = [NH_3] = 1.00$, given

$$K_b = \frac{[NH_4^+][OH^-]}{[NH_3]} = 1.82 \times 10^{-5}$$

For this buffer solution, $K_b = [OH^-] = 1.82 \times 10^{-5}$; so $[H^+] = K_w/[OH^-] = 5.50 \times 10^{-10}$. From (9.7) and the K-values for EDTA there results $D'_{\text{EDTA}} = 9.05 \times 10^{-21}$. Then, from (9.6),

$$F_4 = \frac{8.22 \times 10^{-22}}{9.05 \times 10^{-21}} = 9.08 \times 10^{-2}$$

From (9.2) and data in Table 9-1,

$$D_M = 1 + (151)(1.00) + (2.69 \times 10^4)(1.00)^2 + (5.48 \times 10^6)(1.00)^3 + (5.00 \times 10^8)(1.00)^4 = 5.06 \times 10^8$$

whence $F_{0M} = 1/D_M = 1.98 \times 10^{-9}$. Therefore, from (9.8) and Table 9-2,

$$K'_{\text{form}} = (1.98 \times 10^{-9})(9.08 \times 10^{-2})(3.20 \times 10^{16}) = 5.75 \times 10^6$$

EXAMPLE 9.5 What is the concentration of Zn^{2+} if 1.38×10^{-4} mol of Zn^{2+} and 0.0500 mol of Na_2H_2Y are added to 1.00 L of the solution of Example 9.4?

Since $Zn^{2+} \equiv M$ is the limiting reactant, $[MY] = 1.38 \times 10^{-4}$ and $C'_{\text{EDTA}} = 0.0500 - (1.38 \times 10^{-4}) = 4.99 \times 10^{-2}$. From Example 9.3,

$$5.75 \times 10^6 = K'_{\text{form}} = \frac{[MY]}{C'_M C'_{\text{EDTA}}} = 1.38 \times 10^{-4}/C'_M(4.99 \times 10^{-2})$$

or $C'_M = 4.81 \times 10^{-10}$. Finally, from Example 9.4, $F_{0M} = 1.98 \times 10^{-9}$, so that

$$[Zn^{2+}] \equiv [M] = F_{0M}C'_M = (1.98 \times 10^{-9})(4.81 \times 10^{-10}) = 9.52 \times 10^{-19}$$

Solved Problems

COMMON-ION EFFECT

9.1 For AgCl, $S° = 3.0 \times 10^{-7}$, $K_{sp} = 1.78 \times 10^{-10}$, $K_2 = 66.7$, and $K_3 = 1.0$. What is the solubility, S, of AgCl(s) in 1.00×10^{-4} M NaCl when (a) only the K_{sp}-equilibrium is considered? (b) all four equilibria are considered?

(a) Assuming the [Cl$^-$] produced when AgCl(s) dissolves is smaller than 1% of 10^{-4},

$$[Ag^+] = S = \frac{K_{sp}}{[Cl^-]} = \frac{1.78 \times 10^{-10}}{1.00 \times 10^{-4}} = 1.78 \times 10^{-6}$$

(b) By (1) of Example 9.1,

$$S = (3.0 \times 10^{-7}) + \frac{1.78 \times 10^{-10}}{1.00 \times 10^{-4}}$$
$$+ (66.7)(3.0 \times 10^{-7})(1.00 \times 10^{-4}) + (66.7)(1.0)(3.0 \times 10^{-7})(1.00 \times 10^{-4})^2 = 2.08 \times 10^{-6}$$

9.2 Repeat Problem 9.1 for 1.00×10^{-3} M NaCl.

(a) $$S \equiv \frac{1.78 \times 10^{-10}}{1.00 \times 10^{-3}} = 1.78 \times 10^{-7}$$

(b) $$S = (3.0 \times 10^{-7}) + (1.78 \times 10^{-7}) + (66.7)(3.0 \times 10^{-7})(1.00 \times 10^{-3})$$
$$+ (66.7)(1.0)(3.0 \times 10^{-7})(1.00 \times 10^{-3})^2 = 5.0 \times 10^{-7}$$

9.3 Repeat Problem 9.1 for 1.00×10^{-2} M NaCl.

(a) $$S = \frac{1.78 \times 10^{-10}}{1.00 \times 10^{-2}} = 1.78 \times 10^{-8}$$

(b) $$S = (3.0 \times 10^{-7}) + (1.78 \times 10^{-8}) + (66.7)(3.0 \times 10^{-7})(1.00 \times 10^{-2})$$
$$+ (66.7)(1.0)(3.0 \times 10^{-7})(1.00 \times 10^{-2})^2 = 5.2 \times 10^{-7}$$

9.4 Repeat Problem 9.1 for 1.00×10^{-1} M NaCl.

(a) $$S = \frac{1.78 \times 10^{-10}}{0.100} = 1.78 \times 10^{-9}$$

(b) $$S = (3.0 \times 10^{-7}) + (1.78 \times 10^{-9}) + (66.7)(3.0 \times 10^{-7})(1.00 \times 10^{-1})$$
$$+ (66.7)(1.00)(3.0 \times 10^{-7})(0.100)^2 = 2.5 \times 10^{-6}$$

9.5 Find the value of [Cl$^-$] that minimizes the solubility in Problem 9.1(b), and compute this minimum solubility.

The minimum S corresponds to the [Cl$^-$] for which the derivative of (1), Example 9.1, is zero:

$$\frac{dS}{d[Cl^-]} = -\frac{K_{sp}}{[Cl^-]^2} + K_2 S° + 2K_2 K_3 S°[Cl^-] = 0$$

or

$$-\frac{1.78 \times 10^{-10}}{[Cl^-]^2} + (2.0 \times 10^{-5}) + (4.0 \times 10^{-5})[Cl^-] = 0$$

Solving this cubic equation results in [Cl$^-$] $= 3.0 \times 10^{-3}$. Substitution in (1) of Example 9.1 gives

$$S_{min} = (3.0 \times 10^{-7}) + \frac{1.78 \times 10^{-10}}{3.0 \times 10^{-3}}$$

$$+ (66.7)(3.0 \times 10^{-7})(3.0 \times 10^{-3}) + (66.7)(1.0)(3.0 \times 10^{-7})(3.0 \times 10^{-3})^2 = 4.2 \times 10^{-7}$$

COMPLEXATION

9.6   A solution has an equilibrium NH_4OH concentration of $0.800\ M$ and a total dissolved silver concentration of $0.100\ M$. What is the maximum $[Cl^-]$ that can be present without $AgCl(s)$ forming? $K_{sp} = 1.78 \times 10^{-10}$.

From Table 9-1, $K_{s1} = 2.34 \times 10^3$ and $K_{s2} = 1.62 \times 10^7$. We have $[NH_3] = [L] = 0.800$, and (9.2) gives $D_M = 1.04 \times 10^7$. With $C_M = 0.100$, (9.3) gives

$$[Ag^+] = [M] = \frac{C_M}{D_M} = 9.62 \times 10^{-9}$$

Therefore,

$$[Cl^-]_{max} = \frac{K_{sp}}{[Ag^+]} = 1.85 \times 10^{-2}$$

9.7   Concentrated NH_4OH was added to 1.00 L of $0.200\ M$ $CuSO_4$ until $[NH_3] = 0.600$. All of the copper was in solution, mostly in the form of $Cu(NH_3)_n^{2+}$ complexes. What were the concentrations of the various copper species and how much NH_4OH ($K_b = 1.82 \times 10^{-5}$) was added?

The total cupric concentration was $C_M = 0.200$. From Table 9-1,

$$K_{s1} = 2.04 \times 10^4 \qquad K_{s2} = 9.55 \times 10^7 \qquad K_{s3} = 1.05 \times 10^{11} \qquad K_{s4} = 2.10 \times 10^{13}$$

Since $[NH_3] = [L] = 0.600$, (9.2) gives $D_M = 2.74 \times 10^{12}$. Then, by (9.3),

$$[Cu^{2+}] = [M] = \frac{C_M}{D_M} = 7.30 \times 10^{-14}$$

$$[Cu(NH_3)^{2+}] = [ML] = \frac{C_M K_{s1}[L]}{D_M} = 8.93 \times 10^{-10}$$

$$[Cu(NH_3)_2^{2+}] = [ML_2] = \frac{C_M K_{s2}[L]^2}{D_M} = 2.51 \times 10^{-6}$$

$$[Cu(NH_3)_3^{2+}] = [ML_3] = \frac{C_M K_{s3}[L]^3}{D_M} = 1.66 \times 10^{-3}$$

$$[Cu(NH_3)_4^{2+}] = [ML_4] = \frac{C_M K_{s4}[L]^4}{D_M} = 0.199$$

The added NH_4OH appears in three forms: NH_3 (i.e., NH_4OH), NH_4^+, and the $Cu(NH_3)_n^{2+}$ complexes. The coordinates $(C_b, K_b) = (0.600, 1.82 \times 10^{-5})$ in the "base" version of Fig. 5-1 indicate that the water equilibrium can be neglected and that there is less than 1% dissociation of NH_4OH (Assumptions 1b and 2b); hence, we can neglect $[NH_4^+]$. Since the volume is 1.00 L,

$$\text{Total added } NH_4OH = [NH_4OH] + [Cu(NH_3)^{2+}] + 2[Cu(NH_3)_2^{2+}] + 3[Cu(NH_3)_3^{2+}] + 4[Cu(NH_3)_4^{2+}]$$
$$= 1.40 \text{ mol}$$

9.8   If 25.00 mL of $0.100\ M$ $AgNO_3$ is mixed with 25.00 mL of $0.300\ M$ $NaCN$, find $[Ag^+]$, $[Ag(CN)_2^-]$, and $[CN^-]$.

From Table 9-1, the only reaction of importance is $Ag^+ + 2CN^- \rightarrow Ag(CN)_2^-$, with $K_{s2} = 7.08 \times 10^{19}$. After mixing, but before reaction, $[Ag^+] = 0.0500$ and $[CN^-] = 0.150$. There is a stoichiometric excess of CN^- when forming $Ag(CN)_2^-$. After reaction, but before equilibration, the concentrations of principal species are $[Ag(CN)_2^-] = 0.0500$ and $[CN^-] = 0.050$. Since K_{s2} is very large, the shifts to equilibrium will produce very little $[Ag^+]$ and additional $[CN^-]$, so that the equilibrium $[CN^-]$ and $[Ag(CN)_2^-]$ will be the above-given, principal species values. Therefore,

$$7.08 \times 10^{19} = K_{s2} = \frac{[Ag(CN)_2^-]}{[Ag^+][CN^-]^2} = \frac{0.0500}{[Ag^+](0.050)^2} \qquad \text{or} \qquad [Ag^+] = 2.8 \times 10^{-19}$$

9.9 If 25.00 mL of 0.150 M $AgNO_3$ is mixed with 25.00 mL of 0.100 M NaCN, find $[Ag^+]$, $[Ag(CN)_2^-]$, and $[CN^-]$. Neglect hydrolysis of CN^-.

After mixing, but before reaction, $[Ag^+] = 0.0750$ and $[CN^-] = 0.0500$. There is a stoichiometric excess of Ag^+ when forming $Ag(CN)_2^-$. After reaction, but before equilibration, the concentrations of principal species are $[Ag^+] = 0.0500$ and $[Ag(CN)_2^-] = 0.0250$. Since K_{s2} is so large, the shift to equilibrium results in the production of very little CN^-, and the calculated concentrations of the principal species will be the equilibrium concentrations. Therefore,

$$7.08 \times 10^{19} = K_{s2} = \frac{[Ag(CN)_2^-]}{[Ag^+][CN^-]^2} = \frac{0.0250}{(0.0500)[CN^-]^2} \qquad \text{or} \qquad [CN^-] = 8.40 \times 10^{-11}$$

9.10 How many moles of $AgCl(s)$ ($K_{sp} = 1.78 \times 10^{-10}$) will dissolve in 1.00 L of 0.100 M NaCN? What are the concentrations of the various ions? Neglect $[AgCl(aq)]$ and higher complexes between Ag^+ and Cl^-.

Since K_{s2} for the reaction $Ag^+ + 2CN^- \rightarrow Ag(CN)_2^-$ is extremely large, as shown in Table 9-1, almost all of the silver in solution will be in the form of $Ag(CN)_2^-$, and the principal, overall reaction will be

$$AgCl(s) + 2CN^- \rightarrow Ag(CN)_2^- + Cl^- \tag{1}$$

with equilibrium constant

$$K = (K_{sp})(K_{s2}) = (1.78 \times 10^{-10})(7.08 \times 10^{19}) = 1.26 \times 10^{10}$$

(see Section 3.3). As K is large, very little CN^- will remain uncomplexed provided there is an excess supply of $AgCl(s)$. Therefore,

$$[Ag(CN)_2^-] = \frac{0.100}{2} = 0.0500$$

and, from (1), $[Cl^-]$ also equals 0.0500. From

$$1.26 \times 10^{10} = \frac{[Ag(CN)_2^-][Cl^-]}{[CN^-]^2} = \frac{(0.0500)^2}{[CN^-]^2}$$

we obtain $[CN^-] = 4.45 \times 10^{-7}$; from $K_{sp} = [Ag^+][Cl^-] = [Ag^+](0.0500)$, $[Ag^+] = 3.56 \times 10^{-9}$.

9.11 What minimum concentration of $Na_2S_2O_3$ is needed so that 0.200 mol of $AgCl(s)$ will dissolve in 1.00 L? $K_{sp}(AgCl) = 1.78 \times 10^{-10}$. Neglect $[AgCl(aq)]$ and higher complexes between Ag^+ and Cl^-.

Since $[Cl^-]$ must equal 0.200,

$$[M] \equiv [Ag^+]_{max} = \frac{K_{sp}}{[Cl^-]} = \frac{1.78 \times 10^{-10}}{0.200} = 8.90 \times 10^{-10}$$

The total silver in solution is $C_M = 0.200$. Therefore, from (9.3),

$$D_M = \frac{C_M}{[M]} = \frac{0.200}{8.90 \times 10^{-10}} = 2.25 \times 10^8$$

From (9.2) for D_M and from the data in Table 9-1,

$$D_M = 2.25 \times 10^8 = 1 + (6.61 \times 10^8)[L] + (2.89 \times 10^{13})[L]^2 + (1.42 \times 10^{14})[L]^3$$

Solving this cubic for [L] (using, for example, the graphical intersection method of Section 1.4), we obtain $[L] \equiv [S_2O_3^{2-}] = 2.76 \times 10^{-3}$. Then, from (9.3),

$$[AgS_2O_3^-] \equiv [ML] = \frac{C_M K_{s1}[L]}{D_M} = \frac{(0.200)(6.61 \times 10^8)(2.76 \times 10^{-3})}{2.25 \times 10^8} = 1.60 \times 10^{-3}$$

$$[Ag(S_2O_3)_2^{3-}] \equiv [ML_2] = \frac{C_M K_{s2}[L]^2}{D_M} = \frac{(0.200)(2.89 \times 10^{13})(2.76 \times 10^{-3})^2}{2.25 \times 10^8} = 0.196$$

$$[Ag(S_2O_3)_3^{5-}] \equiv [ML_3] = \frac{C_M K_{s3}[L]^3}{D_M} = \frac{(0.200)(1.42 \times 10^{14})(2.76 \times 10^{-3})^3}{2.25 \times 10^8} = 2.65 \times 10^{-3}$$

Therefore, the required minimum $[Na_2S_2O_3]$ is $[L] + [ML] + 2[ML_2] + 3[ML_3] = 0.404$.

EDTA COMPLEXES

9.12 If 0.0200 mol of $CaCl_2$ and 0.0300 mol of Na_2H_2Y ($Y \equiv EDTA$) are added to 1.00 L of a solution buffered to pH 9.400, determine $[Ca^{2+}]$, $[CaY^{2-}]$, and $[Y^{4-}]$.

From Table 9-2, $K_{form} = 5.01 \times 10^{10}$ for the formation of CaY^{2-}. Since there is a stoichiometric excess of EDTA, the concentrations of the principal species are $[CaY^{2-}] = 0.0200$ and $C'_{EDTA} = 0.0100$. From the pH, $[H^+] = 3.98 \times 10^{-10}$; (9.7) then gives $D'_{EDTA} = 6.77 \times 10^{-21}$ and (9.6) gives

$$[Y^{4-}] = \frac{C'_{EDTA}(8.22 \times 10^{-22})}{D'_{EDTA}} = 1.21 \times 10^{-1}$$

Therefore, from (9.4),

$$[Ca^{2+}] \equiv [M] = \frac{[CaY^{2-}]}{K_{form}[Y^{4-}]} = \frac{0.0200}{(5.01 \times 10^{10})(1.21 \times 10^{-1})} = 3.30 \times 10^{-12}$$

9.13 20.00 mL of 0.0200 M $CaCl_2$ is titrated with 0.0200 M Na_2H_2Y ($Y \equiv EDTA$); both solutions are buffered to pH 9.400. Find pCa upon the addition of the following volumes of Na_2H_2Y: (a) 0 mL, (b) 5.00 mL, (c) 18.00 mL, (d) 19.90 mL, (e) 20.00 mL, (f) 21.00 mL.

(a) $pCa = -\log[Ca^{2+}] = -\log 0.0200 = 1.699$

(b) Since K_{form} is very large (Table 9-2), almost all of the EDTA (which is present in a limiting stoichiometric amount) reacts with Ca^{2+}. Thus, $(0.0200)(5.00) = 0.100$ mmol of EDTA reacts to produce 0.100 mmol of CaY^{2-}, giving

$$[Ca^{2+}] = \frac{\{(0.0200)(20.00) - 0.100\} \text{ mmol}}{25.00 \text{ mL}} = 1.20 \times 10^{-2} \qquad \text{or} \qquad pCa = 1.921$$

(c) As in (b), EDTA is the limiting reagent, and the $(0.0200)(18.00) = 0.360$ mmol of EDTA results in 0.360 mmol of CaY^{2-}.

$$[Ca^{2+}] = \frac{\{(0.0200)(20.00) - 0.360\} \text{ mmol}}{38.00 \text{ mL}} = 1.05 \times 10^{-3} \qquad \text{or} \qquad pCa = 2.978$$

(d) As in (b) and (c), EDTA is the limiting reagent, and the $(0.0200)(19.90) = 0.398$ mmol of EDTA results in 0.398 mmol of CaY^{2-}.

$$[Ca^{2+}] = \frac{\{(0.0200)(20.00) - 0.398\} \text{ mmol}}{39.90 \text{ mL}} = 5.01 \times 10^{-5} \qquad \text{or} \qquad pCa = 4.300$$

(e) This is the stoichiometric equivalence point in the titration, and

$$[CaY^{2-}] = \frac{(0.0200)(20.00) \text{ mmol}}{40.00 \text{ mL}} = 1.00 \times 10^{-2}$$

represents the only principal species (except for Cl^- and Na^+); Ca^{2+} and Y^{4-} will be present only in very small amounts. When S mol/L of CaY^{2-} dissociates, there will result $[Ca^{2+}] = S$ and an EDTA concentration $C'_{EDTA} = S$. From (9.6), $[Y^{4-}] = 8.22 \times 10^{-22} C'_{EDTA}/D'_{EDTA}$ and, from Problem 9.12 (where the pH also was 9.400), $D'_{EDTA} = 6.77 \times 10^{-21}$; hence, $[Y^{4-}] = 0.121 S$ and

$$5.01 \times 10^{10} = K_{form} = \frac{[CaY^{2-}]}{[Ca^{2+}][Y^{4-}]} = \frac{1.00 \times 10^{-2}}{(S)(0.121\,S)}$$

Solving, $S = [Ca^{2+}] = 1.28 \times 10^{-6}$, or pCa = 5.891.

(f) This is past the equivalence point in the titration; Ca^{2+} is the limiting reagent, so that

$$[CaY^{2-}] = \frac{(0.0200)(20.00)\ \text{mmol}}{41.00\ \text{mL}} = 9.76 \times 10^{-3}$$

The remaining EDTA concentration is

$$C'_{EDTA} = \frac{\{(0.0200)(21.00) - (0.0200)(20.00)\}\ \text{mmol}}{41.00\ \text{mL}} = 4.88 \times 10^{-4}$$

From (9.6), $[Y^{4-}] = F_4 C'_{EDTA} = (0.121)(4.88 \times 10^{-4}) = 5.90 \times 10^{-5}$, the pH, and with it F_4, being the same as in (e). Therefore,

$$5.01 \times 10^{10} = K_{form} = \frac{[CaY^{2-}]}{[Ca^{2+}][Y^{4-}]} = \frac{9.76 \times 10^{-3}}{[Ca^{2+}](5.90 \times 10^{-5})}$$

whence $[Ca^{2+}] = 3.30 \times 10^{-9}$, or pCa = 8.482.

CONDITIONAL FORMATION CONSTANTS

9.14 Evaluate the conditional formation constant for the reaction

$$Cd^{2+} + Y^{4-} \rightarrow CdY^{2-}$$

in a pH 9.260 buffer solution that has $[NH_4^+] = [NH_3] = 1.00$.

For this solution, $[H^+] = 5.50 \times 10^{-10}$. From (9.7) and the K-values for EDTA, $D'_{EDTA} = 9.05 \times 10^{-21}$; hence,

$$F_4 = \frac{8.22 \times 10^{-22}}{D'_{EDTA}} = 9.08 \times 10^{-2}$$

From (9.2),

$$D_M = 1 + (447)(1.00) + (5.63 \times 10^4)(1.00)^2 + (1.55 \times 10^6)(1.00)^3 + (1.32 \times 10^7)(1.00)^4 = 1.48 \times 10^7$$

from which

$$F_{0M} = \frac{1}{D_M} = 6.76 \times 10^{-8}$$

Therefore, from (9.8) and Table 9-2,

$$K'_{form} = (6.76 \times 10^{-8})(9.08 \times 10^{-2})(3.89 \times 10^{16}) = 2.39 \times 10^8$$

9.15 30.00 mL of $4.00 \times 10^{-3}\ M$ $CdCl_2$ is titrated with $6.00 \times 10^{-3}\ M$ Na_2H_2Y; both solutions are buffered to pH 9.260, with $[NH_4^+] = [NH_3] = 1.00$. Find pCd upon the addition of the following volumes of Na_2H_2Y: (a) 0 mL, (b) 5.00 mL, (c) 19.00 mL, (d) 20.00 mL, (e) 21.00 mL, (f) 25.00 mL.

(a) From (9.3), $[Cd^{2+}] \equiv [M] = C_M/D_M$. From Problem 9.14, $D_M = 1.48 \times 10^7$, so that

$$[Cd^{2+}] = \frac{4.00 \times 10^{-3}}{1.48 \times 10^7} = 2.70 \times 10^{-10} \qquad \text{or} \qquad pCd = 9.568$$

(b) Since K_{form} is very large and EDTA is the limiting reagent, the $(6.00 \times 10^{-3})(5.00) = 3.00 \times 10^{-2}$ mmol of added EDTA results in

$$[CdY^{2-}] = \frac{3.00 \times 10^{-2}\ \text{mmol}}{35.00\ \text{mL}} = 8.57 \times 10^{-4}$$

and

$$C_M = \frac{\{(4.00 \times 10^{-3})(30.00) - (3.00 \times 10^{-2})\}\ \text{mmol}}{35.00\ \text{mL}} = 2.57 \times 10^{-3}$$

From (9.3) and Problem 9.14,

$$[Cd^{2+}] \equiv [M] = \frac{C_M}{D_M} = \frac{2.57 \times 10^{-3}}{1.48 \times 10^7} = 1.74 \times 10^{-10} \qquad \text{or} \qquad pCd = 9.760$$

(c)   The $(6.00 \times 10^{-3})(19.00) = 0.114$ mmol of added EDTA results in

$$[CdY^{2-}] = \frac{0.114 \text{ mmol}}{49.00 \text{ mL}} = 2.33 \times 10^{-3}$$

and

$$C_M = \frac{\{(4.00 \times 10^{-3})(30.00) - 0.114\} \text{ mmol}}{49.00 \text{ mL}} = 1.22 \times 10^{-4}$$

From (9.3) and Problem 9.19,

$$[Cd^{2+}] \equiv [M] = \frac{C_M}{D_M} = \frac{1.22 \times 10^{-4}}{1.48 \times 10^7} = 8.24 \times 10^{-12} \qquad \text{or} \qquad pCd = 11.084$$

(d)   This is the stoichiometric equivalence point, and

$$[CdY^{2-}] = \frac{(4.00 \times 10^{-3})(30.00) \text{ mmol}}{50.00 \text{ mL}} = 2.40 \times 10^{-3}$$

When x mol/L of CdY^{2-} dissociates, there will result $C_M' = C_{EDTA}' = x$. Therefore, from (9.8) and Problem 9.14, the conditional formation constant is

$$K_{form}' = 2.39 \times 10^8 = \frac{[CdY^{2-}]}{C_M' C_{EDTA}'} = \frac{[CdY^{2-}]}{x^2}$$

Assuming that $x \ll 2.40 \times 10^{-3}$ (i.e., a very small % dissociation of CdY^{2-}), there results

$$2.39 \times 10^8 = \frac{2.40 \times 10^{-3}}{x^2} \qquad \text{or} \qquad x = 3.17 \times 10^{-6}$$

(the assumption is justified). From Problem 9.14, $F_{0M} = 6.76 \times 10^{-8}$, and so

$$[Cd^{2+}] \equiv [M] = F_{0M} C_M' = F_{0M} x = 2.14 \times 10^{-13}$$

or pCd = 12.669. [The reader should compare this evaluation of pCd via K_{form}' with the evaluation of pCa in Problem 9.13(e), where K_{form} was used.]

(e)   This volume corresponds to an excess of EDTA. Therefore,

$$[CdY^{2-}] = \frac{(4.00 \times 10^{-3})(30.00) \text{ mmol}}{51.00 \text{ mL}} = 2.35 \times 10^{-3}$$

$$C_{EDTA}' = \frac{\{(6.00 \times 10^{-3})(21.00) - (4.00 \times 10^{-3})(30.00)\} \text{ mmol}}{51.00 \text{ mL}} = 1.18 \times 10^{-4}$$

From (9.8) and Problem 9.14,

$$2.38 \times 10^8 = K_{form}' = \frac{[CdY^{2-}]}{C_M' C_{EDTA}'} = \frac{2.35 \times 10^{-3}}{C_M'(1.18 \times 10^{-4})} \qquad \text{or} \qquad C_M' = 8.33 \times 10^{-8}$$

Then, using Problem 9.14,

$$[Cd^{2+}] \equiv [M] = F_{0M} C_M' = 5.63 \times 10^{-15} \qquad \text{or} \qquad pCd = 14.249$$

(f)   EDTA is again in excess. Proceeding as in (e):

$$[CdY^{2-}] = \frac{(4.00 \times 10^{-3})(30.00) \text{ mmol}}{55.00 \text{ mL}} = 2.18 \times 10^{-3}$$

$$C_{EDTA}' = \frac{\{(6.00 \times 10^{-3})(25.00) - (4.00 \times 10^{-3})(30.00)\} \text{ mmol}}{55.00 \text{ mL}} = 5.45 \times 10^{-4}$$

$$2.39 \times 10^8 = K_{form}' = \frac{2.18 \times 10^{-3}}{C_M'(5.45 \times 10^{-4})} \qquad \text{or} \qquad C_M' = 1.67 \times 10^{-8}$$

$$[Cd^{2+}] \equiv [M] = F_{0M} C_M' = 1.13 \times 10^{-15} \qquad \text{or} \qquad pCd = 14.947$$

Supplementary Problems

COMMON-ION EFFECT

9.16 Pb^{2+} forms four different complexes with I^-: PbI^+, $PbI_2(aq)$, PbI_3^-, and PbI_4^{2-}, Give the equation for the solubility, S, of $PbI_2(s)$ $(K_{sp} = 7.10 \times 10^{-9})$ as a function of $[I^-]$.
    Ans. $S = (K_{sp}/[I^-]^2) + (K_1 K_{sp}/[I^-]) + K_1 K_2 K_{sp} + K_1 K_2 K_3 K_{sp}[I^-] + K_1 K_2 K_3 K_4 K_{sp}[I^-]^2$, where each K_n is a stepwise formation constant.

9.17   Using Problem 9.16, with $K_1 = 18.2$, $K_2 = 34.7$, $K_3 = 4.17$, and $K_4 = 3.16$, determine the solubility of $PbI_2(s)$ in a 0.400 M NaI solution.   *Ans.* $S = 2.18 \times 10^{-5}$ M

COMPLEXATION

9.18   What will be the concentrations of the copper species if 1.50 mmol of $CuSO_4$ is added to 1.00 L of 1.00 M NH_4OH. Assume that $[NH_3] \equiv [L] = 1.00$.
    Ans. $[Cu^{2+}] = 7.11 \times 10^{-17}$, $[Cu(NH_3)^{2+}] = 1.45 \times 10^{-12}$, $[Cu(NH_3)_2^{2+}] = 6.79 \times 10^{-9}$, $[Cu(NH_3)_3^{2+}] = 7.46 \times 10^{-6}$, $[Cu(NH_3)_4^{2+}] = 1.49 \times 10^{-3}$

9.19   A solution has $[NH_4OH] \equiv [L] = 0.600$ and a total dissolved silver concentration of 3.25×10^{-3} M. How large can $[I^-]$ be without formation of any $AgI(s)$? $K_{sp} = 8.31 \times 10^{-17}$.   *Ans.* 1.49×10^{-7}

9.20   25.00 mL of 0.200 M $AgNO_3$ is mixed with 25.00 mL of (*a*) 0.300 M, (*b*) 0.400 M, (*c*) 0.500 M, NaCN. Neglecting hydrolysis CN^-, compute the equilibrium $[Ag^+]$, $[Ag(CN)_2^-]$, and $[CN^-]$.
    Ans.   (*a*) 0.025, 0.0750, 2.06×10^{-10}; (*b*) 7.07×10^{-8}, 0.100, 1.41×10^{-7}; (*c*) 5.65×10^{-19}, 0.100, 0.0500

EDTA COMPLEXES

9.21   Find F_4 for EDTA in a solution buffered to pH 8.400.   *Ans.* 1.36×10^{-2}

9.22   25.00 mL of 0.0200 M $MnCl_2$ is titrated with 0.0250 M Na_2H_2Y (Y $\equiv$ EDTA); both solutions are buffered to pH 8.400 (refer to Problem 9.21). Determine pMn after the addition of (*a*) 5.00 mL, (*b*) 18.00 mL, (*c*) 20.00 mL, (*d*) 25.00 mL, (*e*) 35.00 mL, of Na_2H_2Y.
    Ans.   (*a*) 1.903; (*b*) 2.93; (*c*) 6.834; (*d*) 11.111; (*e*) 11.588

CONDITIONAL FORMATION CONSTANTS

9.23   Calculate D'_{EDTA}, D_M, and K'_{form} for the reaction $Cu^{2+} + Y^{4-} \rightarrow CuY^{2-}$ (Y $\equiv$ EDTA) in a pH 9.000 buffer solution that has $[NH_4^+] = 1.000$ and $[NH_3] = 0.549$.   *Ans.* 2.32×10^{-15}, 1.93×10^{12}, 1.13

9.24   Find $[Cu^{2+}]$ in a solution prepared by mixing 25.00 mL of 0.200 M $CuSO_4$ with 25.00 mL of 0.200 M Na_2H_2Y (Y $\equiv$ EDTA), if each solution was buffered to pH 9.000 and at equilibrium had $[NH_3] = 0.549$ (cf. Problem 9.23).   *Ans.* 1.55×10^{-13}

Chapter 10

Electrochemical Cells

10.1 GALVANIC CELLS

There are two types of electrochemical cells, both of which involve oxidation-reduction (redox) reactions. *Galvanic cells* can spontaneously release electric energy; batteries are examples of this type. *Electrolytic cells*, on the other hand, must be fed electric energy from an outside voltage source; the electrolysis of water would involve a cell of this type. For either type, the electrode at which *oxidation* occurs is defined as the *anode*, and the electrode at which *reduction* occurs is defined as the *cathode*.

In the spontaneous redox reaction characteristic of a galvanic cell, electrons are released at the site of oxidation. Therefore, *in galvanic cells, the anode is negative and the cathode is positive*. The opposite is true in electrolytic cells, where the external voltage must force electrons in at the reduction site in order for the redox reaction to take place. We will be concerned in this chapter solely with galvanic cells.

We will use a standard notation for cells, which sometimes will indicate that the cell reaction is nonspontaneous as written. This does not imply that the cell is electrolytic rather than galvanic; it means simply that the cell connections must be reversed to obtain the proper spontaneous reaction.

10.2 CELL THERMODYNAMICS

The free-energy change associated with an electrochemical process in a cell at constant pressure and temperature is given by

$$\Delta G_{P,T} = -n\mathscr{F}\mathscr{E} \tag{10.1}$$

Here, n is the number of moles of electrons transferred in the redox reaction (i.e., the number of equivalents of reaction); $\mathscr{E}$ is the voltage of the cell; and $\mathscr{F}$ is the Faraday constant, as given in Table 1-3. If $\mathscr{E} > 0$, then $\Delta G_{P,T} < 0$ and the cell reaction as written is spontaneous. If $\mathscr{E} < 0$, $\Delta G_{P,T} > 0$ and the reverse of the written cell reaction is spontaneous. If $\mathscr{E} = \Delta G_{P,T} = 0$ (e.g., a dead battery), the cell reaction is at equilibrium.

For the hypothetical reaction that involves each substance in its standard state,

$$\Delta G_{P,T}^{\circ} = -n\mathscr{F}\mathscr{E}^{\circ} \tag{10.2}$$

Substituting (10.1) and (10.2), along with the numerical values of $\mathscr{F}$ and R and room temperature $T = 298$ K, into (3.10) yields, after conversion to common logarithms, the *Nernst equation*:

$$\mathscr{E} = \mathscr{E}^{\circ} - \frac{0.0592}{n} \log Q \tag{10.3}$$

or, analogous to (3.12),

$$\mathscr{E} = -\frac{0.0592}{n} \log \frac{Q}{K} \tag{10.4}$$

The Nernst equation is the basic relation between the cell voltage and the chemical activities of the cell substances.

It is not possible to measure the voltage of an isolated reduction half-reaction or an isolated oxidation half-reaction. However, if a single half-reaction is chosen as a reference point, it is possible to determine the voltage of any other half-reaction relative to the reference reaction. By agreement, the standard-state reduction of H^+ at 25 °C has a voltage of exactly zero; this is the meaning of the entry 0.000 in the twentieth line of Table 10-1.

166

Table 10-1

Reduction Half-Reaction	$_rE°$, V	Reduction Half-Reaction	$_rE°$, V
$Ag^+ + e^- \to Ag(s)$	+0.800	$Mn^{2+} + 2e^- \to Mn(s)$	−1.18
$AgCl(s) + e^- \to Ag(s) + Cl^-$	+0.222	$MnO_4^- + e^- \to MnO_4^{2-}$	+0.564
$AgBr(s) + e^- \to Ag(s) + Br^-$	+0.073	$MnO_4^- + 4H^+ + 3e^- \to MnO_2(s) + 2H_2O$	+1.695
$AgI(s) + e^- \to Ag(s) + I^-$	−0.151	$MnO_4^- + 8H^+ + 5e^- \to Mn^{2+} + 4H_2O$	+1.51
$Ag_2O(s) + H_2O + 2e^- \to 2Ag(s) + 2OH^-$	+0.342	$Pb^{2+} + 2e^- \to Pb(s)$	−0.126
$Ag_2S(s) + 2e^- \to 2Ag(s) + S^{2-}$	−0.71	$PbSO_4(s) + 2e^- \to Pb(s) + SO_4^{2-}$	−0.356
$Cd^{2+} + 2e^- \to Cd(s)$	−0.403	$Sn^{2+} + 2e^- \to Sn(s)$	−0.136
$Ce^{4+} + e^- \to Ce^{3+}$ (1 M H_2SO_4)	+1.44	$Sn^{4+} + 2e^- \to Sn^{2+}$	+0.154
$Cr_2O_7^{2-} + 14H^+ + 6e^- \to 2Cr^{3+} + 7H_2O$	+1.33	$Ti^{3+} + e^- \to Ti^{2+}$	−0.368
$Cr^{3+} + e^- \to Cr^{2+}$	−0.41	$Tl^{3+} + 2e^- \to Tl^+$	+1.25
$Cr^{2+} + 2e^- \to Cr(s)$	−0.91	$Tl^{3+} + 2e^- \to Tl^+$ (1 M $HClO_4$)	+1.26
$Co^{3+} + e^- \to Co^{2+}$	+1.842	$Tl^+ + e^- \to Tl(s)$	−0.336
$Co^{2+} + 2e^- \to Co(s)$	−0.277	$UO_2^{2+} + 4H^+ + 2e^- \to U^{4+} + 2H_2O$	+0.334
$Cu^{2+} + 2e^- \to Cu(s)$	+0.337	$UO_2^+ + 4H^+ + e^- \to U^{4+} + 2H_2O$	+0.55
$Cu^{2+} + e^- \to Cu^+$	+0.153	$U^{4+} + e^- \to U^{3+}$	−0.61
$Fe^{3+} + e^- \to Fe^{2+}$	+0.771	$VO_2^+ + 2H^+ + e^- \to VO^{2+} + H_2O$	+1.000
$Fe^{3+} + e^- \to Fe^{2+}$ (1 M H_2SO_4)	+0.68	$VO^{2+} + 2H^+ + e^- \to V^{3+} + H_2O$	+0.361
$Fe^{3+} + e^- \to Fe^{2+}$ (1 M $HClO_4$)	+0.767	$V^{3+} + e^- \to V^{2+}$	−0.255
$Fe^{2+} + 2e^- \to Fe(s)$	−0.440	$V^{2+} + 2e^- \to V(s)$	−1.18
$2H^+ + 2e^- \to H_2(g)$	0.000	$Zn^{2+} + 2e^- \to Zn(s)$	−0.763
$Hg^{2+} + 2e^- \to Hg(l)$	+0.854	$Zn(NH_3)_4^{2+} + 2e^- \to Zn(s) + 4NH_3$	−1.04
$2Hg^{2+} + 2e^- \to Hg_2^{2+}$	+0.919		

We shall distinguish between *full-cell voltages* (corresponding to full redox reactions, whose equations contain no e^--terms) and *half-cell voltages* (corresponding to cell half-reactions) by writing $\mathcal{E}$ for the former and E for the latter. If standard-state processes are involved, then the symbols will be $\mathcal{E}°$ and $E°$, respectively. In addition, if a half-cell voltage is for a reduction half-reaction, we will use the symbol $_rE$, whereas the voltage for an oxidation half-reaction will be symbolized as $_oE$. Table 10-1 lists some standard-state reduction potentials, $_rE°$, at 25 °C.

10.3 COMBINATION OF HALF-CELL REACTIONS

When two electrochemical reactions, both occurring at the same pressure and temperature, are added, their $\Delta G_{P,T}$-values add (by the first law of thermodynamics). This may or may not mean that their voltages also add. We consider three general cases.

A Half-Reaction Is Doubled (Tripled, Etc.)

Using a prime to refer to the sum of k identical half-reactions, we have from (*10.1*):

$$-(kn)\mathcal{F}E' = \Delta G'_{P,T} = k(\Delta G_{P,T}) = k(-n\mathcal{F}E)$$

whence $E' = E$.

Two Different Half-Reactions Are Added, Yielding a Third

In this case, (10.1) gives:

$$-n_3 \mathscr{F} E_3 = \Delta G_3 = \Delta G_1 + \Delta G_2 = -n_1 \mathscr{F} E_1 - n_2 \mathscr{F} E_2$$

where $n_3 = n_1 + n_2$ if the component half-reactions are both oxidation or both reduction, and $n_3 = |n_1 - n_2|$ if one is oxidation and the other is reduction. Hence,

$$E_3 = \frac{n_1}{n_3} E_1 + \frac{n_2}{n_3} E_2 \qquad (10.5)$$

that is, the voltage of the resultant half-reaction is a weighted sum of the voltages of the component half-reactions.

An Oxidation Half-Reaction and a Reduction Half-Reaction Combine to Give a Balanced Redox Reaction (Involving No Free Electrons)

Let the redox reaction (voltage $\mathscr{E}_3$) be expressible as k_o times the oxidation half-reaction plus k_r times the reduction half-reaction. By the above, the voltages of the two summands are $_oE_1$ (the voltage of a *single* oxidation half-reaction) and $_rE_2$ (the voltage of a *single* reduction half-reaction). Then, by (10.1),

$$-n_3 \mathscr{F} \mathscr{E}_3 = \Delta G_3 = \Delta G_1 + \Delta G_2$$
$$= -(k_o n_o) \mathscr{F}_o E_1 - (k_r n_r) \mathscr{F}_r E_2$$

But, since the redox reaction is balanced,

$$k_o n_o = k_r n_r = n_3$$

and so

$$\mathscr{E}_3 = {}_oE_1 + {}_rE_2 = -{}_rE_1 + {}_rE_2 \qquad (10.6)$$

In this case, the component voltages simply add. The second form of (10.6) follows from the fact that the $\Delta G_{P,T}$-values for a forward and for a backward reaction differ only in algebraic sign. This form is convenient in that it involves only reduction potentials, which may often be read directly from tables.

10.4 CELL NOTATION

In the conventional diagram of a cell, the *oxidation* half-cell is shown on the *left* and the *reduction* half-cell is shown on the *right*. Thus, a cell involving the zinc oxidation half-reaction,

$$Zn(s) \rightarrow Zn^{2+}(x\,M) + 2e^-$$

and the silver reduction half-reaction,

$$Ag^+(y\,M) + e^- \rightarrow Ag(s)$$

would be depicted as:

$$Zn(s) \mid Zn^{2+}(x\,M), Ag^+(y\,M) \mid Ag(s)$$

where vertical lines denote phase boundaries between solid zinc and the solution and between the solution and solid silver. Commas are used to separate species existing in the same phase.

If a half-cell reaction involves only ions, then some type of metal contact with that half-cell is necessary. This metal electrode is usually a platinum wire, as in the following cell, which has

$$Sn^{2+} \rightarrow Sn^{4+} + 2e^-$$

as the oxidation half-reaction and

$$Ce^{4+} + e^- \rightarrow Ce^{3+}$$

as the reduction half-reaction:

$$Pt \mid Sn^{2+} (a\ M),\ Sn^{4+} (b\ M),\ Ce^{4+} (c\ M),\ Ce^{3+} (d\ M) \mid Pt$$

One problem with both of the indicated cells is that the redox reaction will begin as soon as the reagents and metals are put into contact with one another, thus making it difficult to measure the voltage of the reaction at the specified reactant concentrations. To prevent such reaction from taking place, a *salt bridge* is interposed between the oxidation half-cell and the reduction half-cell. The salt bridge is usually a solution of KCl in a tube that joins the two half-cells and is indicated in the shorthand cell notation as a pair of vertical lines. For instance, the previous cell with a salt bridge would be shown as

$$Pt \mid Sn^{2+} (a\ M),\ Sn^{4+} (b\ M) \parallel Ce^{4+} (c\ M),\ Ce^{3+} (d\ M) \mid Pt$$

10.5 THE S.C.E. CELL

One of the most useful half-cells is the *saturated calomel electrode* (S.C.E.). It is based on the reduction of mercury (II) in the form of $Hg_2Cl_2(s)$ while present in a saturated KCl solution:

$$Hg_2Cl_2(s) + 2K^+ (sat\ KCl) + 2e^- \xrightarrow{25\ °C} 2Hg(l) + 2KCl(s) \qquad (_rE = +0.242\ V) \qquad (10.7)$$

Since all chemical species are constant in the S.C.E., it has a fixed value of $_rE$ at 25 °C. (This is *not* $_rE°$, since it is not a standard-state set of concentrations.)

Solved Problems

HALF-CELL POTENTIALS

10.1 Use Table 10-1 to calculate the standard-state voltage for the reduction half-reaction

$$Cr^{3+} + 3e^- \rightarrow Cr(s)$$

This reaction is the sum of the two reduction half-reactions

$$\begin{aligned} Cr^{3+} + e^- &\rightarrow Cr^{2+} \qquad (_rE_1° = -0.41\ V) \\ Cr^{2+} + 2e^- &\rightarrow Cr(s) \qquad (_rE_2° = -0.91\ V) \end{aligned}$$

Then, by (10.5), $_rE_3° = [(1)(-0.41) + (2)(-0.91)]/3 = -0.74\ V$.

10.2 Use Table 10-1 to calculate the standard-state voltage for the reduction half-reaction

$$MnO_4^- + 8H^+ + 7e^- \rightarrow Mn(s) + 4H_2O$$

This reaction is the sum of the two reduction half-reactions

$$\begin{aligned} MnO_4^- + 8H^+ + 5e^- &\rightarrow Mn^{2+} + 4H_2O \qquad (_rE_1° = \ \ 1.51\ V) \\ Mn^{2+} + 2e^- &\rightarrow Mn(s) \qquad (_rE_2° = -1.18\ V) \end{aligned}$$

Then, by (10.5), $_rE_3° = [(5)(1.51) + (2)(-1.18)]/7 = +0.741\ V$.

10.3   Use Table 10-1 to calculate the standard-state voltage for the reduction half-reaction

$$VO_2^+ + 4H^+ + 5e^- \rightarrow V(s) + 2H_2O$$

$$
\begin{array}{ll}
VO_2^+ + 2H^+ + e^- \rightarrow VO^{2+} + H_2O & (_rE_1^\circ = 1.000 \text{ V}) \\
VO^{2+} + 2H^+ + e^- \rightarrow V^{3+} + H_2O & (_rE_2^\circ = 0.361 \text{ V}) \\
V^{3+} + e^- \rightarrow V^{2+} & (_rE_3^\circ = -0.255 \text{ V}) \\
V^{2+} + 2e^- \rightarrow V(s) & (_rE_4^\circ = -1.18 \text{ V})
\end{array}
$$

Then, by *(10.5)*, $_rE_5^\circ = [(1)(1.000) + (1)(0.361) + (1)(-0.255) + (2)(-1.18)]/5 = -0.251$ V.

10.4   Use Table 10-1 to calculate the standard-state voltage for the reduction half-reaction

$$MnO_4^{2-} + 4H^+ + 2e^- \rightarrow MnO_2(s) + 2H_2O$$

This reaction may be written as the sum of a reduction half-reaction and an oxidation half-reaction,

$$
\begin{array}{ll}
MnO_4^- + 4H^+ + 3e^- \rightarrow MnO_2(s) + 2H_2O & (_rE_1^\circ = 1.695 \text{ V}) \\
MnO_4^{2-} \rightarrow MnO_4^- + e^- & (_oE_2^\circ = -0.564 \text{ V})
\end{array}
$$

Then, by *(10.5)*, $_rE_3^\circ = [(3)(1.695) + (1)(-0.564)]/2 = 2.261$ V.

EQUILIBRIUM CONSTANTS

10.5   Use Table 10-1 to calculate the equilibrium constant for

(*a*) $PbSO_4(s) \rightarrow Pb^{2+} + SO_4^{2-}$ (*b*) $Sn^{2+} + 2Fe^{3+} \rightarrow Sn^{4+} + 2Fe^{2+}$

From the Nernst equation, *(10.3)*, we have the following general relation between the standard-state potential and the equilibrium constant:

$$\mathscr{E}^\circ = \frac{0.0592}{n} \log K \tag{10.8}$$

(*a*)   The reaction can be written as the sum of the oxidation and reduction half-reactions

$$
\begin{array}{ll}
Pb(s) \rightarrow Pb^{2+} + 2e^- & (_oE^\circ = +0.126 \text{ V}) \\
PbSO_4(s) + 2e^- \rightarrow Pb(s) + SO_4^{2-} & (_rE^\circ = -0.356 \text{ V})
\end{array}
$$

whence, by *(10.6)*, $\mathscr{E}^\circ = 0.126 - 0.356 = -0.230$ V. Then, by *(10.8)*,

$$-0.230 = \frac{0.0592}{2} \log K_{sp} \qquad \text{or} \qquad K_{sp} = 1.7 \times 10^{-8}$$

(*b*)   The reaction can be written as the sum of the oxidation and reduction half-reactions

$$
\begin{array}{ll}
Sn^{2+} \rightarrow Sn^{4+} + 2e^- & (_oE^\circ = -0.154 \text{ V}) \\
2Fe^{3+} + 2e^- \rightarrow 2Fe^{2+} & (_rE^\circ = +0.771 \text{ V})
\end{array}
$$

whence, by *(10.6)*, $\mathscr{E}^\circ = -0.154 + 0.771 = 0.617$ V. Then, by *(10.8)*,

$$0.617 = \frac{0.0592}{2} \log K \qquad \text{or} \qquad K = 7 \times 10^{20}$$

10.6   According to Table 9-1, the equilibrium constant for $Ag^+ + 2CN^- \rightarrow Ag(CN)_2^-$ is $K = 7.08 \times 10^{19}$. Find the standard-state reduction potential for the half-reaction

$$Ag(CN)_2^- + e^- \rightarrow Ag(s) + 2CN^-$$

Using Table 10-1 and *(10.8)*, we have:

Half-Reaction	Voltage
$Ag(s) + 2CN^- \rightarrow Ag(CN)_2^- + e^-$	$_oE_1^\circ = x$
$Ag^+ + e^- \rightarrow Ag(s)$	$_rE_2^\circ = 0.800$ V

$$Ag^+ + 2CN^- \rightarrow Ag(CN)_2^- \qquad \mathscr{E}_3^\circ = (0.0592/1)\log(7.08 \times 10^{19}) = 1.18 \text{ V}$$

By (10.6), $1.18 = x + 0.800$, or $x = 0.38$ V. Hence,

$$_rE_1^\circ = -x = -0.38 \text{ V}$$

10.7 Use data of Table 10-1 to calculate the solubility product of $AgOH(s)$. In water, $AgOH(s)$ dissociates to $Ag_2O(s)$ and H_2O, so that the solubility-product reaction is

$$\tfrac{1}{2}Ag_2O(s) + \tfrac{1}{2}H_2O \rightarrow Ag^+ + OH^-$$

This reaction can be given as the sum of the two half-reactions

$$Ag(s) \rightarrow Ag^+ + e^- \qquad (_oE^\circ = -0.800 \text{ V})$$
$$\tfrac{1}{2}Ag_2O(s) + \tfrac{1}{2}H_2O + e^- \rightarrow Ag(s) + OH^- \qquad (_rE^\circ = +0.342 \text{ V})$$

From (10.6) and (10.8),

$$\mathscr{E}^\circ = -0.800 + 0.342 = -0.458 = \frac{0.0592}{1}\log K_{sp}$$

or $K_{sp} = 1.8_3 \times 10^{-8}$.

10.8 Use data of Table 10-1 to calculate the summed formation constant K_{s4} for $Zn(NH_3)_4^{2+}$.

The summed formation reaction $Zn^{2+} + 4NH_3 \rightarrow Zn(NH_3)_4^{2+}$ can be decomposed into the two half-reactions

$$Zn(s) + 4NH_3 \rightarrow Zn(NH_3)_4^{2+} + 2e^- \qquad (_oE^\circ = +1.04 \text{ V})$$
$$Zn^{2+} + 2e^- \rightarrow Zn(s) \qquad (_rE^\circ = -0.763 \text{ V})$$

From (10.6) and (10.8),

$$\mathscr{E}^\circ = 1.04 - 0.763 = 0.28 = \frac{0.0592}{2}\log K_{s4}$$

or $K_{s4} = 2._9 \times 10^9$.

10.9 Use data of Table 10-1 to calculate the equilibrium constant for the reaction

$$3Sn(s) + 2Cr_2O_7^{2-} + 28H^+ \rightarrow 3Sn^{4+} + 4Cr^{3+} + 14H_2O$$

The reaction may be obtained as the reduction half-reaction

$$2Cr_2O_7^{2-} + 28H^+ + 12e^- \rightarrow 4Cr^{3+} + 14H_2O \qquad (_rE^\circ = 1.33 \text{ V})$$

plus the sum of the two oxidation half-reactions

$$3Sn(s) \rightarrow 3Sn^{2+} + 6e^- \qquad (_oE_1^\circ = +0.136 \text{ V})$$
$$3Sn^{2+} \rightarrow 3Sn^{4+} + 6e^- \qquad (_oE_2^\circ = -0.154 \text{ V})$$

By (10.5), $_oE_3^\circ = [6(0.136) + 6(-0.154)]/12 = -0.009$ V for the overall oxidation half-reaction. Hence, the standard-state potential of the redox reaction is given by (10.6) as

$$\mathscr{E}^\circ = -0.009 + 1.33 = 1.32 \text{ V}$$

and, from (10.8),

$$1.32 = \frac{0.0592}{12}\log K \qquad \text{or} \qquad K = 10^{268}$$

10.10 Use data of Table 10-1 to calculate the equilibrium concentrations of all ions in an ideal solution prepared by mixing 25.00 mL of 0.0400 M Fe^{2+} with 25.00 mL of 0.0600 M Ce^{6+}. Both solutions are also 1 M H_2SO_4.

The redox reaction, $Fe^{2+} + Ce^{4+} \rightarrow Fe^{3+} + Ce^{3+}$, can be given as the sum of the half-reactions

$$Fe^{2+} \rightarrow Fe^{3+} + e^- \qquad (_oE^\circ = -0.68 \text{ V})$$
$$Ce^{4+} + e^- \rightarrow Ce^{3+} \qquad (_rE^\circ = +1.44 \text{ V})$$

From (10.6) and (10.8),

$$\mathscr{E}^\circ = -0.68 + 1.44 = 0.76 = \frac{0.0592}{1} \log K$$

so that $K = 6._9 \times 10^{12}$. The $(0.0400)(25.00) = 1.00$ mmol of Fe^{2+} is less than the $(0.0600)(25.00) = 1.50$ mmol of Ce^{4+}, and since K is so large, virtually all of the Fe^{2+} will react. Writing the equilibrium concentrations as

$$[Fe^{2+}] = x \qquad [Ce^{4+}] = \frac{1.50 - 1.00}{50.00} = 0.010 \qquad [Fe^{3+}] = [Ce^{3+}] = \frac{1.00}{50.00} = 0.0200$$

we have

$$K = 6._9 \times 10^{12} = \frac{[Fe^{3+}][Ce^{3+}]}{[Fe^{2+}][Ce^{4+}]} = \frac{(0.0200)^2}{x(0.010)}$$

or $x = [Fe^{2+}] = 5._8 \times 10^{-15}$.

10.11 Use data of Table 10-1 to calculate the equilibrium concentrations of all ions in an ideal solution prepared by mixing 25.00 mL of 0.100 M Tl^+ with 25.00 mL of 0.200 M Co^{3+}.

The redox reaction, $Tl^+ + 2Co^{3+} \rightarrow Tl^{3+} + 2Co^{2+}$, can be given as the sum of the half-reactions

$$Tl^+ \rightarrow Tl^{+3} + 2e^- \qquad (_oE^\circ = -1.25 \text{ V})$$
$$2Co^{3+} + 2e^- \rightarrow 2Co^{2+} \qquad (_rE^\circ = +1.842 \text{ V})$$

From (10.6) and (10.8),

$$\mathscr{E}^\circ = -1.25 + 1.842 = +0.59 = \frac{0.0592}{2} \log K$$

so that $K = 10^{20}$. The $(0.100)(25.00) = 2.50$ mmol of Tl^+ and the $(0.200)(25.00) = 5.00$ mmol of Co^{3+} are in stoichiometric ratio. Since K is very large, the reactants will be converted almost entirely to products; at equilibrium we will have:

$$[Tl^{3+}] = \frac{2.50}{50.00} = 0.0500 \qquad [Co^{2+}] = \frac{5.00}{50.00} = 0.100 \qquad [Tl^+] = x \qquad [Co^{3+}] = 2x$$

where x is very small as compared to 0.0500 or 0.100. Thus,

$$K = 10^{20} = \frac{[Tl^{3+}][Co^{2+}]^2}{[Tl^+][Co^{3+}]^2} = \frac{(0.0500)(0.100)^2}{(x)(2x)^2}$$

Solving, $x = 10^{-8} = [Tl^+]$ and $[Co^{3+}] = 2x = 2(10^{-8})$, but the 2 is *not* a significant figure.

10.12 Using the value $K = 10^{20}$ (Problem 10.11), calculate the equilibrium concentrations of all ions in a solution prepared by mixing 25.00 mL of 0.100 M Tl^+ with 25.00 mL of 0.100 M Co^{3+}. Include corrections for non-ideality; activity coefficients are

$$f_{Tl^+} = 0.772 \qquad f_{Tl^{3+}} = 0.114 \qquad f_{Co^{2+}} = 0.427 \qquad f_{Co^{3+}} = 0.131$$

The initial solution contains 2.50 mmol each of Tl^+ and Co^{3+}, so that the Tl^+ is in stoichiometric excess. Since K is very large, there are formed 1.25 mmol of Tl^{3+} and 2.50 mmol of Co^{2+}. The equilibrium concentrations will be:

$$[Tl^{3+}] = \frac{1.25}{50.00} = 0.0250 \qquad [Co^{2+}] = \frac{2.50}{50.00} = 0.0500 \qquad [Tl^+] = \frac{2.50 - 1.25}{50.00} = 0.0250$$

and $[Co^{3+}] = x$.

$$10^{20} = \frac{a_{Tl^{3+}}(a_{Co^{2+}})^2}{a_{Tl^+}(a_{Co^{3+}})^2} = \frac{(0.114)(0.0250)\{(0.427)(0.0500)\}^2}{(0.772)(0.0250)\{(0.131)(x)\}^2}$$

Solving, $x = 6(10^{-12})$, where the 6 is *not* a significant figure.

CELL VOLTAGES

10.13 Use data of Table 10-1 to calculate the voltage, $\mathscr{E}$, of the cell

$$Fe(s) \mid Fe^{2+}(0.300M) \parallel Sn^{2+}(0.500M) \mid Sn(s)$$

The cell involves the oxidation of $Fe(s)$ and the reduction of Sn^{2+}, the cell reaction being

$$Fe(s) + Sn^{2+} \to Fe^{2+} + Sn(s)$$

From (*10.6*) and Table 10-1,

$$\mathscr{E}° = {}_oE°_{Fe/Fe^{2+}} + {}_rE°_{Sn^{2+}/Sn} = +0.440 - 0.136 = 0.304 \text{ V}$$

Therefore, by (*10.3*),

$$\mathscr{E} = \mathscr{E}° - \frac{0.0592}{2} \log \frac{[Fe^{2+}]}{[Sn^{2+}]} = 0.304 - \frac{0.0592}{2} \log \frac{0.300}{0.500} = 0.311 \text{ V}$$

10.14 Use data of Table 10-1 to calculate the voltage, $\mathscr{E}$, of the cell

$$Mn(s) \mid Mn^{2+}(0.400M) \parallel Ag^+(0.150M) \mid Ag(s)$$

The cell involves the oxidation of $Mn(s)$ and the reduction of Ag^+, the cell reaction being

$$Mn(s) + 2Ag^+ \to Mn^{2+} + 2Ag(s)$$

From (*10.6*) and Table 10-1,

$$\mathscr{E}° = {}_oE°_{Mn/Mn^{2+}} + {}_rE°_{Ag/Ag} = 1.18 + 0.800 = 1.98 \text{ V}$$

Therefore, by (*10.3*),

$$\mathscr{E} = \mathscr{E}° - \frac{0.0592}{2} \log \frac{[Mn^{2+}]}{[Ag^+]^2} = 1.98 - \frac{0.0592}{2} \log \frac{0.400}{(0.150)^2} = 1.94 \text{ V}$$

10.15 Use data of Table 10-1 to calculate the voltage, $\mathscr{E}$, of the cell

$$Pb(s) \mid PbSO_4(s) \mid Na_2SO_4(0.300M) \parallel Cr_2O_7^{2-}(0.160M), Cr^{3+}(0.270M), H^+(0.400M) \mid Pt$$

The cell involves the oxidation of $Pb(s)$ and the reduction of $Cr_2O_7^{2-}$, the overall reaction being

$$3Pb(s) + 3SO_4^{2-} + Cr_2O_7^{2-} + 14H^+ \to 3PbSO_4(s) + 2Cr^{3+} + 7H_2O$$

From (*10.6*) and Table 10-1,

$$\mathscr{E}° = {}_oE°_{Pb/PbSO_4} + {}_rE°_{Cr_2O_7^{2-}/Cr^{3+}} = +0.356 + 1.33 = 1.69 \text{ V}$$

Therefore, by (*10.3*),

$$\mathscr{E} = \mathscr{E}° - \frac{0.0592}{6} \log \frac{[Cr^{3+}]^2}{[SO_4^{2-}]^3[Cr_2O_7^{2-}][H^+]^{14}} = 1.69 - \frac{0.0592}{6} \log \frac{(0.270)^2}{(0.300)^3(0.160)(0.400)^{14}} = 1.62 \text{ V}$$

10.16 Use data of Table 10-1 to calculate the voltage, $\mathscr{E}$, of the cell

$$Ag(s)\,|\,Ag^+(0.0350M)\,\|\,Cr^{3+}(1.25\times 10^{-6}M)\,|\,Cr(s)$$

The ionic strength of the solution is 0.100; include corrections for non-ideality.

The cell involves the oxidation of $Ag(s)$ and the reduction of Cr^{3+}, according to the redox reaction

$$3Ag(s)+Cr^{3+}\to 3Ag^+ + Cr(s)$$

From (10.6), Table 10-1, and Problem 10.1,

$$\mathscr{E}^\circ = {}_oE^\circ_{Ag/Ag^+} + {}_rE^\circ_{Cr^{3+}/Cr} = -0.800 - 0.74 = -1.54\text{ V}$$

From Tables 3-2 and 3-3, $f_{Ag^+} = 0.754$ and $f_{Cr^{3+}} = 0.179$. Therefore, by (10.3),

$$\mathscr{E} = \mathscr{E}^\circ - \frac{0.0592}{3}\log\frac{(a_{Ag^+})^3}{a_{Cr^{3+}}} = -1.54 - \frac{0.0592}{3}\log\frac{\{(0.0350)(0.754)\}^3}{(1.25\times 10^{-6})(0.179)} = -1.58\text{ V}$$

Since the voltage is negative, it is the reverse reaction that is spontaneous.

10.17 Use data of Table 10-1 to calculate the voltage, $\mathscr{E}$, of the cell

$$Pt\,|\,Cr^{3+}(0.125M),\,Cr_2O_7^{2-}(0.200M),\,H^+(0.600M)\,\|\,MnO_4^-(0.150M),\,Mn^{2+}(0.400M),\,H^+(0.250M)\,|\,Pt$$

The cell involves the oxidation of Cr^{3+} and the reduction of MnO_4^-. However, H^+ appears in both the oxidation and reduction half-reactions, and since the H^+ concentrations differ in the two cell compartments, it is necessary to keep track of the H^+ in writing the overall cell reaction.

Half-Reaction	Voltage
$10Cr^{3+} + 35H_2O \to 5Cr_2O_7^{2-} + 70H^+(0.600M) + 30e^-$	${}_oE^\circ = -1.33$ V
$6MnO_4^- + 48H^+(0.250M) + 30e^- \to 6Mn^{2+} + 24H_2O$	${}_rE^\circ = +1.51$ V

$$6MnO_4^- + 10Cr^{3+} + 11H_2O + 48H^+(0.250M) \to 6Mn^{2+} + 5Cr_2O_7^{2-} + 70H^+(0.600M) \qquad \mathscr{E}^\circ = +0.18\text{ V}$$

By (10.3),

$$\mathscr{E} = \mathscr{E}^\circ - \frac{0.0592}{30}\log\frac{[Cr_2O_7^{2-}]^5\,{}_o[H^+]^{70}\,[Mn^{2+}]^6}{[Cr^{3+}]^{10}\,[MnO_4^-]^6\,{}_r[H^+]^{48}}$$

$$= 0.18 - \frac{0.0592}{30}\log\frac{(0.200)^5(0.600)^{70}(0.400)^6}{(0.125)^{10}(0.150)^6(0.250)^{48}} = 0.14\text{ V}$$

10.18 Use data of Table 10-1 to calculate the voltage, $\mathscr{E}$, of the cell

$$Mn(s)\,|\,Mn(OH)_2(s)\,|\,Mn^{2+}(x\ M),\,OH^-(1.00\times 10^{-4}M)\,\|\,Cu^{2+}(0.675M)\,|\,Cu(s)$$

given that $K_{sp} = 1.9\times 10^{-13}$ for $Mn(OH)_2(s)$.

The cell involves the oxidation of $Mn(s)$ and the reduction of Cu^{2+}, according to the redox reaction

$$Mn(s)+Cu^{2+}\to Mn^{2+}+Cu(s)$$

From (10.6) and Table 10-1,

$$\mathscr{E}^\circ = {}_oE^\circ_{Mn/Mn^{2+}} + {}_rE^\circ_{Cu^{2+}/Cu} = +1.18 + 0.337 = 1.52\text{ V}$$

Also, from

$$K_{sp} = 1.9\times 10^{-13} = [Mn^{2+}][OH^-]^2 = (x)(1.00\times 10^{-4})^2$$

there results $x = [Mn^{2+}] = 1.9\times 10^{-5}$. Then (10.3) gives

$$\mathscr{E} = \mathscr{E}^\circ - \frac{0.0592}{2}\log\frac{[Mn^{2+}]}{[Cu^{2+}]} = 1.52 - \frac{0.0592}{2}\log\frac{1.9\times 10^{-5}}{0.675} = 1.66\text{ V}$$

10.19 Use data in Table 10-1 to calculate the voltage, $\mathscr{E}$, of the cell

$$Ag(s) \mid AgIO_3(s) \mid Ag^+(x\ M), HIO_3(0.300M) \parallel Zn^{2+}(0.175M) \mid Zn(s)$$

if $K_{sp} = 3.02 \times 10^{-8}$ for $AgIO_3(s)$ and $K_a = 0.162$ for HIO_3.

The cell involves the oxidating of $Ag(s)$ and the reduction of Zn^{2+}, with the overall cell reaction

$$2Ag(s) + Zn^{2+} \rightarrow 2Ag^+ + Zn(s)$$

Therefore,

$$\mathscr{E}^\circ = {}_oE^\circ_{Ag/Ag^+} + {}_rE^\circ_{Zn^{2+}/Zn} = -0.800 - 0.763 = -1.563\ V$$

Considering the dissociation of HIO_3, we see that the point $(C_a, K_a) = (0.300, 0.162)$ lies in Region 1 of Fig. 5-1, whence

$$[H^+]^2 + 0.162[H^+] - 4.86 \times 10^{-2} = 0$$

The solution to this quadratic is $[H^+] = 0.154$. Since the water equilibrium can be neglected, $[H^+] = [IO_3^-] = 0.154$, and

$$x = [Ag^+] = \frac{K_{sp}}{[IO_3^-]} = \frac{3.02 \times 10^{-8}}{0.154} = 1.96 \times 10^{-7}$$

Now, from (10.3),

$$\mathscr{E} = \mathscr{E}^\circ - \frac{0.0592}{2}\log\frac{[Ag^+]^2}{[Zn^{2+}]} = -1.563 - \frac{0.0592}{2}\log\frac{(1.96 \times 10^{-7})^2}{0.175} = -1.188\ V$$

Since $\mathscr{E}$ is negative, it is the reverse reaction that is spontaneous.

10.20 The voltage of the cell

$$Pb(s) \mid PbSO_4(s) \mid NaHSO_4(0.600M) \parallel Pb^{2+}(2.50 \times 10^{-5}M) \mid Pb(s)$$

is $\mathscr{E} = +0.061\ V$. Use Table 10-1 to calculate $K_2 = [H^+][SO_4^{2-}]/[HSO_4^-]$, the dissociation constant for HSO_4^-.

The cell involves the oxidation of $Pb(s)$ to $PbSO_4(s)$ and the reduction of Pb^{2+} to $Pb(s)$; the cell reaction is

$$Pb^{2+} + SO_4^{2-} \rightarrow PbSO_4(s)$$

From (10.6) and Table 10-1,

$$\mathscr{E}^\circ = {}_oE^\circ_{Pb/PbSO_4} + {}_rE^\circ_{Pb^{2+}/Pb} = +0.356 - 0.126 = 0.230\ V$$

The Nernst equation then gives

$$0.061 = 0.230 - \frac{0.0592}{2}\log\frac{1}{(2.50 \times 10^{-5})[SO_4^{2-}]}$$

so that $[SO_4^{2-}] = 7.8_1 \times 10^{-2}$. Therefore, since there is one H^+ produced for every SO_4^{2-}, $[H^+] = 7.8_1 \times 10^{-2}$ and

$$[HSO_4^-] = 0.600 - (7.8_1 \times 10^{-2}) = 0.522$$

Consequently,

$$K_2 = \frac{(7.8_1 \times 10^{-2})^2}{0.522} = 1.1_7 \times 10^{-2}$$

10.21 The voltage of the cell

$$Zn(s) \mid Zn(CN)_4^{2-}(0.450M), CN^-(2.65 \times 10^{-3}M) \parallel Zn^{2+}(3.84 \times 10^{-4}M) \mid Zn(s)$$

is $\mathscr{E} = +0.099\ V$. Use Table 10-1 to calculate the constant K_{s4} for $Zn^{2+} + 4CN^- \rightarrow Zn(CN)_4^{2-}$, the only $Zn^{2+} + CN^-$ complexation reaction of importance.

The cell involves the oxidation of $Zn(s)$ to Zn^{2+} and also the reduction of Zn^{2+} to $Zn(s)$. Such an arrangement, where the oxidation standard-state reaction is the reverse of the reduction standard-state reaction, is known as a *concentration cell*; such cells obviously have $\mathscr{E}° = 0.000$ V. In the present concentration cell, the half-reactions are:

$$Zn(s) \to Zn^{2+}(x\ M) + 2e^-$$
$$Zn^{2+}(3.84 \times 10^{-4}\ M) + 2e^- \to Zn(s)$$

so that (10.3) becomes:

$$0.099 = 0.000 - \frac{0.0592}{2} \log \frac{x}{3.84 \times 10^{-4}}$$

whence $x = [Zn^{2+}] = 1.7_4 \times 10^{-7}$. Then,

$$K_{s4} = \frac{[Zn(CN)_4{}^{2-}]}{[Zn^{2+}][CN^-]^4} = \frac{0.450}{(1.7_4 \times 10^{-7})(2.65 \times 10^{-3})^4} = 5.2_4 \times 10^{16}$$

10.22 The voltage of the cell

$$Pt \mid H_2(g,\ 608\ torr) \mid HClO(4.34 \times 10^{-3}\ M) \parallel S.C.E.$$

is $\mathscr{E} = +0.532$ V. Use data of Table 10-1 to calculate the dissociation constant for HClO.

The (nonstandard) oxidation half-reaction is

$$H_2(g) \to 2H^+ + 2e^-$$

for which (10.3) and Table 10-1 give

$$_oE = {}_oE° - \frac{0.0592}{2} \log \frac{[H^+]^2}{P_{H_2}}$$
$$= 0.000 - \frac{0.0592}{2} \log \frac{[H^+]^2}{608/760}$$

where the necessary conversion of the gas pressure from torrs to atms was accomplished using Table 1-1. The (nonstandard) reduction half-reaction is given by (10.7); it has $_rE = 0.242$ V. Hence, by (10.6),

$$0.532 = -\frac{0.0592}{2} \log \frac{[H^+]^2}{608/760} + 0.242$$

from which $[H^+] = 1.1_3 \times 10^{-5}$. Consequently, $[ClO^-] = 1.1_3 \times 10^{-5}$,

$$[HClO] = (4.34 \times 10^{-3}) - (1.1_3 \times 10^{-5}) = 4.33 \times 10^{-3}$$

and

$$K_a = \frac{[H^+][ClO^-]}{[HClO]} = 2.9_5 \times 10^{-8}$$

10.23 The voltage of the cell

$$Pt \mid H_2(g,\ 570\ torr) \mid HNO_2(2.28 \times 10^{-3}M) \parallel HCl(0.0250M) \mid AgCl(s) \mid Ag(s)$$

is $\mathscr{E} = +0.493$ V. Using data of Table 10-1, calculate the dissociation constant of HNO_2. Include corrections for non-ideality. The activity coefficients for the oxidation half-cell are $f_{H^+} = 0.854$, $f_{NO_2^-} = 0.807$, and $f_{HNO_2} = 1.000$; the H_2 can be considered an ideal gas. The activity coefficients for the reduction half-cell, which is at a different ionic strength, are $f_{H^+} = 0.835$, $f_{Cl^-} = 0.772$, and $f_{Ag^+} = 0.772$.

The cell involves the oxidation of H_2 to H^+ and the reduction of $AgCl(s)$ to $Ag(s)$, the overall cell reaction being

$$\tfrac{1}{2}H_2(g) + AgCl(s) \to H^+ + Ag(s) + Cl^-$$

where the H^+-term is associated with the oxidation cell. From (10.6) and Table 10-1,

$$\mathscr{E}^\circ = {}_oE^\circ_{H_2/H^+} + {}_rE^\circ_{AgCl/Ag} = 0.000 + 0.222 = 0.222$$

The expression (10.3) for the cell voltage is:

$$\mathscr{E} = \mathscr{E}^\circ - \frac{0.0592}{1} \log \frac{a_{H^+}\, a_{Cl^-}}{P_{H_2}^{1/2}}$$

$$0.493 = 0.222 - \frac{0.0592}{1} \log \frac{[H^+](0.854)(0.0250)(0.772)}{(570/760)^{1/2}}$$

Solving, $[H^+] = 1.3_9 \times 10^{-3}$. Therefore, $[NO_2^-] = 1.3_9 \times 10^{-3}$,

$$[HNO_2] = (2.28 \times 10^{-3}) - (1.3_9 \times 10^{-3}) = 0.8_9 \times 10^{-3}$$

and

$$K_a = (a_{H^+})(a_{NO_2^-})/a_{HNO_2} = \frac{(1.3_9 \times 10^{-3})(0.854)(1.3_9 \times 10^{-3})(0.807)}{(0.8_9 \times 10^{-3})(1.000)} = 1._5 \times 10^{-3}$$

Supplementary Problems

HALF-CELL POTENTIALS

10.24 Use Table 10-1 to find the standard-state voltages for the reduction half-reactions (a) $Co^{3+} + 3e^- \rightarrow Co(s)$, (b) $Sn^{4+} + 4e^- \rightarrow Sn(s)$, (c) $MnO_4^{2-} + 8H^+ + 6e^- \rightarrow Mn(s) + 4H_2O$.
   Ans.   (a) ${}_rE^\circ = +0.429$ V; (b) ${}_rE^\circ = +0.009$ V; (c) ${}_rE^\circ = +0.771$ V

EQUILIBRIUM CONSTANTS

10.25 Use Table 10-1 to calculate the solubility product for (a) $AgBr(s)$, (b) $Ag_2S(s)$.
   Ans.   (a) $5._2 \times 10^{-13}$; (b) $9._7 \times 10^{-52}$

10.26 Use data of Table 10-1 to calculate the equilibrium constant for the reaction

$$3Sn^{4+} + 2MnO_2(s) + 4H_2O \rightarrow 2MnO_4^- + 8H^+ + 3Sn^{2+}$$

   Ans. $K = 10^{-156}$

10.27 Use data of Table 10-1 to calculate the equilibrium constant for the reaction

$$2Cr^{3+} + 3Sn^{2+} \rightarrow 2Cr(s) + 3Sn^{4+}$$

   Ans. $K = 1 \times 10^{-91}$

CELL VOLTAGES

10.28 Use data of Table 10-1 to calculate the voltage, $\mathscr{E}$, of the cell

$$Fe(s) \mid Fe^{2+}(0.325M) \parallel Ag^+(5.50 \times 10^{-4}M) \mid Ag(s)$$

   Ans.   +1.061 V

10.29 Use data of Table 10-1 to calculate the voltage $\mathscr{E}$, of the cell

$$Pt \mid Cu^{2+}(0.175M),\, Cu^+(0.344M) \parallel Sn^{4+}(0.263M),\, Sn^{2+}(0.108M) \mid Pt$$

   Ans.   +0.030 V

10.30 Use data of Table 10-1 to calculate the voltage, $\mathscr{E}$, of the cell

$$Pt \mid H_2(g,\, 0.900 \text{ atm}) \mid HAc(0.350M),\, NaAc(0.675M) \parallel S.C.E.$$

   if $K_a = 1.78 \times 10^{-5}$ for HAc.   *Ans.*   +0.539 V

10.31 Use data of Table 10-1 to calculate the voltage, $\mathscr{E}$, of the cell

$$Ag(s) \,|\, Ag(CN)_2^- \,(0.340M), \, CN^- \,(2.00 \times 10^{-6}M) \,\|\, Ag^+ \,(0.500M) \,|\, Ag(s)$$

given that $K_{s2} = 7.1 \times 10^{19}$ for $Ag^+ + 2CN^- \rightarrow Ag(CN)_2^-$.   *Ans.*   +0.510 V

10.32 The voltage of the cell

$$Pt \,|\, H_2(g, \, 0.850 \text{ atm}) \,|\, NaHSO_3 \,(0.400M), \, Na_2SO_3 \,(6.44 \times 10^{-3}M) \,\|\, Zn^{2+} \,(0.300M) \,|\, Zn(s)$$

is $\mathscr{E} = -0.460$ V. Use data of Table 10-1 to calculate $K_2 = [H^+][SO_3^{2-}]/[HSO_3^-]$.   *Ans.* $6._2 \times 10^{-8}$

10.33 The voltage of the following cell is $\mathscr{E} = +0.982$ V:

$$Pt \,|\, H_2(g, \, 0.580 \text{ atm}) \,|\, HOCN \,(1.30 \times 10^{-3}M) \,\|\, Ag^+ \,(0.800M) \,|\, Ag(s)$$

Evaluate $K_a = [H^+][OCN^-]/[HOCN]$, the dissociation constant for hydrogen cyanate.
Ans. $3.3_4 \times 10^{-4}$

10.34 The cell

$$Cu(s) \,|\, Cu^{2+} \,(0.0725M) \,\|\, CuCl_2^- \,(0.0425M), \, Cl^- \,(0.0250M) \,|\, Cu(s)$$

has $\mathscr{E} = +0.034$ V. Use data of Table 10-1 to determine (*a*) $_rE^\circ$ for $Cu^+ + e^- \rightarrow Cu(s)$, and (*b*) the summed formation constant $K_{2s} = [CuCl_2^-]/[Cu^+][Cl^-]^2$.   *Ans.*   (*a*) +0.521 V; (*b*) $8.6_3 \times 10^4$

10.35 The voltage of the cell

$$Pt \,|\, CuI(s) \,|\, Cu^{2+} \,(0.0525M), \, I^- \,(4.28 \times 10^{-5}M) \,\|\, Cu^{2+} \,(0.0270M) \,|\, Cu(s)$$

is $\mathscr{E} = -0.235$ V. Use Table 10-1 to calculate $_rE^\circ$ for $Cu^{2+} + I^- + e^- \rightarrow CuI(s)$.   *Ans.*   +0.860 V

Chapter 11

Potentiometric Titrations

11.1 TITRATION OF A SINGLE ION

A redox reaction may be used to determine the concentration of either the oxidizing or reducing ion, provided that:

1. The reaction mixture comes to equilibrium almost instantaneously at all stages in the titration. Sometimes a catalyst must be added to ensure this.

2. The equilibrium constant for the reaction is sufficiently large so that, at the equivalence point, at least 99.9% of the reactants have been converted to products. In view of (10.8), this requires that $\mathscr{E}°$ be at least about +0.2 to +0.4 V.

3. If the concentrations of more than one ionic species are being determined together, the $\mathscr{E}°$-values for the separate redox reactions must differ from each other by at least 0.2 V; this is necessary for distinctness of the equivalence points in the titration data.

4. The electrochemical cell is arranged so that the redox reaction occurs in only one half-cell compartment. The other half-cell should be a constant-voltage electrode such as the S.C.E.

Requirement 4 is based on the fact that, if the redox titration represented the full cell, the cell voltage $\mathscr{E}$ would be zero at all stages in the titration, since the redox reaction would be at equilibrium at all times. A typical cell will be of the form

$$\text{S.C.E.} \parallel \textit{redox titration mixture} \mid \text{Pt} \qquad (11.1)$$

The S.C.E. (or similar constant-voltage electrode) is called the *reference electrode*; the platinum (or similar electrode immersed in the redox titration solution) is called the *indicator electrode*. The above cell can be reversed, if desired, so that the reference electrode is the cathode rather than the anode (see Section 10.1).

From (10.6), the full-cell voltage will be

$$\mathscr{E}_{cell} = {}_oE_{\text{S.C.E.}} + {}_rE_{reac} = (-0.242 \text{ V}) + {}_rE_{reac} \qquad (11.2)$$

In other words, the cell voltage is, to within a subtractive constant, equal to the reduction potential of the redox titration mixture. Consider now the redox reaction, with voltage $\mathscr{E}_{reac}$, as the sum of an oxidation half-reaction, with voltage ${}_oE_x$, and a reduction half-reaction, with voltage ${}_rE_y = {}_rE_{reac}$. By (10.6)

$$\mathscr{E}_{reac} = {}_oE_x + {}_rE_y$$

But, by Requirement 1 above, $\mathscr{E}_{reac} = 0$; hence, ${}_rE_y = -{}_oE_x$ and

$$_rE_{reac} = {}_rE_y = {}_rE_x \qquad (11.3)$$

The generalization of (11.3) is that *all half-reactions in equilibrium in the titration mixture must have a common reduction potential at any stage in the titration.* This condition is of importance in choosing a color redox titration indicator. It also tells us that if the concentrations of several ions are being determined in a potentiometric titration, then, at any stage, the reduction potential of the titrant half-reaction will equal the half-cell reduction potential of each ion.

By (11.2) and (11.3), the cell voltage may be found from either ${}_rE_x$ or ${}_rE_y$, *whichever choice makes for the simpler calculation.*

EXAMPLE 11.1 What is the voltage of the redox titration cell (11.1) when 5.00 mL of 0.200 M Ce^{4+} has been added in the titration of 20.00 mL of 0.100 M Sn^{2+}? The titration reaction is

$$Sn^{2+} + 2Ce^{4+} \rightarrow Sn^{4+} + 2Ce^{3+}$$

Since 20.00 mL of Ce^{4+} is required to reach the equivalence point, only

$$\frac{5.00}{20.00} = 25\%$$

of the Sn^{2+} has been oxidized to Sn^{4+}, leaving 75% in the form of Sn^{2+}. Therefore, $[Sn^{2+}]/[Sn^{4+}] = 75\%/25\% = 3$, and the simplest calculation of the cell voltage will be based on the Sn^{4+} reduction half-reaction. Thus, from (11.2), (10.3), and Table 10-1:

$$\mathscr{E}_{cell} = -0.242 + {}_rE_{Sn^{4+}/Sn^{2+}}$$

$$= -0.242 + \left(0.154 - \frac{0.0592}{2} \log 3\right) = -0.102 \text{ V}$$

EXAMPLE 11.2   What is the voltage of the cell of Example 11.1 when 25.00 mL of Ce^{4+} has been added?

This is 5.00 mL in excess of the stoichiometric amount of Ce^{4+}, so that $[Ce^{3+}]/[Ce^{4+}] = 20.00/5.00 = 4$. The simplest calculation of the cell voltage will be based on the Ce^{4+} reduction half-reaction. From (11.2), (10.3), and Table 10-1:

$$\mathscr{E}_{cell} = -0.242 + {}_rE_{Ce^{4+}/Ce^{3+}}$$

$$= -0.242 + \left(1.44 - \frac{0.0592}{1} \log 4\right) = 1.16 \text{ V}$$

EXAMPLE 11.3   What is the voltage of the cell of Example 11.1 at the equivalence point in the titration (where 20.00 mL of Ce^{4+} has been added)?

Since the balanced redox reaction is $2Ce^{4+} + Sn^{2+} \rightarrow 2Ce^{3+} + Sn^{4+}$, and since the only source of products is via this reaction, the identity

$$[Ce^{3+}] = 2[Sn^{4+}]$$

holds at any stage. At the stoichiometric equivalence point, there is very little of either reactant (because, for the reaction,

$$\mathscr{E}^\circ = {}_oE^\circ_{Sn^{2+}/Sn^{4+}} + {}_rE^\circ_{Ce^{4+}/Ce^{3+}} = -0.154 + 1.44 = +1.29 \text{ V}$$

which, by (10.8), corresponds to $K = 3.81 \times 10^{43}$). However, what little there is, will be present in a stoichiometric ratio, so that we have the additional identity

$$[Ce^{4+}] = 2[Sn^{2+}]$$

The simplest calculation for the cell voltage exploits both identities. We express $\mathscr{E}_{cell}$ two ways, as in Examples 11.1 and 11.2:

$$\mathscr{E}_{cell} = -0.242 + \left(0.154 - \frac{0.0592}{2} \log \frac{[Sn^{2+}]}{[Sn^{4+}]}\right)$$

$$\mathscr{E}_{cell} = -0.242 + \left(1.44 - \frac{0.0592}{1} \log \frac{[Ce^{3+}]}{[Ce^{4+}]}\right)$$

Adding twice the first equation to the second (i.e., adding the expressions for $n\mathscr{E}_{cell}$) gives

$$3\mathscr{E}_{cell} = 3(-0.242) + 2(0.154) + 1.44 - 0.0592 \log \left(\frac{[Sn^{2+}]}{[Ce^{4+}]} \frac{[Ce^{3+}]}{[Sn^{4+}]}\right)$$

$$= 1.022 - 0.0592 \log \left(\tfrac{1}{2} \cdot 2\right) = 1.022$$

from which $\mathscr{E}_{cell} = 0.341$ V.

11.2   TITRATION OF MIXTURES

When more than one ion is being determined in a potentiometric titration, the redox reaction with the largest $\mathscr{E}^\circ_{reac}$ will occur first, followed in sequence by reactions with lesser $\mathscr{E}^\circ$-values.

The cell voltage at an intermediate equivalence point is most easily calculated by applying the method of Example 11.3 to the redox reaction that proceeds to the greatest extent (i.e., has the largest $\mathscr{E}^\circ$-value) at that point.

11.3 REDOX TITRATION INDICATORS

Just as colored weak acids serve as color indicators in acid-base titrations, colored redox ions can serve as indicators in redox titrations. One group of redox indicators has a half-reaction of the type

$$\text{ox.} + ne^- \rightarrow \text{red.} \tag{11.4}$$

where $n = 2$ is the most common case, and ox. and/or red. are colored. The reduction potential is:

$$_rE_{\text{ox./red.}} = {_rE_{\text{ox./red.}}^\circ} - \frac{0.0592}{n}\log\frac{[\text{red.}]}{[\text{ox.}]} \tag{11.5}$$

so that, in general, the principal color change occurs when [red.] = [ox.] and

$$_rE_{\text{ox./red.}} = {_rE_{\text{ox./red.}}^\circ}$$

The indicator, although present in a very small concentration as compared to the concentrations of the principal species involved in the titration reaction, is also in equilibrium with the half-reactions of the titration reaction. Therefore (Section 11.1), the proper redox indicator of the type (11.4) will be the one with $_rE_{\text{ox./red.}}^\circ$ closest in value to $_rE_{\text{reac}}$ at the equivalence point of the titration.

However, most redox indicators are of the type

$$\text{ox.} + 2H^+ + 2e^- \rightarrow \text{red.} \tag{11.6}$$

and have the reduction potential

$$_rE_{\text{ox./red.}} = {_rE_{\text{ox./red.}}^\circ} - \frac{0.0592}{2}\log\frac{[\text{red.}]}{[\text{ox.}][H^+]^2} \tag{11.7}$$

so that at the point of maximum color change, where [red.] = [ox.],

$$_rE_{\text{ox./red.}} = {_rE_{\text{ox./red.}}^\circ} - 0.0592\,\text{pH} \tag{11.8}$$

This type of indicator allows for greater choice in that the pH of the titration solution can be adjusted so that $_rE_{\text{ox./red.}}$ is as close as possible to the equivalence-point $_rE_{\text{reac}}$ of the titration solution.

Solved Problems

TITRATION OF A SINGLE ION

11.1 20.00 mL of 0.120 M Cu^+ is titrated with 0.100 M Ce^{4+}. What is the voltage of the cell (11.1) when the amount of added Ce^{4+} is (a) 5.00 mL? (b) 12.00 mL? (c) 22.00 mL? The redox titration reaction is:

$$Cu^+ + Ce^{4+} \rightarrow Cu^{2+} + Ce^{3+}$$

From Table 10-1 and (10.6),

$$\mathscr{E}_{\text{reac}}^\circ = {_oE_{Cu^+/Cu^{2+}}^\circ} + {_rE_{Ce^{4+}/Ce^{3+}}^\circ} = -0.153 + 1.44 = +1.29\text{ V}$$

so that, from (10.8), $K = 6.2 \times 10^{21}$, and the reaction can be assumed to go to completion. By (1.3), with $n_o = n_r = 1$, it requires 24.00 mL of Ce^{4+} to reach the equivalence point. In all three cases, the added Ce^{4+} is less than this amount, so the cell voltage is most easily calculated from the reduction potential of the copper half-reaction:

$$\mathscr{E}_{\text{cell}} = {_oE_{\text{S.C.E.}}} + {_rE_{Cu^{2+}/Cu^+}^\circ} - \frac{0.0592}{1}\log\frac{[Cu^+]}{[Cu^{2+}]}$$

$$= -0.242 + 0.153 - \frac{0.0592}{1}\log\frac{[Cu^+]}{[Cu^{2+}]}$$

(a)   The 5.00 mL of Ce^{4+} will convert 5.00/24.00 of the Cu^+ to Cu^{2+}, so that

$$\frac{[Cu^+]}{[Cu^{2+}]} = \frac{19.00/24.00}{5.00/24.00} = 3.80$$

and $\mathscr{E}_{cell} = -0.123$ V.

(b)   The 12.00 mL of Ce^{4+} (which is halfway to the equivalence point) will convert 12.00/24.00 of the Cu^+ to Cu^{2+}, so that $[Cu^+]/[Cu^{2+}] = 1.000$; the log term vanishes and $\mathscr{E}_{cell} = -0.089$ V.

(c)   The 22.00 mL of Ce^{4+} will convert 22.00/24.00 of the Cu^+ to Cu^{2+}, so that

$$\frac{[Cu^+]}{[Cu^{2+}]} = \frac{2.00/24.00}{22.00/24.00} = 0.0909$$

and $\mathscr{E}_{cell} = -0.027$ V.

11.2   What is the voltage of the cell of Problem 11.1 at the equivalence point in the titration?

The simplest expression for the equivalence-point cell voltage is obtained by adding the equations based on the Ce and Cu half-reactions,

$$\mathscr{E}_{cell} = {}_oE_{S.C.E.} + {}_rE^\circ_{Ce^{4+}/Ce^{3+}} - 0.0592 \log \frac{[Ce^{3+}]}{[Ce^{4+}]}$$

and

$$\mathscr{E}_{cell} = {}_oE_{S.C.E.} + {}_rE^\circ_{Cu^{2+}/Cu^+} - 0.0592 \log \frac{[Cu^+]}{[Cu^{2+}]}$$

and making use of the relations $[Cu^{2+}] = [Ce^{3+}]$ (valid throughout the titration) and $[Cu^+] = [Ce^{4+}]$ (valid at the equivalence point) to eliminate the logarithmic term in the sum. The result is:

$$\mathscr{E}_{cell} = {}_oE_{S.C.E.} + \frac{{}_rE^\circ_{Ce^{4+}/Ce^{3+}} + {}_rE^\circ_{Cu^{2+}/Cu^+}}{2}$$

$$= -0.242 + \frac{1.44 + 0.153}{2} = +0.555 \text{ V}$$

11.3   Find the voltages of the cell of Problem 11.1 after the addition of (a) 26.00 mL, (b) 48.00 mL, and (c) 60.00 mL, of Ce^{4+}.

These volumes are all in excess of that needed to reach the equivalence point. Hence, the simplest expression for the cell voltage is the one based on the Ce half-reaction:

$$\mathscr{E}_{cell} = {}_oE_{S.C.E.} + {}_rE^\circ_{Ce^{4+}/Ce^{3+}} - \frac{0.0592}{1} \log \frac{[Ce^{3+}]}{[Ce^{4+}]}$$

$$= -0.242 + 1.44 - \frac{0.0592}{1} \log \frac{[Ce^{3+}]}{[Ce^{4+}]}$$

(a)   The first 24.00 mL of Ce^{4+} is converted to Ce^{3+}, so that

$$\frac{[Ce^{3+}]}{[Ce^{4+}]} = \frac{24.00}{2.00} = 12.0$$

and $\mathscr{E}_{cell} = 1.13$ V.

(b)   This volume is precisely twice that needed to reach the equivalence point; hence, $[Ce^{3+}] = [Ce^{4+}]$, the log term is zero, and $\mathscr{E}_{cell} = +1.20$ V.

(c)

$$\frac{[Ce^{3+}]}{[Ce^{4+}]} = \frac{24.00}{36.00} = 0.6667$$

and $\mathscr{E}_{cell} = 1.21$ V.

11.4   List the reduction potentials of the titration solution in Problems 11.1, 11.2, and 11.3 for the specified volumes of Ce^{4+}.

By (11.2), $_rE_{reac} = \mathscr{E}_{cell} + (0.242\ V)$. Thus:

vol Ce^{4+}, mL	5.00	12.00	22.00	24.00	26.00	48.00	60.00
$_rE_{reac}$, V	0.119	0.153	0.215	0.797	1.37	1.44	1.45

The corresponding titration curve is sketched in Fig. 11-1(a).

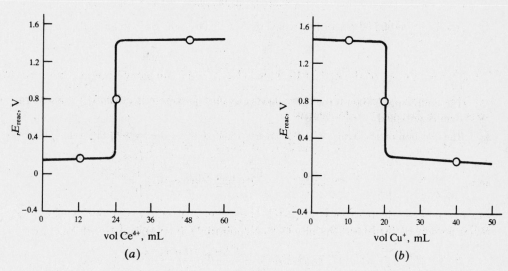

Fig. 11-1

11.5   Sketch the titration curve when 24.00 mL of 0.100 M Ce^{4+} is titrated with 0.120 M Cu^+.

Now the reducing agent, Cu^+, is the titrant, rather than the oxidizing agent, Ce^{4+}, which was the titrant in Problems 11.1–11.3. Therefore, the titration curve should look like the reverse of Fig. 11-1(a). By (1.3), with $n_o = n_r = 1$, it requires 20.00 mL of Cu^+ to reach the equivalence point. When 10.00 mL of Cu^+ is added, which is halfway to the equivalence point, $[Ce^{3+}] = [Ce^{4+}]$, whence

$$\log\frac{[Ce^{3+}]}{[Ce^{4+}]} = 0 \qquad \text{and} \qquad _rE_{reac} = {_rE^\circ_{Ce^{4+}/Ce^{3+}}} = +1.44\ V$$

When 40.00 mL of Cu^+ is added, which is twice the volume needed to reach the equivalence point, $[Cu^+] = [Cu^{2+}]$, whence

$$\log\frac{[Cu^+]}{[Cu^{2+}]} = 0 \qquad \text{and} \qquad _rE_{reac} = {_rE^\circ_{Cu^{2+}/Cu^+}} = +0.153\ V$$

The equivalence-point reduction potential is the same as in Problem 11.2,

$$_rE_{reac} = 0.555 + 0.242 = 0.797\ V$$

These results yield Fig. 11-1(b).

11.6   20.00 mL of 0.100 M Sn^{2+} is titrated with 0.200 M Fe^{3+}, according to the redox reaction

$$Sn^{2+} + 2Fe^{3+} \rightarrow Sn^{4+} + 2Fe^{2+}$$

Find the voltages of the cell (11.1) when (a) 5.00 mL, (b) 10.00 mL, and (c) 15.00 mL, of Fe^{3+} are added.

From Table 10-1 and (10.6),

$$\mathscr{E}^\circ_{reac} = {}_oE^\circ_{Sn^{2+}/Sn^{4+}} + {}_rE^\circ_{Fe^{3+}/Fe^{2+}} = -0.154 + 0.771 = +0.617 \text{ V}$$

so that, from (10.8), $K = 7._0 \times 10^{20}$, and the reaction can be assumed to go to completion. From (1.3), 20.00 mL of Fe^{3+} is required to reach the equivalence point. The volumes added here are all less than this, so that the cell voltage is most easily calculated using the expression based on the Sn half-reaction:

$$\mathscr{E}_{cell} = {}_oE_{S.C.E.} + {}_rE^\circ_{Sn^{4+}/Sn^{2+}} - \frac{0.0592}{2} \log \frac{[Sn^{2+}]}{[Sn^{4+}]}$$

$$= -0.242 + 0.154 - \frac{0.0592}{2} \log \frac{[Sn^{2+}]}{[Sn^{4+}]}$$

(a) The addition of 5.00 mL of Fe^{3+} converts 5.00/20.00 of the Sn^{2+} to Sn^{4+}; thus,

$$\frac{[Sn^{2+}]}{[Sn^{4+}]} = \frac{15.00/20.00}{5.00/20.00} = 3.00$$

and $\mathscr{E}_{cell} = -0.102$ V.

(b) The addition of 10.00 mL of Fe^{3+} converts half of the Sn^{2+} to Sn^{4+}; hence, $[Sn^{2+}] = [Sn^{4+}]$, the log term is zero, and $\mathscr{E}_{cell} = -0.088$ V.

(c) The addition of 15.00 mL of Fe^{3+} converts 15.00/20.00 of the Sn^{2+} to Sn^{4+}; thus,

$$\frac{[Sn^{2+}]}{[Sn^{4+}]} = \frac{5.00/20.00}{15.00/20.00} = 0.333$$

and $\mathscr{E}_{cell} = -0.074$ V.

11.7 What is the cell voltage at the equivalence point in the titration of Problem 11.6?

As in Example 11.3, add the two expressions for $n\mathscr{E}_{cell}$:

$$2\mathscr{E}_{cell} = 2\,{}_oE_{S.C.E.} + 2\,{}_rE^\circ_{Sn^{4+}/Sn^{2+}} - 0.0592 \log \frac{[Sn^{2+}]}{[Sn^{4+}]}$$

$$1\mathscr{E}_{cell} = {}_oE_{S.C.E.} + {}_rE^\circ_{Fe^{3+}/Fe^{2+}} - 0.0592 \log \frac{[Fe^{2+}]}{[Fe^{3+}]}$$

$$3\mathscr{E}_{cell} = 3\,{}_oE_{S.C.E.} + 2\,{}_rE^\circ_{Sn^{4+}/Sn^{2+}} + {}_rE^\circ_{Fe^{3+}/Fe^{2+}} - 0.0592 \log \frac{[Sn^{2+}][Fe^{2+}]}{[Sn^{4+}][Fe^{3+}]}$$

At any stage in the titration, the products are formed in a ratio such that $[Fe^{2+}] = 2[Sn^{4+}]$. At the equivalence point, $[Fe^{3+}] = 2[Sn^{2+}]$. When these identities are inserted in the above equation, the log term reduces to zero and

$$\mathscr{E}_{cell} = {}_oE_{S.C.E.} + \frac{2\,{}_rE^\circ_{Sn^{2+}/Sn^{4+}} + {}_rE^\circ_{Fe^{3+}/Fe^{2+}}}{3}$$

$$= -0.242 + \frac{(2)(0.154) + 0.771}{3} = +0.118 \text{ V}$$

11.8 Obtain the cell voltages in the titration of Problem 11.6 after (a) 30.00 mL, (b) 40.00 mL, and (c) 50.00 mL, of Fe^{3+} have been added.

These volumes are all in excess of that required for the stoichiometric equivalence point, and the cell reaction based on the iron half-reaction is the simplest to use.

$$\mathscr{E}_{cell} = {}_oE_{S.C.E.} + {}_rE^\circ_{Fe^{3+}/Fe^{2+}} - \frac{0.0592}{1} \log \frac{[Fe^{2+}]}{[Fe^{3+}]}$$

$$= -0.242 + 0.771 - 0.0592 \log \frac{[Fe^{2+}]}{[Fe^{3+}]}$$

(a) 30.00 mL of Fe^{3+} is 10.00 mL in excess of what is needed for the stoichiometric equivalence point. Hence,

$$\frac{[Fe^{2+}]}{[Fe^{3+}]} = \frac{20.00}{10.00} = 2.00$$

and $\mathscr{E}_{cell} = 0.511$ V.

(b) 40.00 mL is 20.00 mL in excess of the equivalence volume, and

$$\frac{[Fe^{2+}]}{[Fe^{3+}]} = \frac{20.00}{20.00} = 1.00$$

Hence, the log term is zero and $\mathscr{E}_{cell} = 0.529$ V.

(c) 50.00 mL is 30.00 mL in excess of the equivalence volume;

$$\frac{[Fe^{2+}]}{[Fe^{3+}]} = \frac{20.00}{30.00} = 0.666$$

and $\mathscr{E}_{cell} = 0.539$ V.

11.9 25.00 mL of 0.150 M Ti^{2+} (buffered to pH = 4.000) is titrated with 0.125 M UO_2^{2+} (also buffered to pH = 4.000), according to the reaction

$$2Ti^{2+} + UO_2^{2+} + 4H^+ \rightarrow U^{4+} + 2Ti^{3+} + 2H_2O$$

Find the voltages of the cell (11.1) after (a) 5.00 mL, (b) 7.50 mL, and (c) 14.00 mL, of UO_2^{2+} have been added.

From Table 10-1 and (10.6), $\mathscr{E}^{\circ}_{reac} = 0.368 + 0.334 = 0.702$, so that $K = 5._2 \times 10^{23}$ and the reaction can be assumed to go to completion. From (1.3), with $n_o = 1$ and $n_r = 2$, it requires 15.0 mL to reach the equivalence point; hence, Ti^{2+} is still present and the cell equation based on the titanium half-reaction is the simplest to solve.

$$\mathscr{E}_{cell} = {}_oE_{S.C.E.} + {}_rE^{\circ}_{Ti^{3+}/Ti^{2+}} - \frac{0.0592}{1} \log \frac{[Ti^{2+}]}{[Ti^{3+}]}$$

$$= -0.242 - 0.368 - 0.0592 \log \frac{[Ti^{2+}]}{[Ti^{3+}]}$$

(a) When 5.00 mL is added, 5.00/15.00 of the Ti^{2+} is oxidized, so that

$$\frac{[Ti^{2+}]}{[Ti^{3+}]} = \frac{10.00/15.00}{5.00/15.00} = 2.00$$

and $\mathscr{E}_{cell} = -0.628$ V.

(b) When 7.50 mL is added, half of the Ti^{2+} is oxidized, $[Ti^{2+}]/[Ti^{3+}] = 1.00$, the log term is zero, and $\mathscr{E}_{cell} = -0.610$ V.

(c) When 14.00 mL is added, 14.00/15.00 of the Ti^{2+} is oxidized, so that

$$\frac{[Ti^{2+}]}{[Ti^{3+}]} = \frac{1.00/15.00}{14.00/15.00} = \frac{1.00}{14.00}$$

and $\mathscr{E}_{cell} = -0.542$ V.

11.10 What is the voltage of the cell of Problem 11.9 at the equivalence point in the titration? (The only reaction of importance is the original redox titration reaction; other oxidation states of uranium are unimportant.)

Addition of the two expressions for $n\mathscr{E}_{cell}$ furnished by the two titration half-reactions results in

$$3\mathscr{E}_{cell} = 3 {}_o E_{S.C.E.} + {}_r E^\circ_{Ti^{3+}/Ti^{2+}} + 2 {}_r E^\circ_{UO_2^{2+}/U^{4+}} - 0.0592 \log \frac{[Ti^{2+}][U^{4+}]}{[Ti^{3+}][UO_2^{2+}][H^+]^4}$$

At any stage in the titration, $[Ti^{3+}] = 2[U^{4+}]$, whereas, at the equivalence point, $[Ti^{2+}] = 2[UO_2^{2+}]$. Inserting these identities in the log term, we obtain

$$\mathscr{E}_{cell} = {}_o E_{S.C.E.} + \tfrac{1}{3}({}_r E^\circ_{Ti^{3+}/Ti^{2+}} + 2 {}_r E^\circ_{UO_2^{2+}/U^{4+}}) - \tfrac{1}{3}(0.0592) \log \frac{1}{[H^+]^4}$$

$$= -0.242 + \tfrac{1}{3}[-0.368 + 2(0.334)] - \tfrac{1}{3}(0.0592)(4)(4.000) = -0.458 \text{ V}$$

11.11 Give the voltages of the cell of Problem 11.9 when the amounts of added UO_2^{2+} are (a) 20.00 mL, (b) 30.00 mL, and (c) 40.00 mL.

The simplest expression for $\mathscr{E}_{cell}$ will be the one based on the uranium half-reaction, since the given volumes all exceed the equivalence-point volume, 15.00 mL.

$$\mathscr{E}_{cell} = {}_o E_{S.C.E.} + {}_r E^\circ_{UO_2^{2+}/U^{4+}} - \frac{0.0592}{2} \log \frac{[U^{4+}]}{[UO_2^{2+}][H^+]^4}$$

$$= -0.242 + 0.334 - \frac{0.0592}{2} \log \frac{[U^{4+}]}{[UO_2^{2+}](1.00 \times 10^{-4})^4}$$

(a) This is 5.00 mL past the equivalence point:

$$\frac{[U^{4+}]}{[UO_2^{2+}]} = \frac{15.00}{5.00} = 3.00$$

so that $\mathscr{E}_{cell} = -0.396$ V.

(b) This is 15.00 mL past the equivalence point:

$$\frac{[U^{4+}]}{[UO_2^{2+}]} = \frac{15.00}{15.00} = 1.000$$

and $\mathscr{E}_{cell} = -0.382$ V.

(c) This is 25.00 mL past the equivalence point:

$$\frac{[U^{4+}]}{[UO_2^{2+}]} = \frac{15.00}{25.00} = 0.6000$$

and $\mathscr{E}_{cell} = -0.375$ V.

11.12 Check Problem 11.10 by separate calculations of $\mathscr{E}_{cell}$ from the two half-cell reactions.

The concentrations of principal species can be calculated as follows: At the equivalence point, the original $(0.150)(25.00) = 3.75$ mmol of Ti^{2+} is converted to Ti^{3+}; it is present in a volume of 40.00 mL, so that

$$[Ti^{3+}] = \frac{3.75}{40.00} = 0.0938$$

The $(0.125)(15.00) = 1.87_5$ mmol of UO_2^{2+} is converted to U^{4+}, which also is present in a volume of 40.00 mL; thus

$$[U^{4+}] = \frac{1.87_5}{40.00} = 0.0469$$

Write $[UO_2^{2+}] = x$; then $[Ti^{2+}] = 2x$. Knowing K from Problem 11.9, we have:

$$K = 5._2 \times 10^{23} = \frac{[U^{4+}][Ti^{3+}]^2}{[UO_2^{2+}][Ti^{2+}]^2[H^+]^4} = \frac{(0.0469)(0.0938)^2}{(x)(2x)^2(1.00 \times 10^{-4})^4}$$

from which $x = [UO_2^{2+}] = 1._{26} \times 10^{-4}$ and $2x = [Ti^{2+}] = 2._{52} \times 10^{-4}$.

Using the expression based on titanium (see Problem 11.9),

$$\mathcal{E}_{cell} = -0.242 - 0.368 - 0.0592 \log \frac{2._{52} \times 10^{-4}}{0.0938} = -0.45_8 \text{ V}$$

Using the expression based on uranium (see Problem 11.11),

$$\mathcal{E}_{cell} = -0.242 + 0.334 - \frac{0.0592}{2} \log \frac{0.0469}{(1._{26} \times 10^{-4})(1.00 \times 10^{-4})^4} = -0.458 \text{ V}$$

11.13 25.00 mL of 0.240 M Cu^+ is titrated with 0.0500 M $Cr_2O_7^{2-}$, using the cell (*11.1*). Both solutions are buffered to pH = 2.000. What are the cell voltages when (*a*) 5.00 mL, (*b*) 10.00 mL, and (*c*) 15.00 mL, of $Cr_2O_7^{2-}$ are added? The redox reaction is

$$Cr_2O_7^{2-} + 6Cu^+ + 14H^+ \rightarrow 2Cr^{3+} + 6Cu^{2+} + 7H_2O$$

From Table 10-1 and (*10.6*),

$$\mathcal{E}_{reac}^\circ = {}_oE_{Cu^+/Cu^{2+}}^\circ + {}_rE_{Cr_2O_7^{2-}/Cr^{3+}}^\circ = -0.153 + 1.33 = +1.18 \text{ V}$$

From (*10.8*), $K = 10^{119}$, so that the reaction can be assumed to go to completion. From (*1.3*), with $n_o = 1$ and $n_r = 6$, 20.00 mL of $Cr_2O_7^{2-}$ is required to reach the equivalence point. The volumes here are all less than 20.00 mL, so it is simplest to use the expression based on the copper half-reaction:

$$\mathcal{E}_{cell} = {}_oE_{S.C.E.} + {}_rE_{Cu^{2+}/Cu}^\circ - \frac{0.0592}{1} \log \frac{[Cu^+]}{[Cu^{2+}]}$$

$$= -0.242 + 0.153 - 0.0592 \log \frac{[Cu^+]}{[Cu^{2+}]}$$

(*a*) The 5.00 mL of $Cr_2O_7^{2-}$ oxidizes 5.00/20.00 of the Cu^+ to Cu^{2+}, so that

$$\frac{[Cu^+]}{[Cu^{2+}]} = \frac{15.00/20.00}{5.00/20.00} = 3.00$$

and $\mathcal{E}_{cell} = -0.117$ V.

(*b*) The 10.00 mL of $Cr_2O_7^{2-}$ converts half of the Cu^+ to Cu^{2+}; hence, $[Cu^+]/[Cu^{2+}] = 1.000$, the log term is zero, and $\mathcal{E}_{cell} = -0.089$ V.

(*c*) The 15.00 mL of $Cr_2O_7^{2-}$ converts 15.00/20.00 of the Cu^+ to Cu^{2+}, so that

$$\frac{[Cu^+]}{[Cu^{2+}]} = \frac{5.00/20.00}{15.00/20.00} = \frac{1.00}{3.00}$$

and $\mathcal{E}_{cell} = -0.061$ V.

11.14 What is the cell voltage at the equivalence point in the titration of Problem 11.13?

Take the sum of the $n\mathcal{E}_{cell}$-expressions furnished by the two half-reactions:

$$6\mathcal{E}_{cell} = 6\,{}_oE_{S.C.E.} + 6\,E_{Cr_2O_7^{2-}/Cr^{3+}}^\circ - 0.0592 \log \frac{[Cr^{3+}]^2}{[Cr_2O_7^{2-}][H^+]^{14}}$$

$$1\mathcal{E}_{cell} = {}_oE_{S.C.E.} + {}_rE_{Cu^{2+}/Cu^+}^\circ - 0.0592 \log \frac{[Cu^+]}{[Cu^{2+}]}$$

$$\overline{7\mathcal{E}_{cell} = 7\,{}_oE_{S.C.E.} + 6\,{}_rE_{Cr_2O_7^{2-}/Cr^{3+}}^\circ + {}_rE_{Cu^{2+}/Cu^+}^\circ - 0.0592 \log \frac{[Cr^{3+}]^2[Cu^+]}{[Cr_2O_7^{2-}][Cu^{2+}][H^+]^{14}}}$$

At any stage in the titration, $[Cu^{2+}] = 3[Cr^{3+}]$; also, at the equivalence point, $[Cu^+] = 6[Cr_2O_7^{2-}]$ is valid. Substituting these in the summed expression leads to

$$\mathcal{E}_{cell} = {}_oE_{S.C.E.} + \frac{6\,{}_rE_{Cr_2O_7^{2-}/Cr^{3+}}^\circ + {}_rE_{Cu^{2+}/Cu^+}^\circ}{7} - \frac{0.0592}{7} \log \frac{2[Cr^{3+}]}{[H^+]^{14}}$$

Since the balanced chromium half-reaction is unsymmetrical in that $1Cr_2O_7^{2-} \rightarrow 2Cr^{3+}$, an extra $[Cr^{3+}]$ and the factor 2 remain after canceling. [This and $2Hg^{2+} + 2e^- \rightarrow Hg_2^{2+}$ are the only common half-reactions for which extra terms of this sort remain in the equivalence-point calculation.] Since $(0.0500)(20.00) = 1.00$ mmol of $Cr_2O_7^{2-}$ was added at the equivalence point, there is $2(1.00) = 2.00$ mmol of Cr^{3+} product, and

$$[Cr^{3+}] = \frac{2.00}{45.00} = 0.0444$$

Therefore,

$$\mathscr{E}_{cell} = -0.242 + \frac{6(1.33) + 0.153}{7} - \frac{0.0592}{7} \log \frac{2(0.0444)}{(1.00 \times 10^{-2})^{14}} = 0.69_2 = 0.69_2 \text{ V}$$

11.15 Determine the cell voltages in the titration of Problem 11.13 when the amounts of added $Cr_2O_7^{2-}$ are (a) 28.00 mL, (b) 40.00 mL, and (c) 52.00 mL.

The simplest expression for the cell voltage is

$$\mathscr{E}_{cell} = {}_oE_{S.C.E.} + {}_rE^{\circ}_{Cr_2O_7^{2-}/Cr^{3+}} - \frac{0.0592}{6} \log \frac{[Cr^{3+}]^2}{[Cr_2O_7^{2-}][H^+]^{14}}$$

$$= -0.242 + 1.33 - \frac{0.0592}{6} \log \frac{[Cr^{3+}]^2}{[Cr_2O_7^{2-}](1.00 \times 10^{-2})^{14}}$$

Of the added $Cr_2O_7^{2-}$, $(0.0500)(20.00) = 1.00$ mmol is converted to $2Cr^{3+}$, so that there is 2.00 mmol of Cr^{3+}.

(a) The total volume is 53.00 mL, so that

$$[Cr^{3+}] = \frac{2.00}{53.00} = 0.0377 \qquad [Cr_2O_7^{2-}] = \frac{(0.0500)(28.00 - 20.00)}{53.00} = 7.55 \times 10^{-3}$$

and $\mathscr{E}_{cell} = 0.81_9$ V.

(b) Since the total volume is 65.00 mL

$$[Cr^{3+}] = \frac{2.00}{65.00} = 0.0308 \qquad [Cr_2O_7^{2-}] = \frac{(0.0500)(40.00 - 20.00)}{65.00} = 0.0154$$

and $\mathscr{E}_{cell} = 0.82_4$ V.

(c) Since the total volume is 77.00 mL,

$$[Cr^{3+}] = \frac{2.00}{77.00} = 0.0260 \qquad [Cr_2O_7^{2-}] = \frac{(0.0500)(52.00 - 20.00)}{77.00} = 0.0208$$

and $\mathscr{E}_{cell} = 0.82_6$ V.

11.16 20.00 mL of 0.100 M Ag^+ is titrated with 0.125 M Cl^- using the cell

$$\text{S.C.E.} \parallel redox\ titration \mid AgCl(s) \mid Ag(s)$$

The redox titration is $Ag^+ + Cl^- \rightarrow AgCl(s)$, which is the sum of the two half-reactions

$$Ag(s) + Cl^- \rightarrow AgCl(s) + e^- \qquad Ag^+ + e^- \rightarrow Ag(s)$$

Find (a) $\mathscr{E}^{\circ}_{reac}$ and (b) the equivalence-point volume.

(a) From Table 10-1 and (10.6), $\mathscr{E}^{\circ}_{reac} = {}_oE^{\circ}_{Ag/AgCl} + {}_rE^{\circ}_{Ag^+/Ag} = -0.222 + 0.800 = +0.578$ V.

(b) From (1.3), with $n_o = n_r = 1$, the equivalence-point volume is 16.0 mL.

11.17 What is the reduction potential of the titration solution of Problem 11.16 when the amount of added Cl^- is (a) 5.00 mL?   (b) 8.00 mL?   (c) 14.00 mL?

Prior to the equivalence point, appreciable Ag^+ will still remain in solution, and the simplest calculation will make use of the $Ag(s)$ reduction potential:

$$_rE_{reac} = {}_rE_{Ag^+/Ag} = {}_rE^\circ_{Ag^+/Ag} - \frac{0.0592}{1} \log \frac{1}{[Ag^+]}$$

$$= +0.800 + 0.0592 \log [Ag^+]$$

Since we started with $(0.100)(20.00) = 2.00$ mmol of Ag^+, we have:

(a) $$[Ag^+] = \frac{2.00 - (0.125)(5.00)}{25.00} = 0.0550 \qquad _rE_{reac} = 0.725 \text{ V}$$

(b) $$[Ag^+] = \frac{2.00 - (0.125)(8.00)}{28.00} = 0.0357 \qquad _rE_{reac} = 0.714 \text{ V}$$

(c) $$[Ag^+] = \frac{2.00 - (0.125)(14.00)}{34.00} = 0.0073_5 \qquad _rE_{reac} = 0.674 \text{ V}$$

11.18 What is the reduction potential of the titration solution of Problem 11.16 at the equivalence point?

Add the expressions for n_rE_{reac} as given by the two half-reactions:

$$1_rE_{reac} = {}_rE^\circ_{Ag^+/Ag} - 0.0592 \log \frac{1}{[Ag^+]}$$

$$1_rE_{reac} = {}_rE^\circ_{AgCl/Ag} - 0.0592 \log \frac{[Cl^-]}{1}$$

$$\overline{\rule{0pt}{1em}\hspace{6cm}}$$

$$2_rE_{reac} = {}_rE^\circ_{Ag^+/Ag} + {}_rE^\circ_{AgCl/Ag} - 0.0592 \log \frac{[Cl^-]}{[Ag^+]}$$

At the stoichiometric equivalence point, $[Ag^+] = [Cl^-]$, the log term is zero, and

$$_rE_{reac} = \frac{0.800 + 0.222}{2} = 0.511 \text{ V}$$

11.19 What is the reduction potential of the titration solution of Problem 11.16 when the amount of added Cl^- is (a) 20.00 mL?   (b) 32.00 mL?   (c) 45.00 mL?

These volumes are all past the equivalence point and involve excess Cl^-, so that the calculation is most easily performed using the reduction potential based on $AgCl(s)$:

$$_rE_{reac} = {}_rE_{AgCl/Ag} = {}_rE^\circ_{AgCl/Ag} - \frac{0.0592}{1} \log \frac{[Cl^-]}{1}$$

$$= +0.222 - 0.0592 \log [Cl^-]$$

Since $(0.125)(16.00) = 2.00$ mmol of Cl^- is consumed in the reaction with Ag^+, we have:

(a) $$[Cl^-] = \frac{(0.125)(20.00) - 2.00}{40.00} = 0.012_5 \qquad _rE_{reac} = 0.334 \text{ V}$$

(b) $$[Cl^-] = \frac{(0.125)(32.00) - 2.00}{52.00} = 0.0385 \qquad _rE_{reac} = 0.306 \text{ V}$$

(c) $$[Cl^-] = \frac{(0.125)(45.00) - 2.00}{65.00} = 0.0558 \qquad _rE_{reac} = 0.296 \text{ V}$$

TITRATION OF MIXTURES

11.20 A solution containing Fe^{2+} and Tl^+ is titrated with Ce^{4+}, both solutions also being 1 M $HClO_4$; take $_rE°$ for Ce^{4+} to be 1.70 V. (*a*) Which ion is oxidized first? (*b*) Do the reactions go to completion? (*c*) Are separate equivalence points observed?

The redox reactions are:

$$Ce^{4+} + Fe^{2+} \rightarrow Ce^{3+} + Fe^{3+} \tag{I}$$

$$2Ce^{4+} + Tl^+ \rightarrow 2Ce^{3+} + Tl^{3+} \tag{II}$$

(*a*) From Table 10-1 and (*10.6*),

$$\mathscr{E}_I° = {_o}E°_{Fe^{2+}/Fe^{3+}} + {_r}E°_{Ce^{4+}/Ce^{3+}} = -0.767 + 1.70 = +0.93 \text{ V}$$

$$\mathscr{E}_{II}° = {_o}E°_{Tl^+/Tl^{3+}} + {_r}E°_{Ce^{4+}/Ce^{3+}} = -1.26 + 1.70 = +0.44 \text{ V}$$

Corresponding to the larger standard-state potential, the Fe^{2+} is oxidized first.

(*b*) As $\mathscr{E}_I°$ and $\mathscr{E}_{II}°$ each exceed +0.2 V, the K-values are very large, and both reactions may be assumed to go to completion.

(*c*) Since $\mathscr{E}_I° - \mathscr{E}_{II}° = 0.49 > 0.2$ V and each $\mathscr{E}°$ exceeds 0.2 V, there is sufficient difference between the two reactions for separate equivalence points to be seen.

11.21 A solution containing Sn^{2+} and Cu^+ is titrated with Fe^{3+}. (*a*) Which ion is oxidized first? (*b*) Do the reactions go to completion? (*c*) Are separate equivalence points observed?

The redox reactions are:

$$2Fe^{3+} + Sn^{2+} \rightarrow 2Fe^{2+} + Sn^{4+} \tag{I}$$

$$Fe^{3+} + Cu^+ \rightarrow Fe^{2+} + Cu^{2+} \tag{II}$$

(*a*) From Table 10-1 and 10.6,

$$\mathscr{E}_I° = {_o}E°_{Sn^{2+}/Sn^{4+}} + {_r}E°_{Fe^{3+}/Fe^{2+}} = -0.154 + 0.771 = +0.617 \text{ V}$$

$$\mathscr{E}_{II}° = {_o}E°_{Cu^+/Cu^{2+}} + {_r}E°_{Fe^{3+}/Fe^{2+}} = -0.153 + 0.771 = +0.618 \text{ V}$$

Since $\mathscr{E}_{II}°$ is only slightly greater than $\mathscr{E}_I°$, at any stage in the titration there will be slightly more oxidation of Cu^+ than of Sn^{2+}.

(*b*) Since $\mathscr{E}_I°$ and $\mathscr{E}_{II}°$ are both greater than 0.2 V, the K-values are very large, and both reactions may be assumed to go to completion.

(*c*) Since $\mathscr{E}_{II}° - \mathscr{E}_I° = 0.001 \ll 0.2$ V, *separate* equivalence points are not observed. However, since each $\mathscr{E}°$ exceeds 0.2 V, a single equivalence point corresponding to the sum of the two redox reactions will be seen.

11.22 A mixture of Tl^{3+} and Fe^{3+} is titrated with Sn^{2+}. (*a*) Which ion is reduced first? (*b*) Do the reactions go to completion? (*c*) Are separate equivalence points observed?

The redox reactions are:

$$Sn^{2+} + Tl^{3+} \rightarrow Sn^{4+} + Tl^+ \tag{I}$$

$$Sn^{2+} + 2Fe^{3+} \rightarrow Sn^{4+} + 2Fe^{2+} \tag{II}$$

(*a*) From Table 10-1 and (*10.6*),

$$\mathscr{E}_I° = {_o}E°_{Sn^{2+}/Sn^{4+}} + {_r}E°_{Tl^{3+}/Tl^+} = -0.154 + 1.25 = 1.10 \text{ V}$$

$$\mathscr{E}_{II}° = {_o}E°_{Sn^{2+}/Sn^{4+}} + {_r}E°_{Fe^{3+}/Fe^{2+}} = -0.154 + 0.771 = 0.617 \text{ V}$$

Since $\mathscr{E}_I° > \mathscr{E}_{II}°$, and Tl^{3+} is reduced first.

(*b*) Since $\mathscr{E}_I°$ and $\mathscr{E}_{II}°$ each exceed 0.2 V, the K-values are very large, and both reactions may be assumed to go to completion.

(*c*) Since $\mathscr{E}_I° - \mathscr{E}_{II}° = 0.48 > 0.2$ V and each $\mathscr{E}°$ is greater than 0.2 V, separate equivalence points are observed.

11.23 12.00 mL of a solution that is 0.100 M Fe^{2+} and 0.150 M Tl^+ is titrated with 0.120 M Ce^{4+} using the cell (11.1); both solutions are also 1 M $HClO_4$ (see Problem 11.20). What is the cell voltage when the amount of added Ce^{4+} is (a) 2.00 mL?   (b) 5.00 mL?   (c) 9.00 mL?

From (1.3), it will require 10.00 mL of Ce^{4+} to reach the equivalence point for reaction I and an additional 30.00 mL (40.00 mL total) to reach the equivalence point for reaction II. The specified volumes are prior to the first equivalence point, so that not all Fe^{2+} is oxidized. Therefore, the simplest expression is the cell equation based on the iron half-reaction:

$$\mathscr{E}_{cell} = {}_oE_{S.C.E.} + {}_rE^\circ_{Fe^{3+}/Fe^{2+}} - \frac{0.0592}{1} \log \frac{[Fe^{2+}]}{[Fe^{3+}]}$$

$$= -0.242 + 0.767 - 0.0592 \log \frac{[Fe^{2+}]}{[Fe^{3+}]}$$

(a) 2.00 mL of Ce^{4+} converts 2.00/10.00 of the Fe^{2+} to Fe^{3+}; hence,

$$\frac{[Fe^{2+}]}{[Fe^{3+}]} = \frac{8.00/10.00}{2.00/10.00} = 4.00$$

and $\mathscr{E}_{cell} = 0.489$ V.

(b) 5.00 mL of Ce^{4+} converts half of the Fe^{2+} to Fe^{3+}. Thus, $[Fe^{2+}] = [Fe^{3+}]$, the log term is zero, and $\mathscr{E}_{cell} = 0.525$ V.

(c) 9.00 mL of Ce^{4+} converts 9.00/10.00 of the Fe^{2+} to Fe^{3+}; hence,

$$\frac{[Fe^{2+}]}{[Fe^{3+}]} = \frac{1.00/10.00}{9.00/10.00} = 0.111$$

and $\mathscr{E}_{cell} = 0.582$ V.

11.24 What is the voltage of the cell in Problem 11.23 at the first equivalence point in the titration?

The first equivalence point occurs when 10.00 mL of Ce^{4+} (1.20 mmol) has been added. The total volume of the solution is 22.00 mL and the concentrations of the principal species are:

$$[Ce^{3+}] = \frac{1.20}{22.00} = 0.0545 \qquad [Fe^{3+}] = \frac{1.20}{22.00} = 0.0545 \qquad [Tl^+] = \frac{(0.150)(12.00)}{22.00} = 0.0818$$

The concentrations of the other species, Ce^{4+}, Fe^{2+}, and Tl^{3+}, are each very small. There are two possible reactions involving the principal species:

$$Fe^{3+} + Ce^{3+} \rightarrow Fe^{2+} + Ce^{4+} \tag{III}$$
$$2Fe^{3+} + Tl^+ \rightarrow 2Fe^{2+} + Tl^{3+} \tag{IV}$$

Reaction III is simply the reverse of reaction I of Problem 11.20; it has $\mathscr{E}^\circ_{III} = -\mathscr{E}^\circ_I = -0.93$ V. For reaction IV,

$$\mathscr{E}^\circ_{IV} = {}_oE^\circ_{Tl^+/Tl^{3+}} + {}_rE^\circ_{Fe^{3+}/Fe^{2+}} = -1.26 + 0.767 = -0.49 \text{ V}$$

Thus, neither reaction proceeds to any appreciable extent [by (10.8), $K_{III} = 10^{-16}$, $K_{IV} = 10^{-17}$]. However, since $\mathscr{E}^\circ_{IV} > \mathscr{E}^\circ_{III}$, and since they differ by more than 0.2 V, reaction IV will proceed much further than reaction III. Since reaction IV is the predominant reaction, the products of IV must be in a stoichiometric ratio, $[Fe^{2+}] = 2[Tl^{3+}]$. Therefore, the simplest calculation of the cell voltage will involve adding the $n\mathscr{E}_{cell}$-expressions based on iron and thallium:

$$1\mathscr{E}_{cell} = {}_oE^\circ_{S.C.E.} + {}_rE^\circ_{Fe^{3+}/Fe^{2+}} - 0.0592 \log \frac{[Fe^{2+}]}{[Fe^{3+}]}$$

$$2\mathscr{E}_{cell} = 2{}_oE^\circ_{S.C.E.} + 2{}_rE^\circ_{Tl^{3+}/Tl^+} - 0.0592 \log \frac{[Tl^+]}{[Tl^{3+}]}$$

$$\rule{11cm}{0.4pt}$$

$$3\mathscr{E}_{cell} = 3{}_oE^\circ_{S.C.E.} + {}_rE^\circ_{Fe^{3+}/Fe^{2+}} + 2{}_rE^\circ_{Tl^{3+}/Tl^+} - 0.0592 \log \frac{[Fe^{2+}][Tl^+]}{[Fe^{3+}][Tl^{3+}]}$$

Substituting $[Fe^{2+}] = 2[Tl^{3+}]$ and the calculated $[Tl^+]$ and $[Fe^{3+}]$, we obtain

$$\mathscr{E}_{cell} = -0.242 + \frac{0.767 + 2(1.26)}{3} - \frac{0.0592}{3} \log \frac{2(0.0818)}{0.0545} = 0.84_4 \text{ V}$$

11.25 Find the cell voltages in Problem 11.23 when the amounts of added Ce^{4+} are (a) 12.50 mL, (b) 25.00 mL, and (c) 39.00 mL.

These volumes correspond to the second titration step, where Tl^+ is converted to Tl^{3+}. The simplest expression for $\mathscr{E}_{cell}$ is based on the thallium half-reaction:

$$\mathscr{E}_{cell} = {}_oE_{S.C.E.} + {}_rE^{\circ}_{Tl^{3+}/Tl^+} - \frac{0.0592}{2} \log \frac{[Tl^+]}{[Tl^{3+}]}$$

$$= -0.242 + 1.26 - \frac{0.0592}{2} \log \frac{[Tl^+]}{[Tl^{3+}]}$$

Since the first 10.00 mL of added Ce^{4+} was consumed in oxidizing the Fe^{2+}, the additional volumes of Ce^{4+} used in oxidizing Tl^+ are 2.50, 15.00, and 29.00 mL.

(a) The 2.50 mL of additional Ce^{4+} converts 2.50/30.00 of the Tl^+ to Tl^{3+}, so that

$$\frac{[Tl^+]}{[Tl^{3+}]} = \frac{27.50/30.00}{2.50/30.00} = 11.0$$

and $\mathscr{E}_{cell} = 0.99$ V.

(b) The 15.00 mL of additional Ce^{4+} converts half of the Tl^+ to Tl^{3+}, so that $[Tl^+] = [Tl^{3+}]$, the log term is zero, and $\mathscr{E}_{cell} = 1.02$ V.

(c) The 29.00 mL of additional Ce^{4+} converts 29.00/30.00 of the Tl^+ to Tl^{3+}, so that

$$\frac{[Tl^+]}{[Tl^{3+}]} = \frac{1.00/30.00}{29.00/30.00} = 0.0345$$

and $\mathscr{E}_{cell} = 1.06$ V.

11.26 What is the cell voltage at the second equivalence point in the titration of Problem 11.23?

The second equivalence point occurs when a total of 40.00 mL Ce^{4+} has been added. The total volume is 52.00 mL, and the concentrations of principal species are:

$$[Ce^{3+}] = \frac{(0.120)(40.00)}{52.00} = 0.0923 \qquad [Fe^{3+}] = \frac{(0.100)(12.00)}{52.00} = 0.0231 \qquad [Tl^{3+}] = \frac{(0.150)(12.00)}{52.00} = 0.0346$$

The only reactions are the reverses of reactions I and II of Problem 11.20:

$$Ce^{3+} + Fe^{3+} \rightarrow Ce^{4+} + Fe^{2+} \qquad\qquad\qquad\qquad\qquad\qquad (III)$$
$$2Ce^{3+} + Tl^{3+} \rightarrow 2Ce^{4+} + Tl^+ \qquad\qquad\qquad\qquad\qquad (IV)$$

Thus $\mathscr{E}^{\circ}_{III} = -0.93$ V and $\mathscr{E}^{\circ}_{IV} = -0.44$ V. Since $\mathscr{E}^{\circ}_{IV}$ is algebraically the greater and $\mathscr{E}^{\circ}_{IV} - \mathscr{E}^{\circ}_{III} > 0.2$ V, reaction IV is the principal equilibrium at the second equivalence point. The simplest expression for the cell voltage will involve the sum of the $n\mathscr{E}_{cell}$-expressions based on the half-reactions for Ce^{4+} and Tl^+:

$$1\mathscr{E}_{cell} = {}_oE_{S.C.E.} + {}_rE^{\circ}_{Ce^{4+}/Ce^{3+}} - 0.0592 \log \frac{[Ce^{3+}]}{[Ce^{4+}]}$$

$$2\mathscr{E}_{cell} = 2{}_oE_{S.C.E.} + 2{}_rE^{\circ}_{Tl^{3+}/Tl^+} - 0.0592 \log \frac{[Tl^+]}{[Tl^{3+}]}$$

_____

$$3\mathscr{E}_{cell} = 3{}_oE_{S.C.E.} + {}_rE^{\circ}_{Ce^{4+}/Ce^{3+}} + 2{}_rE^{\circ}_{Tl^{3+}/Tl^+} - 0.0592 \frac{[Ce^{3+}][Tl^+]}{[Ce^{4+}][Tl^{3+}]}$$

The stoichiometry of reaction IV requires that $[Ce^{4+}] = 2[Tl^+]$. Inserting this, and the values

$$[Ce^{3+}] = 0.0923 \qquad [Tl^{3+}] = 0.0346$$

in the expression for $3\mathscr{E}_{cell}$ results in

$$\mathscr{E}_{cell} = -0.242 + \frac{1.70 + 2(1.26)}{3} - \frac{0.0592}{3} \log \frac{0.0923}{2(0.0346)} = 1.16_3 \text{ V}$$

11.27 Determine the cell voltages in Problem 11.23 when the total amounts of added Ce^{4+} are (a) 44.00 mL, (b) 80.00 mL, and (c) 95.00 mL.

The simplest expression to use is the cell equation based on the cerium half-reaction:

$$\mathscr{E}_{cell} = {}_oE_{S.C.E.} + {}_rE^\circ_{Ce^{4+}/Ce^{3+}} - \frac{0.0592}{1} \log \frac{[Ce^{3+}]}{[Ce^{4+}]}$$

$$= -0.242 + 1.70 - \frac{0.0592}{1} \log \frac{[Ce^{3+}]}{[Ce^{4+}]}$$

The first 40.00 mL of added Ce^{4+} is converted to Ce^{3+}.

(a) $\qquad \dfrac{[Ce^{3+}]}{[Ce^{4+}]} = \dfrac{40.00}{4.00} = 10.0 \qquad \mathscr{E}_{cell} = 1.40 \text{ V}$

(b) $\qquad \dfrac{[Ce^{3+}]}{[Ce^{4+}]} = \dfrac{40.00}{40.00} = 1.00 \qquad \mathscr{E}_{cell} = 1.46 \text{ V}$

(c) $\qquad \dfrac{[Ce^{3+}]}{[Ce^{4+}]} = \dfrac{40.00}{55.00} = 0.7273 \qquad \mathscr{E}_{cell} = 1.46_6 = 1.47 \text{ V}$

11.28 16.00 mL of a solution that is $0.0750\ M$ Sn^{2+} and $0.100\ M$ Fe^{2+} is titrated with $0.100\ M$ Ce^{4+}; each solution is also $1\ M$ $HClO_4$; take $_rE^\circ$ for Ce^{4+} as 1.70 V. Sketch the titration curve for the reduction potential of the titration solution, $_rE_{reac}$, as a function of added Ce^{4+}.

The two redox reactions are:

$$Sn^{2+} + 2Ce^{4+} \rightarrow Sn^{4+} + 2Ce^{3+} \tag{I}$$
$$Fe^{2+} + Ce^{4+} \rightarrow Fe^{3+} + Ce^{3+} \tag{II}$$

From Table 10-1 and (10.6),

$$\mathscr{E}^\circ_I = -0.154 + 1.70 = 1.55 \text{ V} \qquad \mathscr{E}^\circ_{II} = -0.767 + 1.70 = +0.93 \text{ V}$$

Since $\mathscr{E}^\circ_I > \mathscr{E}^\circ_{II}$, Sn^{2+} is oxidized first; and, since $\mathscr{E}^\circ_I - \mathscr{E}^\circ_{II} > 0.2$ V, separate equivalence points are seen. From (1.3), 24.00 mL of Ce^{4+} is required to reach the first equivalence point. Halfway to the first equivalence point (after 12.00 mL),

$$[Sn^{2+}] = [Sn^{4+}] \qquad \text{and} \qquad _rE_{reac} = {}_rE^\circ_{Sn^{4+}/Sn^{2+}} = +0.154 \text{ V}$$

From (1.3), an additional 16.00 mL (i.e., 40.00 mL total) is needed to reach the second equivalence point. Thus, the addition of 32.00 mL of Ce^{4+} corresponds to a point halfway through the second titration step:

$$[Fe^{2+}] = [Fe^{3+}] \qquad \text{and} \qquad _rE_{reac} = {}_rE^\circ_{Fe^{3+}/Fe^{2+}} = +0.767 \text{ V}$$

When a total of 80.00 mL of Ce^{4+} is added,

$$[Ce^{3+}] = [Ce^{4+}] \qquad \text{and} \qquad _rE^\circ_{reac} = {}_rE^\circ_{Ce^{4+}/Ce^{3+}} = +1.70 \text{ V}$$

The only two reactions at the first equivalence point are:

$$Sn^{4+} + 2Ce^{3+} \rightarrow Sn^{2+} + 2Ce^{4+} \tag{III}$$
$$Sn^{4+} + 2Fe^{2+} \rightarrow Sn^{2+} + 2Fe^{3+} \tag{IV}$$

From Table 10-1 and (*10.6*),

$$\mathscr{E}^\circ_{\text{III}} = -1.70 + 0.154 = -1.55 \text{ V} \qquad \mathscr{E}^\circ_{\text{IV}} = -0.767 + 0.154 = -0.613 \text{ V}$$

so that IV proceeds to a much greater extent than III. The simplest calculation of $_rE_{\text{reac}}$ involves the sum of redox potentials for Sn^{4+} and Fe^{3+}:

$$3\,_rE_{\text{reac}} = 2\,_rE^\circ_{\text{Sn}^{4+}/\text{Sn}^{2+}} + \,_rE^\circ_{\text{Fe}^{3+}/\text{Fe}^{2+}} - 0.0592 \frac{[\text{Sn}^{2+}][\text{Fe}^{2+}]}{[\text{Sn}^{4+}][\text{Fe}^{3+}]}$$

The equality $[Fe^{3+}] = 2[Sn^{2+}]$ follows from the stoichiometry of IV. Also,

$$[\text{Sn}^{4+}] = \frac{(0.0750)(16.00)}{40.00} = 0.0300 \qquad [\text{Fe}^{2+}] = \frac{(0.100)(16.00)}{40.00} = 0.0400$$

so that

$$_rE_{\text{reac}} = \frac{(2)(0.154) + 0.767}{3} - \frac{0.0592}{3} \log \frac{0.0400}{2(0.0300)} = 0.362 \text{ V}$$

at the first equivalence point.

The principal reaction at the second equivalence point is $Ce^{3+} + Fe^{3+} \rightarrow Ce^{4+} + Fe^{2+}$, so that

$$[\text{Ce}^{4+}] = [\text{Fe}^{2+}] \qquad [\text{Ce}^{3+}] = \frac{(0.100)(40.00)}{56.00} = 0.0714 \qquad [\text{Fe}^{3+}] = \frac{(0.100)(16.00)}{56.00} = 0.0286$$

The sum of the Ce^{4+} and Fe^{3+} redox potentials is

$$2\,_rE_{\text{reac}} = \,_rE^\circ_{\text{Ce}^{4+}/\text{Ce}^{3+}} + \,_rE^\circ_{\text{Fe}^{3+}/\text{Fe}^{2+}} - 0.0592 \log \frac{[\text{Ce}^{3+}][\text{Fe}^{2+}]}{[\text{Ce}^{4+}][\text{Fe}^{3+}]}$$

$$_rE_{\text{reac}} = \frac{1.70 + 0.767}{2} - \frac{0.0592}{2} \log \frac{0.0714}{0.0286} = 1.22 \text{ V}$$

at the second equivalence point. The titration curve is sketched in Fig. 11-2.

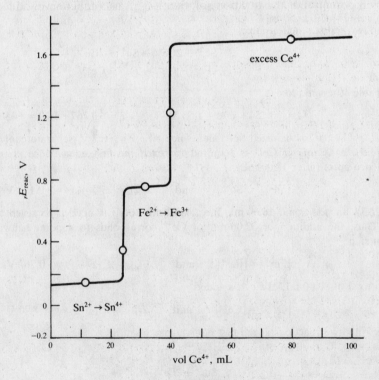

Fig. 11-2

REDOX INDICATORS

11.29 What should be the value of $_rE^\circ_{ox./red.}$ for an indicator of the type (*11.4*), if it is to be used in the titration of Problems 11.6–11.8?

From Problem 11.7, the reduction potential of the solution at the equivalence point is

$$_rE_{reac} = \mathscr{E}_{cell} - {}_oE_{S.C.E.} = 0.118 + 0.242 = 0.360 \text{ V}$$

The redox indicator should have $_rE^\circ_{ox./red.} = 0.360$ V.

11.30 A redox indicator obeys (*11.8*) and has $_rE^\circ_{ox./red.} = +1.12$ V. To what pH should the solutions of Problems 11.1–11.3 be buffered so that this indicator could be used?

From Problem 11.2, the reduction potential at the equivalence point is

$$_rE_{reac} = \mathscr{E}_{cell} - {}_oE_{S.C.E.} = +0.555 + 0.242 = 0.797 \text{ V}$$

From (*11.8*), $0.797 = 1.12 - 0.0592\,\text{pH}$, or pH = 5.5.

Supplementary Problems

TITRATION OF A SINGLE ION

11.31 18.00 mL of 0.125 M Sn^{4+} is titrated with 0.100 M Ti^{2+}, according to the reaction

$$Sn^{4+} + Ti^{2+} \rightarrow Sn^{2+} + Ti^{3+}$$

Find (a) $\mathscr{E}^\circ_{reac}$, (b) the equivalence-point volume of Ti^{2+}.   *Ans.* (a) 0.522 V; (b) 45.0_0 mL

11.32 What is the reduction potential, $_rE_{reac}$, of the solution of Problem 11.31 when the volume of added Ti^{2+} is (a) 15.00 mL? (b) 22.50 mL? (c) 45.00 mL? (d) 60.00 mL? (e) 90.00 mL?
Ans. (a) +0.163 V; (b) +0.154 V; (c) −0.020 V; (d) −0.340 V; (e) −0.368 V

11.33 24.00 mL of 0.100 M Sn^{2+} is titrated with 0.0500 M MnO_4^-; both solutions are buffered to pH = 4.000. The redox titration reaction is

$$5Sn^{2+} + 2MnO_4^- + 16H^+ \rightarrow 5Sn^{4+} + 2Mn^{2+} + 8H_2O$$

Find (a) $\mathscr{E}^\circ_{reac}$, (b) the equivalence-point volume of MnO_4^-.   *Ans.* (a) +1.36 V; (b) 19.2_0 mL

11.34 What is the reduction potential, $_rE_{reac}$, of the solution of Problem 11.33 when the amount of added MnO_4^- is (a) 5.00 mL? (b) 9.60 mL? (c) 19.20 mL? (d) 24.00 mL? (e) 38.40 mL? (f) 50.00 mL?
Ans. (a) 0.141 V; (b) 0.154 V; (c) 0.852 V; (d) 1.12_4 V; (e) 1.13_1 V; (f) 1.13_4 V

11.35 18.00 mL of 0.100 M SO_4^{2-} is titrated with 0.0900 M Pb^{2+}. The titration reaction,

$$Pb^{2+} + SO_4^{2-} \rightarrow PbSO_4(s)$$

is the sum of the half-reactions

$$Pb(s) + SO_4^{2-} \rightarrow PbSO_4(s) + 2e^- \qquad Pb^{2+} + 2e^- \rightarrow Pb(s)$$

Find (a) $\mathscr{E}^\circ_{reac}$, (b) the volume of added Pb^{2+} at the equivalence point.
Ans. (a) +0.230 V; (b) 20.0 mL

11.36 What is the reduction potential, $_rE_{reac}$, of the titration solution of Problem 11.35 when the amount of added Pb^{2+} is (a) 10.00 mL? (b) 20.00 mL? (c) 40.00 mL?
Ans. (a) −0.400 V; (b) −0.241 V; (c) −0.171 V

TITRATION OF MIXTURES

11.37 15.00 mL of a solution that is 0.0800 M Tl^{3+} and 0.120 M Ce^{4+} is titrated with 0.100 M Fe^{2+}, according to the reactions

$$Tl^{3+} + 2Fe^{2+} \rightarrow Tl^+ + 2Fe^{3+} \tag{I}$$
$$Ce^{4+} + Fe^{2+} \rightarrow Ce^{3+} + Fe^{3+} \tag{II}$$

Both solutions are 1 M $HClO_4$; $_rE^\circ = 1.70$ V for Ce^{4+}. (a) Determine $\mathscr{E}_I^\circ$ and $\mathscr{E}_{II}^\circ$. (b) Which ion is reduced first? (c) May reactions I and II be assumed to go to completion? (d) Are separate equivalence points observed? (e) What equivalence volumes of Fe^{2+} are required for each step in the titration?
Ans. (a) $\mathscr{E}_I^\circ = +0.49$ V, $\mathscr{E}_{II}^\circ = +0.93$ V; (b) Ce^{4+}; (c) yes; (d) yes; (e) 18.0_0 mL for the Ce^{4+} and an additional 20.0_0 mL for the Tl^+

11.38 What is the reduction potential, $_rE_{reac}$, of the titration solution of Problem 11.37 when the amount of added Fe^{2+} is (a) 6.00 mL? (b) 9.00 mL? (c) 16.00 mL?   *Ans.*   (a) 1.72 V; (b) 1.70 V; (c) 1.65 V

11.39 What is the reduction potential, $_rE_{reac}$, of the titration solution of Problem 11.37 at the first equivalence point?   *Ans.*   1.41 V

11.40 What is the reduction potential, $_rE_{reac}$, of the titration solution of Problem 11.37 when the total amount of added Fe^{2+} is (a) 20.00 mL? (b) 30.00 mL? (c) 40.00 mL?
Ans.   (a) 1.29 V; (b) 1.26 V; (c) 1.23 V

11.41 What is the reduction potential, $_rE_{reac}$, of the titration solution of Problem 11.37 at the second equivalence point?   *Ans.*   1.10 V

11.42 What is the reduction potential, $_rE_{reac}$, of the titration solution of Problem 11.37 when the total amount of added Fe^{2+} is (a) 45.00 mL? (b) 84.00 mL? (c) 100.00 mL?
Ans.   (a) 0.835 V; (b) 0.767 V; (c) 0.759 V

REDOX INDICATORS

11.43 What should be the value of $_rE^\circ_{ox./red.}$ for an indicator of the type (*11.4*), if it is to be used in the titration of Problems 11.33 and 11.34?   *Ans.*   0.58 V

11.44 If the solutions involved in the titration of Problems 11.33 and 11.34 are both buffered to pH = 4.500, what should be the value of $_rE^\circ_{ox./red.}$ for an indicator obeying (*11.8*)?   *Ans.*   0.85 V

Chapter 12

Phase Separations
and Chromatography

12.1 INTRODUCTION

Analyses of real substances are usually much more difficult than the analyses encountered in a teaching laboratory, because (i) other species may interfere with the analysis and/or (ii) the concentration of the species may be too low to detect. As a result, most real substances require some type of separation and/or preconcentration prior to analysis. Such procedures usually rely on various phase equilibria. One of the simplest separation methods involves selective precipitation. Other methods are based on solid-vapor equilibria as in sublimation, liquid-vapor equilibria as in distillation and fractionation, liquid-liquid equilibria as in solvent extraction, or solid-liquid equilibria as in liquid chromatography or electrolytic deposition. Considered in this chapter are three of the more general methods: *fractional distillation*, *solvent extraction*, and *gas chromatography*.

12.2 FRACTIONAL DISTILLATION

If the substances that comprise an ideal liquid mixture have different vapor pressures, then it is possible to separate the components of the mixture by fractional distillation. In the ideal mixture, at some temperature T, the *partial vapor pressure* of component i is given by $P_i = x_i P^\circ_i$, where x_i is the mole fraction of component i in the liquid phase and P°_i is the vapor pressure of pure i at the temperature T. If the mixture contains only two substances, A and B, then:

$$P_A = x_A P^\circ_A \qquad P_B = x_B P^\circ_B \qquad (x_A + x_B = 1) \tag{12.1}$$

If the *vapor* mixture is also ideal, so that each component obeys the ideal gas law, and if y_A and y_B denote the mole fractions of A and B in the vapor phase, then:

$$\frac{y_A}{y_B} = \frac{P_A}{P_B} = \frac{x_A P^\circ_A}{x_B P^\circ_B} \equiv \alpha \frac{x_A}{x_B} \tag{12.2}$$

where we have set $\alpha \equiv P^\circ_A / P^\circ_B$ and have supposed the designations A and B chosen such that $P^\circ_A > P^\circ_B$, making $\alpha > 1.00$. From (12.2), $y_A/y_B > x_A/x_B$; i.e., the vapor phase is enriched in A, the more volatile component.

If a series of n distillation-condensation steps occurs (as in a fractionation column), and if α can be assumed constant over the temperature range of the column, then (12.2) generalizes to

$$\frac{y_A}{y_B} = \alpha^{n+1} \frac{x_A}{x_B} \tag{12.3}$$

where y_A/y_B is the molar (or mole fractional) ratio in the distillate and x_A/x_B is the molar (or mole fractional) ratio in the fractionation pot. The integer n is called the *number of theoretical plates* in the fractionation column.

An approximate formula for α was obtained by A. Rose:

$$\log \alpha = 8.9 \frac{T_{bB} - T_{bA}}{T_{bB} + T_{bA}} \tag{12.4}$$

where T_{bA} and T_{bB} are the normal boiling points on the Kelvin scale.

12.3 SOLVENT EXTRACTIONS

Separation by solvent extraction depends on the differences in solubility of substances in two immiscible solvents. One of the solvents is usually H_2O; the other solvent is then a liquid immiscible with H_2O, such as octanol, diethyl ether, CCl_4, $CHCl_3$, or benzene. The *partition coefficient*, K_p, for a species X is the thermodynamic equilibrium constant for the distribution of X between the two phases. If one of the phases is water, w, and the other is an organic solvent, o, the separation process is diagramed as follows:

$$\left(K_p \equiv \frac{[X]_o}{[X]_w} \right) \tag{12.5}$$

It is not uncommon for X to dimerize or dissociate to some degree in either or both solvents. As a result, a more practical term, the *distribution coefficient*, D_c, is defined as the ratio of the *total* concentrations in the two phases:

$$D_c \equiv \frac{[\text{total concentration of all forms of X}]_o}{[\text{total concentration of all forms of X}]_w} \tag{12.6}$$

If X is an undissociated monomer in both phases, then $D_c = K_p$.

X Dissociates in the Aqueous Phase

If X is an acid, HA, the pictorial representation is

$$HA \underset{K_a}{\rightleftharpoons} H^+ + A^-$$

In this case, $D_c = [HA]_o/([HA]_w + [A^-]_w)$, which can be converted to

$$D_c = \frac{K_p [H^+]_w}{[H^+]_w + K_a} \tag{12.7}$$

Thus, D_c is dependent on the pH of the aqueous phase.

X Dimerizes in the Organic Phase

The dimerization of acetic acid in benzene is a typical example. The pictorial diagram for these systems is:

$$X \rightleftharpoons \tfrac{1}{2}X_2 \quad \text{or} \quad 2X \overset{K_d}{\rightleftharpoons} X_2$$

where $K_d = [X_2]_o/[X]_o^2$ is the equilibrium constant for the dimerization process. The distribution coefficient is

$$D_c = \frac{[X]_o + 2[X_2]_o}{[X]_w} = K_p + 2K_d K_p^2 [X]_w \tag{12.8}$$

Sequential Extraction

Let a single equilibration result in amounts $C_o V_o$ and $C_w V_w$ of all forms of X in the organic and water phases (whose respective volumes are V_o and V_w). Then, defining

$f_o \equiv$ fraction of original X in organic phase

$f_w \equiv 1 - f_o \equiv$ fraction of original X in water phase

$V_r \equiv V_o/V_w \equiv$ volume ratio

we have, using (12.6),

$$f_o = \frac{C_o V_o}{C_o V_o + C_w V_w} = \frac{\dfrac{C_o V_o}{C_w V_w}}{\dfrac{C_o V_o}{C_w V_w} + 1} = \frac{D_c V_r}{D_c V_r + 1} \qquad (12.9)$$

$$f_w = 1 - f_o = \frac{1}{D_c V_r + 1} \qquad (12.10)$$

Note that D_c and V_r are pure numbers; therefore, any convenient concentration unit can be chosen for C_o and C_w, and any convenient volume unit for V_o and V_w.

EXAMPLE 12.1   30.00 mL of a 0.100 M solution of X in CCl₄ is extracted successively with four fresh portions of water, each 10.00 mL in volume. If

$$K_p = \frac{[X]_o}{[X]_w} = 5.00$$

find the distribution of X between the CCl₄ and the fourth portion of water.

Since $V_r = 30.00/10.00 = 3.000$ and $D_c = K_p = 5.00$, (12.9) gives

$$f_o = \frac{(5.00)(3.000)}{(5.00)(3.000) + 1} = 0.937_5$$

Therefore, in four extractions, the fraction of X that remains in the CCl₄ phase is

$$f_o^4 = (0.937_5)^4 = 0.772_5 = 77.3\%$$

At any stage, the amounts of X in water and in CCl₄ are in the ratio f_w/f_o. Hence, the fraction of X in the fourth portion of water is

$$\frac{f_w}{f_o}(f_o^4) = f_w f_o^3 = (1 - f_o) f_o^3 = (0.062_5)(0.937_5)^3 \approx 5.1_5\%$$

12.4 CHROMATOGRAPHIC SEPARATIONS

In both gas chromatography and liquid chromatography, solute molecules having different K_p-values are separated by allowing the molecules to be absorbed/desorbed numerous times while being carried through the column by the moving gaseous or liquid phase. As the principles are similar in both chromatographic systems, we shall consider only gas chromatographic separations.

Figure 12-1 shows an idealized gas chromatogram for the separation of two substances, A and B. Here, t_M is the time required for a nonabsorbing substance to emerge from the column, $(t_R)_A$ is the retention time for substance A and $(t_R)_B$ is the retention time for substance B. The widths of the peaks at their bases, when the peaks are approximated as triangles, are $(t_W)_A$ and $(t_W)_B$. The area of a (nearly triangular) peak is proportional to the number of molecules, and thus to the number of moles, for most eluted substances.

The number of theoretical plates associated with a given peak is

$$N = 16\left(\frac{t_R}{t_W}\right)^2 \qquad (12.11)$$

If the column length is L, the *height-equivalent-theoretical-plate* (HETP or H) is

$$H = \frac{L}{N} \qquad (12.12)$$

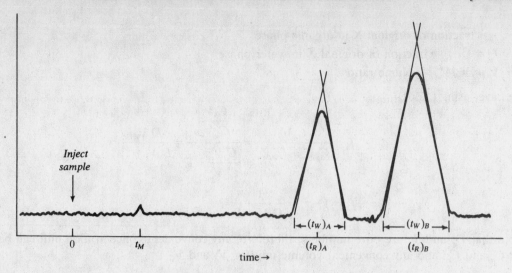

Fig. 12-1

The value of H is a function of gas flow rate, F; it is usually described fairly well by the equation

$$H = \theta + \frac{\beta}{F} + \gamma F \qquad (12.13)$$

where θ, β, and γ are constants for a given column and substance. Ordinarily, L is given in cm and F in cm^3/min, so that the units of θ, β, and γ would be cm, cm^4/min, and min/cm^2, respectively. By differentiation of (12.13) it is found that H is a minimum for $F = \sqrt{\beta/\gamma}$.

The resolution of two adjacent peaks, $\mathscr{R}$, is defined as:

$$\mathscr{R} \equiv \frac{\text{distance between peaks in time units}}{\text{average } t_W \text{ of the two peaks}} \qquad (12.14)$$

For the chromatogram shown in Fig. 12-1,

$$\mathscr{R} = \frac{(t_R)_B - (t_R)_A}{((t_W)_A + (t_W)_B)/2}$$

If $\mathscr{R} \approx 1.0$, a reasonable separation between peaks exists, since there is only 2% cross-contamination of the peaks when they are assumed to be Gaussian in shape. If $\mathscr{R} \geqslant 1.5$, there is less than 1% cross-contamination and the peaks are taken to be completely separated.

According to theory, $\mathscr{R}$ is directly proportional to $\sqrt{N}$ and hence to $\sqrt{L}$. Therefore, to double the value of $\mathscr{R}$ requires that the column length be increased by a factor of 4.

The retention time for a substance depends on the Kelvin temperature T of the gas chromatography column via

$$\log(t_R - t_M) = \frac{\zeta}{T} + \eta \qquad (12.15)$$

where ζ and η are constants for a given column and substance.

For a series of similar compounds (alkanes, alcohols, alkyl bromides, etc.), the order in which the compounds emerge from the column is almost always the order of their normal boiling points. If the column temperature and flow rate are kept constant, the following relationship usually exists for a simple homologous series of compounds (e.g., straight-chain hydrocarbons):

$$\log((t_R)_n - t_M) = \mu n + \nu \qquad (12.16)$$

where μ and ν are constants for the particular column and separation conditions, and n is the number of carbon atoms in the compound. If the flow rate is constant but the column temperature increases linearly with time during the separation, (12.16) is replaced by

$$(t_R)_n - t_M = \mu^* n + \nu^* \qquad (12.17)$$

where μ^* and ν^* are a different pair of constants.

Solved Problems

FRACTIONAL DISTILLATION

12.1 A mixture consisting of 0.700 mole-fraction CCl_4 and 0.300 mole-fraction $CHCl_3$ boils at 70 °C. At this temperature, the vapor pressure of pure CCl_4 is 622 torr, whereas that of $CHCl_3$ is 1019 torr. What is the barometric pressure in the laboratory?

The boiling point is the temperature at which the vapor pressure of the mixture is equal to the barometric pressure. By (12.1), the total vapor pressure of the mixture at 70 °C is

$$P_T = x_{CHCl_3}P^\circ_{CHCl_3} + x_{CCl_4}P^\circ_{CCl_4} = (0.300)(1019) + (0.700)(622) = 741 \text{ torr}$$

12.2 What is the value of α for the $CHCl_3/CCl_4$ mixture of Problem 12.1?

$$\alpha = P^\circ_{CHCl_3}/P^\circ_{CCl_4} = 1019/622 = 1.63_8 = 1.64$$

12.3 The normal boiling point of $CHCl_3$ is 60.9 °C, whereas that of CCl_4 is 76.8 °C. What is the value of α for $CHCl_3/CCl_4$, based on (12.4)?

$T_{bCHCl_3} = 334.1$ K, $T_{bCCl_4} = 350.0$ K, and from (12.4):

$$\log \alpha = \frac{(8.9)(350.0 - 334.1)}{350.0 + 334.1} = 0.20_7 \qquad \text{or} \qquad \alpha = 1.6_1$$

12.4 A mixture of $CHCl_3/CCl_4$ was fractionated in a small column at 70 °C. Analysis showed that the distillate had 96.0 mole % $CHCl_3$ when the equilibrium liquid in the still pot had 30.0 mole % $CHCl_3$. Use the value of α from Problem 12.2 to calculate the number of theoretical plates in the column.

From (12.3),

$$\frac{0.960}{0.040} = 1.63_8{}^{n+1}\left(\frac{0.300}{0.700}\right)$$

$$56._0 = 1.63_8{}^{n+1}$$

$$\log 56._0 = (n + 1) \log 1.63_8$$

$$n + 1 = \frac{1.74_8}{0.214_4} = 8.15$$

or $n = 7.15$.

12.5 How many theoretical plates should there be in the column of Problem 12.4 so that the distillate will be 99.90 mole % $CHCl_3$ when the still pot has 30.0 mole % $CHCl_3$?

From (12.3),

$$\frac{0.9990}{0.0010} = 1.63_8{}^{n+1}\left(\frac{0.300}{0.700}\right)$$

Solving, $n = 14.7$.

12.6 If the column in Problem 12.4 had 22.0 theoretical plates, what composition of distillate would be obtained when the still pot had a mixture with 30.0 mole % $CHCl_3$?

From (12.3),

$$\frac{y_{CHCl_3}}{y_{CCl_4}} = 1.63_8{}^{23.0}\left(\frac{0.300}{0.700}\right) = 3.64 \times 10^4$$

Since this molar ratio is very large, y_{CHCl_3} is very close to 1.00, and it is easier to solve for y_{CCl_4}:

$$\frac{1 - y_{CCl_4}}{y_{CCl_4}} = 3.64 \times 10^4 \qquad \text{or} \qquad y_{CCl_4} = \frac{1}{3.64_{001} \times 10^4} = 0.0000275$$

and $y_{CHCl_3} = 0.9999725$. Therefore, the distillate is 99.99725 mole % $CHCl_3$.

SOLVENT EXTRACTION

12.7   The partition coefficient for I_2 between $CHCl_3$ and H_2O is

$$K_p = \frac{[I_2]_{CHCl_3}}{[I_2]_{H_2O}} = 132$$

What percent of the I_2 dissolved in 100.0 mL of water will remain in the aqueous phase if it is extracted once with 50.0 mL of $CHCl_3$?

$V_r = 50.0/100.0 = 0.500$, and from (12.10):

$$f_w = \frac{1}{(0.500)(132) + 1} = 0.0149$$

i.e., 1.49% of the I_2 remains in the aqueous phase.

12.8   In Problem 12.7, what percent of the I_2 will remain in the water if the aqueous phase is extracted successively with five fresh 10.0 mL portions of $CHCl_3$?

$V_r = 10.0/100.0 = 0.100$, and from (12.10):

$$f_w^5 = \left(\frac{1}{(0.100)(132) + 1}\right)^5 = 0.0704^5 = 1.73 \times 10^{-6}$$

so that only $1.73 \times 10^{-4}\%$ remains in the water phase.

12.9   For a certain weak acid distributed between benzene and water, $K_p = 100$. Calculate the distribution ratio for this acid if the pH of the aqueous phase is 7.000, if $K_a = 1.00 \times 10^{-6}$ for the acid in the water phase, and if the acid is a monomer in the benzene phase.

By (12.7),

$$D_c = \frac{(100)(1.00 \times 10^{-7})}{(1.00 \times 10^{-7}) + (1.00 \times 10^{-6})} = 9.09$$

12.10   A compound X dimerizes in ether, with $K_d = [X_2]_o/[X]_o^2 = 15.8$, but is a monomer in water. If 20.00 mL of 0.100 M X in H_2O is extracted with 50.00 mL of ether and if the concentration of X in the aqueous extract is 0.0250 M, evaluate $K_p = [X]_o/[X]_w$.

Initially there was $(0.100)(20.00) = 2.00$ mmol of X in the aqueous phase; following extraction, there was $(0.0250)(20.00) = 0.500$ mmol. By difference, 1.50 mmol was in the organic phase, so that

$$[\text{all forms of X}]_o = \frac{1.50}{50.00} = 0.0300$$

Therefore, $D_c = 0.0300/0.0250 = 1.20$, and, by (12.8),

$$1.20 = K_p + (2)(15.8)K_p^2(0.0250) = K_p + 0.790\,K_p^2$$

The solution to this quadratic is $K_p = 0.753$.

12.11 A weak acid, HA, of molecular weight 250, has $K_a = 1.00 \times 10^{-5}$ in water. This monomer weak acid dimerizes to some extent in benzene, but the value of

$$K_d = \frac{[H_2A_2]_o}{[HA]_o^2}$$

is unknown. The partition coefficient is $K_p = [HA]_o/[HA]_w = 3.20$. A separatory funnel was filled with 20.00 mL of an aqueous solution of 1.00 M HCl containing 0.2500 g (1.0 mmol) of HA. 20.00 mL of benzene was added to the separatory funnel, the mixture was shaken, the two phases were separated, and water (and HCl) distilled from the water phase, leaving 0.0185 g of HA. The benzene was distilled from the organic phase, leaving 0.2315 g of HA. What is the value of K_d?

To compute D_c, we can in this case measure both concentrations in g:

$$D_c = \frac{0.2315}{0.0185} = 12.5_1$$

Since the $[H^+]$ of the aqueous phase is so high, due to the presence of 1.00 M HCl, negligible HA is dissociated. Therefore, (12.8) applies, giving

$$D_c = 12.5_1 = 3.20 + 2K_d(3.20)^2 \frac{0.0185/250}{20.00 \times 10^{-3}} = 3.20 + 0.0758 K_d$$

whence $K_d = 1.2_3 \times 10^2$.

12.12 A weak acid, H_2A, is a monomer in benzene but dissociates in water, with equilibrium constants K_1 and K_2. Derive a general expression for D_c in terms of

$$K_p = \frac{[H_2A]_o}{[H_2A]_w}$$

K_1, K_2, and $[H^+]_w$.

$$D_c = \frac{[H_2A]_o}{[H_2A]_w + [HA^-]_w + [A^{2-}]_w} = \frac{K_p}{1 + \dfrac{[HA^-]_w}{[H_2A]_w} + \dfrac{[A^{2-}]_w}{[H_2A]_w}}$$

$$= \frac{K_p}{1 + \dfrac{K_1}{[H^+]_w} + \dfrac{K_1 K_2}{[H^+]_w^2}} = \frac{[H^+]_w^2 K_p}{[H^+]_w^2 + K_1[H^+]_w + K_1 K_2}$$

12.13 In Problem 12.12, $K_1 = 4.00 \times 10^{-4}$ and $K_2 = 4.00 \times 10^{-9}$. If 4.00 mmol of H_2A is dissolved in 10.00 mL of a pH 4.000 buffer solution, and if this solution is extracted with 10.00 mL of benzene, then 3.00 mmol of H_2A is found in the benzene layer. Evaluate K_p.

Here,

$$D_c = \frac{3.00}{4.00 - 3.00} = 3.00$$

and the equation derived in Problem 12.12 gives

$$3.00 = \frac{1.00 \times 10^{-8} K_p}{(1.00 \times 10^{-8}) + (4.00 \times 10^{-8}) + (1.6 \times 10^{-12})}$$

or $K_p = 15.0$.

12.14 A 2.00 L sample of water collected downstream from a chemical plant is to be analyzed for an organic chemical X. Because the concentration of X in the river water is too low to allow direct measurement, preconcentration is needed. Fortunately, X is extremely soluble in ether, with $K_p = [X]_o/[X]_w = 500$; X is present only as a monomer in both phases. The above sample is shaken with 10.0 mL of ether and the concentration of X in the ether solution extract is found to be $[X]_o = 2.00 \times 10^{-3}$. What is the concentration of X in the original river water?

From (12.5), $[X]_w = (2.00 \times 10^{-3})/500 = 4.00 \times 10^{-6}$. The amounts of X in ether and in water are then

$$(2.00 \times 10^{-3})(10.0 \times 10^{-3}) = 20.0 \times 10^{-6} \text{ mol} \quad \text{and} \quad (4.00 \times 10^{-6})(2.00) = 8.00 \times 10^{-6} \text{ mol}$$

for a total of 28.0×10^{-6} mol. Hence, the original concentration in the river water sample was

$$\frac{28.0 \times 10^{-6} \text{ mol}}{2.00 \text{ L}} = 1.40 \times 10^{-5} \ M$$

12.15 What fraction of X in Problem 12.14 would remain in the 2.00 L of river water if the water sample were extracted instead with five fresh portions of ether, each of 2.00 mL?

Now $V_r = 2.00/(2.00 \times 10^3) = 1.00 \times 10^{-3}$ and $D_c = K_p = 500$. From (12.10),

$$f_w = \frac{1}{500(1.00 \times 10^{-3}) + 1} = 0.667$$

so that $f_w^5 = (0.667)^5 = 0.132$. [Compare this with the mole fraction

$$\frac{8.00 \times 10^{-6}}{28.0 \times 10^{-6}} = 0.286$$

found in Problem 12.14.]

12.16 Thioacetic acid, HA, has K_a in the range 10^{-3} to 10^{-5} in water. Two 100.0 mL aqueous solutions were prepared, each containing 2.00 mmol of HA. The first solution was buffered to pH 1.000 and was extracted with 50.00 mL of $CHCl_3$; following the extraction, the aqueous solution contained 0.535 mmol of HA. The second solution, which was buffered to pH 4.000, was also extracted with 50.00 mL of $CHCl_3$; following the extraction, the aqueous solution contained 1.35 mmol of HA. Assuming that HA neither dimerizes nor dissociates in $CHCl_3$, but does dissociate in H_2O, evaluate K_a and $K_p = [HA]_o/[HA]_w$.

Equation (12.7) applies to both solutions. For solution 1, there is

$$2.00 - 0.535 = 1.47 \text{ mmol}$$

of HA in the 50.00 mL of $CHCl_3$, and 0.535 mmol of HA in the 100.0 mL of water; thus,

$$D_{c1} = \frac{1.47/50.00}{0.535/100.0} = 5.49_1$$

Similarly, for solution 2,

$$D_{c2} = \frac{0.65/50.00}{1.35/100.0} = 0.96_3$$

Applying (12.7),

$$5.49_1 = \frac{(1.00 \times 10^{-1})K_p}{K_a + (1.00 \times 10^{-1})}$$

$$0.96_3 = \frac{(1.00 \times 10^{-4})K_p}{K_a + (1.00 \times 10^{-4})}$$

and solving simultaneously gives $K_a = 4.7_3 \times 10^{-4}$, $K_p = 5.52$.

12.17 An acid, HA, dimerizes in the organic phase, with

$$K_d = \frac{[H_2A_2]_o}{[HA]_o^2}$$

but dissociates in the water phase, with

$$K_a = \frac{[H^+]_w[A^-]_w}{[HA]_w}$$

If $K_p = [HA]_o/[HA]_w$, derive an expression for D_c as a function of K_d, K_a, K_p, $[H^+]_w$, and $[HA]_w$.

By the basic definition (12.6):

$$D_c = \frac{[HA]_o + 2[H_2A_2]_o}{[HA]_w + [A^-]_w}$$

$$= \frac{K_p[HA]_w + 2K_dK_p^2[HA]_w^2}{[HA]_w + (K_a[HA]_w/[H^+]_w)} = \frac{K_p[H^+]_w + 2K_dK_p^2[HA]_w[H^+]_w}{[H^+]_w + K_a}$$

12.18 Acetic acid has $K_a = 1.78 \times 10^{-5}$ in water and its partition coefficient between benzene and water is

$$K_p = \frac{[HAc]_o}{[HAc]_w} = 7.98 \times 10^{-3}$$

However, acetic acid dimerizes in benzene, with $K_d = [H_2Ac_2]_o/[HAc]_o^2$. Find K_d, if 5.00 mL of 1.000 M HAc that is buffered to pH 5.000 is shaken with 100.0 mL of benzene, the phases are separated, and the water phase proves to be 0.900 M HAc.

To solve the equation derived in Problem 12.17 for K_d, we need to know only the values of D_c and $[HAc]_w$. The original aqueous solution contained $(1.000)(5.00) = 5.00$ mmol of HAc; after extraction with benzene, the aqueous solution contained $(0.900)(5.00) = 4.50$ mmol of HAc. Therefore, 0.50 mmol of HAc was transferred to the benzene phase, and the total acetic acid concentration in that phase $(HAc + H_2Ac_2)$ is $0.50/100.0 = 5.0 \times 10^{-3}$ M. Since the total acetic acid concentration in the water phase $(HAc + Ac^-)$ is 0.900 M,

$$D_c = \frac{5.0 \times 10^{-3}}{0.900} = 5.5_6 \times 10^{-3}$$

Since $[H^+] = 1.00 \times 10^{-5}$ (by buffering), we can use the expression for K_a to calculate $[HAc]_w$:

$$1.78 \times 10^{-5} = \frac{(1.00 \times 10^{-5})[Ac^-]_w}{[HAc]_w} = \frac{(1.00 \times 10^{-5})(0.900 - [HAc]_w)}{[HAc]_w}$$

or $[HAc]_w = 0.324$. Now Problem 12.17 gives

$$5.5_6 \times 10^{-3} = \frac{(7.98 \times 10^{-3})(1.00 \times 10^{-5}) + 2K_d(7.98 \times 10^{-3})^2(0.324)(1.00 \times 10^{-5})}{(1.00 \times 10^{-5}) + (1.78 \times 10^{-5})}$$

Solving, $K_d = 1.8_1 \times 10^2$.

12.19 A water sample contains 3.00×10^{-3} M chloroacetone, C, which is soluble to a limited extent in both water and octanol, with

$$K_{pC} = \frac{[C]_o}{[C]_w} = 1.91$$

The water sample also contains 3.00×10^{-3} M methyl iodide, M, which also dissolves to a limited extent in both water and octanol, with

$$K_{pM} = \frac{[M]_o}{[M]_w} = 49.0$$

In order to measure some of the physical properties of chloroacetone, it is necessary to reduce as much as possible the relative concentration of M. The following procedure was used: 25.00 mL of the (C + M) water sample was shaken with 25.00 mL of octanol, the phases separated, and the octanol extract placed in contact with 25.00 mL of fresh water. This second mixture was shaken, the phases separated, and the water extract placed in contact with 25.00 mL of fresh octanol. This third mixture was shaken, the phases separated, and the octanol extract placed in contact with 25.00 mL of fresh water. This fourth mixture was shaken and the phases separated. By what factor has the ratio $[C]_w/[M]_w$ changed when this last water extract is compared with the original water sample? What fraction of the original amount of C was recovered in this last water extract?

When C is transferred from water to octanol, the fraction of C that will be in the octanol phase is

$$f_{oC} = \frac{D_{cC}}{D_{cC}+1} = \frac{1.91}{2.91} = 0.656$$

since $V_r = 1.000$. Similarly, the fraction of C that will be in the water phase is

$$f_{wC} = \frac{1}{D_{cC}+1} = \frac{1}{2.91} = 0.344$$

Similarly, $f_{oM} = 49.0/50.0 = 0.980$ and $f_{wM} = 1/50.0 = 0.0200$.

Since a term f represents the fraction of a particular species, C or M, that is present in a particular phase, o or w, a sequence of extractions will result in an expression that is the product of f-values. For the extractions described here, the concentration of C in the final water phase is

$$(\text{initial concentration})(f_{oC})(f_{wC})(f_{oC})(f_{wC}) = (3.00 \times 10^{-3})(0.656)^2(0.344)^2 = 1.53 \times 10^{-4} \ M$$

Similarly, the concentration of M in the final water phase is

$$(3.00 \times 10^{-3})(f_{oM})^2(f_{wM})^2 = (3.00 \times 10^{-3})(0.980)^2(0.0200)^2 = 1.15 \times 10^{-6} \ M$$

The ratio $[C]_w/[M]_w$ changes from an original value of 1.00 to

$$\frac{1.53 \times 10^{-4}}{1.15 \times 10^{-6}} = 133$$

i.e., by a factor of 133. However, only $(1.53 \times 10^{-4})/(3.00 \times 10^{-3}) = 0.0510$ of the original C is contained in the final water extract.

CHROMATOGRAPHIC SEPARATIONS

12.20 A 4.00 mL blood sample from a patient suspected of suffering from ketosis (abnormally high ketone levels) was analyzed for acetone by extracting the sample with 25.0 mL of $CHCl_3$. The partitioning for acetone is such that $f_o = 0.970$. When 5.00 μL of the $CHCl_3$ extract was injected in a gas chromatography column, a 70.0 mm^2 acetone peak was observed. When 5.00 μL of a standard, containing 61.1 μg acetone/10.0 mL was injected in the same column, there resulted a 44.0 mm^2 acetone peak. What is the acetone concentration of the blood sample, in units of mg acetone/100 mL blood?

Since the same volume of liquid was injected in the column for both the unknown and the standard, the acetone peak areas are directly proportional to the concentrations of acetone in $CHCl_3$:

$$\frac{61.1}{C} = \frac{44.0}{70.0}$$

$$C = 97.2 \ \frac{\mu g \ \text{acetone}}{10.0 \ \text{mL} \ CHCl_3} = 0.243 \ \frac{mg \ \text{acetone}}{25.0 \ \text{mL} \ CHCl_3}$$

Thus the 4.00 mL of blood contained

$$\frac{0.243}{0.970} = 0.250 \text{ mg acetone}$$

and so 100 mL of blood would contain $25 \times 0.250 = 6.25$ mg acetone.

12.21 Two species, A and B, are separated by isothermal gas chromatography, using a 2.00 m column having 5000 theoretical plates when the flow rate is 15.0 mL/min. A nonabsorbing air peak emerges in 30.0 s; peak A emerges in 5.00 min; peak B emerges in 12.00 min. (*a*) What are the base widths of peaks A and B? (*b*) What fraction of time does B spend in the mobile phase while passing through the column? (*c*) What is the value of H for this column?

(*a*) From (*12.11*), $t_W = 4t_R/\sqrt{N}$; hence,

$$(t_W)_A = \frac{4(300)}{\sqrt{5000}} = 17.0 \text{ s} \qquad (t_W)_B = \frac{12.00}{5.00}(t_W)_A = 40.8 \text{ s}$$

(*b*) The mobile phase for any substance lasts $t_M = 30.0$ s, the time required for the air peak to emerge. Since B spends a total of $(12.00)(60) = 720$ s in the column, the desired fraction is

$$\frac{30.0}{720} = 0.0417$$

(*c*) By (*12.12*),

$$H = \frac{200 \text{ cm}}{5000 \text{ plates}} = 0.0400 \text{ cm/plate}$$

12.22 A mixture of five alkyl iodides is separated by gas chromatography using a column packed with crushed firebrick coated with silicone oil. The column is so heated that its temperature rises linearly during the separation. Data for the peaks obtained in the separation are displayed in Table 12-1. What is the resolution, $\mathscr{R}$, of peaks 4 and 5?

Table 12-1

Peak	Identity	Molecular Weight	t_R, min	t_W, min	Temperature, °C	Chart Area, cm²
1	air peak	—	$t_M = 0.50$	small	55	small
2	CH_3I	141.9	6.60	0.55	100	13.0
3	C_2H_5I	156.0	9.82	1.00	127	12.0
4	iso-C_3H_7I	170.0	11.90	1.04	139	10.0
5	n-C_3H_7I	170.0	13.04	1.08	148	7.2
6	CH_2I_2	267.8	19.10	1.60	193	2.0

From (*12.14*),

$$\mathscr{R} = \frac{13.04 - 11.90}{(1.04 + 1.08)/2} = 1.08$$

12.23 The column used in the separation of Problem 12.22 was 365 cm long. What length of column would be needed so that the resolution of peaks 4 and 5 was $\mathscr{R}' = 1.50$?

Since $\mathscr{R}$ is proportional to $\sqrt{L}$, $\mathscr{R}'/\mathscr{R} = \sqrt{L'/L}$. Thus,

$$\frac{1.50}{1.08} = \sqrt{\frac{L'}{365}} \qquad \text{or} \qquad L' = 704 \text{ cm}$$

12.24 The detector used in the chromatographic analysis of Problem 12.22 had a response that was directly proportional to the number of molecules of alkyl iodide. If 20.0 μg of CH_3I was present in the sample of Problem 12.22, how much C_2H_5I was present?

$$\frac{\text{mols } C_2H_5I}{\text{mols } CH_3I} = \frac{\text{area } C_2H_5I}{\text{area } CH_3I}$$

$$\frac{x/156.0}{20.0/141.9} = \frac{12.0}{13.0}$$

$$x = 20.3 \ \mu\text{g } C_2H_5I$$

12.25 The column used in the analysis of Problem 12.22 separates a homologous series of compounds (such as straight-chain alkyl iodides) in accord with theory. What value of t_R would be expected for $n\text{-}C_4H_9I$?

Table 12-1 gives, for three straight-chain alkyl iodides,

No. C atoms (n)	1	2	3
$(t_R)_n - t_M$, min	6.10	9.32	12.54

Since the column temperature was raised linearly with time, (*12.17*) applies. A least-squares fit of the data (Section 2.9) results in

$$(t_R)_n - t_M = 3.22\,n + 2.88$$

which gives, when $n = 4$,

$$(t_R)_4 = 15.8 + 0.50 = 16.3 \text{ min}$$

12.26 Figure 12-2 is a gas chromatogram obtained for a mixture of straight-chain hydrocarbons: C_nH_{2n+2}. Peak M is due to a nonabsorbing species; peak A results from C_3H_8; peak F is due to $C_{20}H_{42}$. The column, 120 cm in length, is operated isothermally at a carrier-gas flow rate of 50.0 cm^3/min. Data for retention times and widths of peaks are found in Table 12-2. The flame ionization detector employed in the analysis can be assumed to respond in direct proportion to the total carbon content of the eluted hydrocarbon. Find N_A, the number of theoretical plates based on peak A?

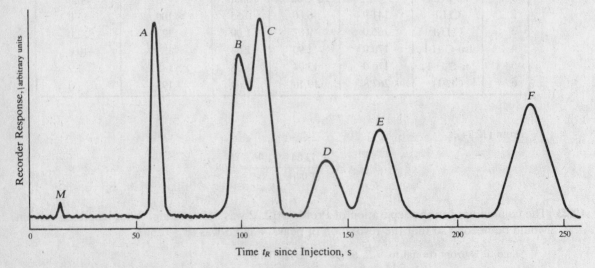

Fig. 12-2

Table 12-2

Peak	M	A	B	C	D	E	F
t_R, s	15	60	100	110	140	165	235
t_W, s	small	9.00	15.00	16.50	21.00	24.75	35.25
Area, cm^2	small	0.900	1.200	1.650	0.610	1.040	1.900

From (12.11), $N_A = 16(60/9.00)^2 = 7.1 \times 10^2$.

12.27 For the separation described in Problem 12.26, measurements of H for different flow rates, F, were fitted by (12.13) with $\theta = 0.0320$ cm, $\beta = 0.150$ cm^4/min, and $\gamma = 8.00 \times 10^{-4}$ min/cm^2. What is the optimum flow rate to use in order to minimize H?

$$F_{\text{opt}} = \sqrt{\frac{\beta}{\gamma}} = \sqrt{\frac{0.150}{8.00 \times 10^{-4}}} = 13.7 \text{ cm}^3/\text{min}$$

12.28 What is the number of theoretical plates for the column of Problem 12.26 when operated at the optimum flow rate calculated in Problem 12.27?

From (12.13),

$$H = 0.0320 + \frac{0.150}{13.7} + (8.00 \times 10^{-4})(13.7) = 0.0539 \text{ cm}$$

From (12.12),

$$N = \frac{120 \text{ cm}}{0.0539 \text{ cm}} = 2.23 \times 10^3 \text{ plates}$$

12.29 What is the formula for the hydrocarbon eluted as peak E in the separation of Problem 12.26?

Since the separation is of a homologous series of compounds and since the temperature is kept constant, (12.16) applies. Values of n are available only for peaks A ($n = 3$) and F ($n = 20$), for which

$$(t_R)_3 - t_M = 45 \text{ s} \qquad (t_R)_{20} - t_M = 220 \text{ s}$$

Using these two pairs of data to solve (12.16), we find $\mu = 4.05 \times 10^{-2}$, $\nu = 1.53_1$. Peak E has $t_R - t_M = 150$ s, so that

$$\log 150 = (4.05 \times 10^{-2})n + 1.53_1$$

Solving, $n = 15._9$, so that peak E is $C_{16}H_{34}$.

12.30 A mixture of alkyl bromides is separated by gas chromatography. The column parameters are: length, 150 cm; temperature, 140 °C; carrier gas, helium; flow rate, 20 cm^3/min; detector, flame ionization. The gas chromatogram shown in Fig. 12-3 was obtained for an unknown mixture; however, it is known that peak F is due to 6.61 μg of n-C_5H_{11}Br. Data for the various peaks are shown in Table 12-3; molecular weights and boiling points of possible alkyl bromides are given in Table 12-4. How short can the column be so that peaks D and E are resolved with only 1% overlap?

With the present column, the resolution of peaks D and E is

$$\mathcal{R} = \frac{1222 - 1029}{(81.5 + 68.6)/2} = 2.57$$

Since $\mathcal{R}$ is proportional to $\sqrt{L}$, and since a separation with 1% overlap corresponds to $\mathcal{R} = 1.50$,

$$\frac{1.50}{2.57} = \sqrt{\frac{L}{150}}$$

and a length $L = 51.0$ cm is adequate to achieve $\mathcal{R} = 1.50$.

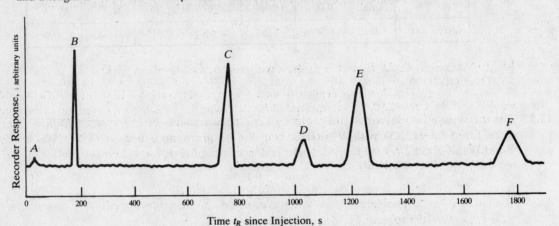

Fig. 12-3

Table 12-3

Peak	A	B	C	D	E	F
t_R, s	$t_M = 25$	177	750	1029	1222	1775
t_W, s	1.7	11.8	54.5	68.6	81.5	118.3
Area, cm²	0.01	18.1	101.1	26.5	98.9	61.0

Table 12-4

Compound	MW	bp, K	10^3/bp, K^{-1}	Compound	MW	bp, K	10^3/bp, K^{-1}
CH_3Br	95	277	3.61	t-C_4H_9Br	137	346	2.89
C_2H_5Br	109	311	3.22	n-$C_5H_{11}Br$	151	401	2.49
n-C_3H_7Br	123	344	2.91	$C_3H_6(CH_3)CH_2Br$	151	393	2.55
i-C_3H_7Br	123	333	3.00	$(CH_3)_2C_3H_5Br$	151	394	2.54
n-C_4H_9Br	137	375	2.67	$(CH_3)_3C_2H_2Br$	151	362	2.76
s-C_4H_9Br	137	364	2.75				

12.31 Determine N, the number of theoretical plates, for peaks B, C, D, E, and F in Problem 12.30. Which, if any, of these peaks is probably due to an unresolved mixture of two alkyl bromides?

From (12.11), $N_B = 16(177/11.8)^2 = 3.60 \times 10^3$. Similarly,

$$N_C = 3.03 \times 10^3 \qquad N_D = 3.60 \times 10^3 \qquad N_E = 3.60 \times 10^3 \qquad N_F = 3.60 \times 10^3$$

Since peak C has a smaller calculated value of N than the other peaks, it most likely arises from an unresolved mixture of two (or more) alkyl bromides (an unresolved peak would have a larger t_W and, hence, a smaller calculated value of N).

12.32 Using mixtures of known alkyl bromides, it was shown that the boiling points bp (in K) and the adjusted retention times

$$t'_R = t_R - t_M$$

(in s) for the alkyl bromide separation of Problem 12.30 are related by the equation

$$\log t'_R = \frac{-956}{bp} + 5.633 \tag{1}$$

Identify peaks B, C, D, E, and F in the chromatogram of Problem 12.33.

Since $(t'_R)_B = 177 - 25 = 152$, (1) gives $10^3/bp_B = 3.61$; thus, from Table 12-4, B is CH_3Br. Similarly, we find that $10^3/bp_C = 2.90$. From Problem 12.31 and Table 12-4, peak C is a mixture—a mixture of n-C_3H_7Br ($10^3/bp = 2.91$) and t-C_4H_9Br ($10^3/bp = 2.89$). Since $(t'_R)_D = 1004$, (1) gives $10^3/bp_D = 2.75$. According to Problem 12.31, peak D is primarily due to a single compound; from Table 12-4, D is either s-C_4H_9Br ($10^3/bp = 2.75$) or $(CH_3)_3C_2H_2Br$ ($10^3/bp = 2.76$). In the same fashion, we identify E as n-C_4H_9Br and F as n-$C_5H_{11}Br$.

12.33 With reference to Problem 12.32, how many μg of the bromide in peak B is present in the chromatogram of Problem 12.30?

B is CH_3Br. According to Problem 12.30, peak F, which is due to n-$C_5H_{11}Br$, arises from 6.61 μg of the bromide. Since a flame ionization detector has a response that is proportional to the number of carbon atoms in the effluent,

$$\frac{\text{peak area } F}{\mu\text{mols C in } F} = \frac{\text{peak area } B}{\mu\text{mols C in } B}$$

$$\frac{61.0}{5(6.61/151)} = \frac{18.1}{1(x/95)}$$

where 151 and 95 are the molecular weights of F and B, respectively, and 5 and 1 are the respective numbers of carbon atoms per molecule. Solving, $x = 6.17$ μg CH_3Br.

Supplementary Problems

FRACTIONAL DISTILLATION

12.34 At 100 °C, toluene has a vapor pressure of 557 torr, whereas benzene has a vapor pressure of 1335 torr. What is the value of α for benzene/toluene at 100 °C?   *Ans.*   2.40

12.35 If a fractionation column has 12 theoretical plates, what is the composition of a distillate of benzene/toluene at 100 °C, if the equilibrium liquid in the still pot is 5.00 mole % benzene in toluene? Refer to Problem 12.34.   *Ans.*   99.9783 mole % benzene

SOLVENT EXTRACTIONS

12.36 The partition coefficient for ethyl iodide, E, between octanol and water is $K_p = [E]_o/[E]_w = 100$. What percent of E present in 50.00 mL of water would remain if the water was extracted with 10.00 mL of octanol?   *Ans.*   4.76%

12.37 What percent of E in Problem 12.36 would remain in the water phase if it was extracted successively with five fresh 10.00 mL portions of octanol?   *Ans.* 2.44×10^{-5}%

12.38 The partition coefficient for oxalic acid, H_2A, between ether and water is $K_p = [H_2A]_o/[H_2A]_w = 0.115$. Dissociation constants for H_2A are $K_1 = 6.46 \times 10^{-2}$ and $K_2 = 6.17 \times 10^{-5}$. 100.0 mL of ether containing some dissolved oxalic acid is extracted with 10.0 mL of water buffered to pH 3.000. What fraction of the oxalic acid is transferred to the water phase? Refer to Problem 12.12.   *Ans.*   98.38%

CHROMATIC SEPARATIONS

12.39 Straight-chain alcohols were separated by isothermal gas chromatography at 105 °C. Table 12-5 contains data for three standards and an unknown mixture. A small nonabsorbing air peak emerged at $t_M = 40.0$ s. What are the identities of A, B, and C, based on a plot of $\log(t_R - t_M)$ versus the number of carbon atoms for the three standards?   *Ans.* A is C_2H_5OH, B is n-C_3H_7OH, and C is n-$C_5H_{11}OH$.

Table 12-5

Compound	Standards			Unknown		
	C_2H_5OH	n-C_4H_9OH	n-$C_5H_{11}OH$	A	B	C
Volume, μL	5.00	5.00	5.00	?	?	?
t_R, s	84.0	209.1	364.5	82.7	132.4	366.2
t_W, s	22.8	33.7	59.1	21.9	25.5	60.2
Peak Area, cm^2	1.16	1.18	1.13	0.725	1.68	1.23

12.40 Find the values of N for peaks A, B, and C of Problem 12.39.
        Ans. $N_A = 228$, $N_B = 431$, $N_C = 592$

12.41 Determine the volume of each alcohol in the unknown mixture of Problem 12.39 from the average chromatography response for the standards.
        Ans.   3.13 μL C_2H_5OH, 7.26 μL n-C_3H_7OH, 5.32 μL n-$C_5H_{11}OH$

12.42 When 5.00 μL of i-C_4H_9OH was injected in the column of Problem 12.39, it emerged at $t_R = 173.3$ s, with $t_W = 24.2$ s. If the column was 100 cm long, how long a column would be needed so that i-C_4H_9OH was separated from n-C_4H_9OH with $\mathcal{R} = 1.50$?   *Ans.*   147 cm

Chapter 13

Spectral Measurements

13.1 UNITS OF MEASUREMENT

Analytical measurements embrace a very large span of spectral energies. Indeed, the shortest-wavelength photons have energies around 10^{12} times as great as those of the longest-wavelength photons. It is therefore convenient to use different units of energy, wavelength, or frequency for the different regions of the electromagnetic spectrum. In most cases, as shown in Fig. 13-1, the unit chosen for a region is the one that results in values close to 1.0 for the region. Some of the units are true energy units; others are directly or inversely proportional to energy, as given by the expressions below.

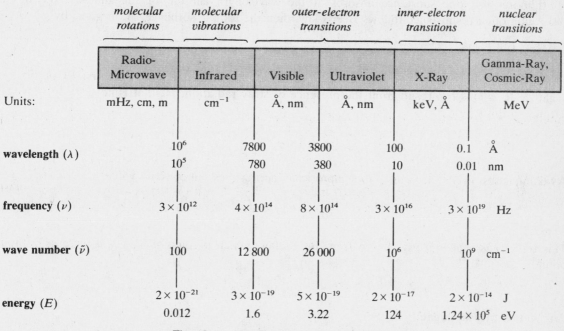

Fig. 13-1. The Electromagnetic Spectrum

Photon energy, E, is related to photon frequency, ν, by the equation

$$E = h\nu \tag{13.1}$$

where h is the Planck constant. The wavelength, λ, of a photon is related to its frequency by

$$\nu = \frac{c}{\lambda} \tag{13.2}$$

where c is the velocity of light. For E in joules, ν in hertz, and λ in meters (the SI units), h and c have the numerical values shown in Table 1-3.

Combination of (13.1) and (13.2) yields

$$E = \frac{hc}{\lambda} = hc\tilde{\nu} \tag{13.3}$$

213

where $\bar{\nu} \equiv 1/\lambda$ is the *wave number* of the photon, commonly given in cm^{-1}. Using Table 1-3, we can rewrite (13.3) in the convenient "mixed" form

$$E \text{ (eV)} = \frac{1239.8}{\lambda \text{ (nm)}} = \frac{\bar{\nu} \text{ (cm}^{-1})}{8066} \tag{13.4}$$

13.2 BEER'S LAW

Beer's law, (13.5), relates the incident power, P_o, of monochromatic radiation and the transmitted power, P, emerging from an analyte. Defining *transmittance* $T \equiv P/P_o$ and *absorbance* $A \equiv -\log T$, we have:

$$A = \epsilon b C \tag{13.5}$$

Here, ϵ is the *absorptivity* of the analyte (at the given incident wavelength λ), b is the absorbing path length, and C is the concentration of the analyte. These are commonly given in L$\cdot$mol$^{-1}\cdot$cm^{-1}, cm, and M, respectively. Note that A is dimensionless.

If more than one substance absorbs at the wavelength chosen for the analysis, and if the absorption of each substance follows Beer's law, then the total absorbance, A_t, is given by

$$A_t = \sum A_i = b \sum \epsilon_i C_i \tag{13.6}$$

Deviations from Beer's law arise from chemical reactions and other factors. In particular, negative deviations can result from the presence of stray radiation, P_s, that is not absorbed by the sample; in this case,

$$A = -\log \frac{P + P_s}{P_o + P_s} \tag{13.7}$$

As C increases, P decreases and A asymptotically approaches the constant value

$$A_{C \to \infty} = -\log \frac{P_s}{P_o + P_s} \tag{13.8}$$

This represents a negative deviation from Beer's law, which predicts that A will increase without bound.

Precision of Measurements

Sometimes the principal cause of error in a concentration measurement by absorption is the uncertainty in determining the transmittance. The uncertainty in T might be due to errors in reading a transmittance scale or to electrical noise. In either case, if the uncertainty in T is the fixed value σ_T, then the propagation-of-error expression (2.10) applied to Beer's law in the form

$$C = -\frac{1}{\epsilon b} \log T$$

yields

$$\frac{\sigma_C}{\bar{C}} = -0.434 \frac{\sigma_T}{\bar{T} \log \bar{T}} \tag{13.9}$$

According to (13.9), $\sigma_C/\bar{C}$ (the relative error in the concentration) is fairly flat in the range $0.13 < \bar{T} < 0.63$, which corresponds to $0.9 > \bar{A} > 0.2$, and becomes very large outside this range (see Fig. 13-2). Consequently, many experimenters try to keep within this absorbance interval.

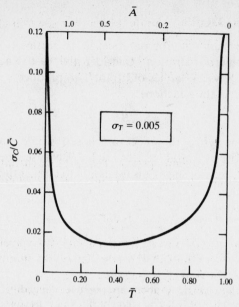

Fig. 13-2

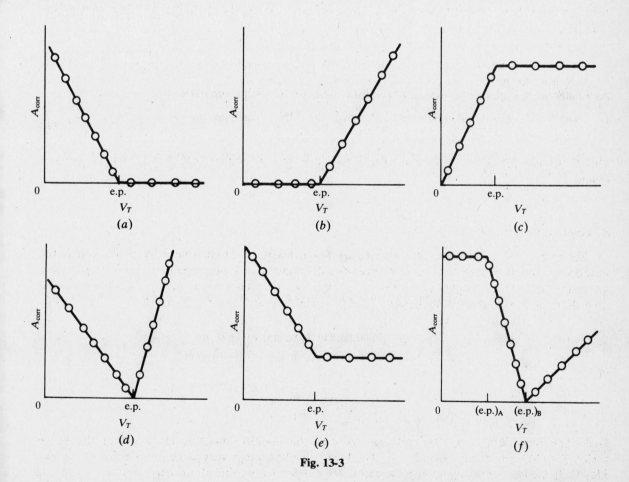

Fig. 13-3

13.3 PHOTOMETRIC TITRATIONS

If one or more of the reactants and/or products in a reaction absorbs significantly, then it is possible to perform a photometric titration. Consider the case where the process involves the reaction of α moles of sample S with β moles of titrant T to produce γ moles of products P (where P represents all products formed in the reaction):

$$\alpha S + \beta T \rightarrow \gamma P$$

As is usual in quantitative analyses, the reaction chosen would have a large K-value so that the equilibrium would be displaced far to the right. Because the addition of titrant dilutes the concentrations of reactants and products, it is necessary to correct the measured absorbance via the expression

$$A_{corr} = A_{meas} \frac{V_S + V_T}{V_S} \qquad (13.10)$$

where V_S is the initial volume of sample S and V_T is the volume of added titrant T.

EXAMPLE 13.1   A number of photometric titrations are possible, depending on which of the reactants and/or products absorb at the wavelength chosen for measurement. Some of the possibilities are shown in Fig. 13-3.

(a) **Only S absorbs.** Then, along with the concentration of S, the absorbance decreases, until the equivalence point (e.p.) is reached.

(b) **Only T absorbs.** Then the absorbance will occur only after reaching the equivalence point, when excess T becomes available.

(c) **Only P absorbs.** Then the amount of product and the absorbance together increase until the equivalence point is reached, after which the absorbance remains constant.

(d) **Both S and T absorb.** Then the absorbance first decreases, reaching zero at the equivalence point, after which it increases because of excess T. The diagram is drawn for the case $\beta \epsilon_T > \alpha \epsilon_S$, so that the slope following the equivalence point is greater than the slope prior to the equivalence point.

(e) **Both S and at least one P absorb** (and $\alpha \epsilon_S \neq \gamma \epsilon_P$). The diagram illustrates the case $\alpha \epsilon_S > \gamma \epsilon_P$.

(f) Mixtures of substances that have different absorptivities (or whose products have different absorptivities) can also be titrated photometrically if there is a difference in their reactivity with the titrant. In the diagram, only species B and the titrant T absorb, but species A reacts before species B.

Solved Problems

13.1   Characterize a 250.0 nm photon in (a) μm, (b) cm, (c) cm^{-1}, (d) Å, (e) Hz, (f) ergs, (g) eV, (h) MeV.

(a) $(250.0 \text{ nm})(10^{-3} \ \mu\text{m/nm}) = 0.2500 \ \mu\text{m}$

(b) $(250.0 \text{ nm})(10^{-7} \ \text{cm/nm}) = 2.500 \times 10^{-5} \text{ cm}$

(c) $\tilde{\nu} = \dfrac{1}{2.500 \times 10^{-5} \text{ cm}} = 4.000 \times 10^4 \text{ cm}^{-1}$

(d)   Since 1 nm = 10 Å, 250.0 nm = 2500 Å.

(e)   From (13.2),

$$\nu = \frac{c}{\lambda} = \frac{2.998 \times 10^8 \text{ m/s}}{250.0 \times 10^{-9} \text{ m}} = 1.199 \times 10^{15} \text{ Hz}$$

(f)   Since 1 erg = 10^{-7} J, Table 1-3 and (13.1) give

$$E = h\nu = (6.626 \times 10^{-27} \text{ erg} \cdot \text{s})(1.199 \times 10^{15} \text{ s}^{-1}) = 7.945 \times 10^{-12} \text{ erg}$$

(g)   From (13.4),

$$E = \frac{1239.8}{250.0} = 4.959 \text{ eV}$$

(h) $(4.959 \text{ eV})(10^{-6} \text{ MeV/eV}) = 4.959 \times 10^{-6} \text{ MeV}$

13.2   A $4.00 \times 10^{-4}\,M$ solution of aniline in H_2O has absorbance $A = 0.504$ at 280 nm, when measured in a 1.00 cm cell. Find the transmittance of a $1.50 \times 10^{-3}\,M$ solution of aniline in water when measured at the same wavelength but in a 0.500 cm cell.

From (13.5), $0.504 = \epsilon(1.00)(4.00 \times 10^{-4})$, so that $\epsilon = 1.26 \times 10^3 \text{ L} \cdot \text{mol}^{-1} \cdot \text{cm}^{-1}$. For the second solution, (13.5) now gives

$$-\log T = (1.26 \times 10^3)(0.500)(1.50 \times 10^{-3}) = 0.945$$

whence $T = 0.114$.

13.3   The ultraviolet absorbances of a series of nine aqueous standards having different NO_3^- contents were determined at 220 nm using a 1.00 cm cell; the data are:

NO_3^-, mg/mL	0	0.0040	0.015	0.025	0.035	0.040	0.050	0.060	0.070
A	0	0.097	0.208	0.347	0.450	0.553	0.620	0.668	0.688

Eight samples of river water taken downstream from a chemical plant had an average absorbance of 0.642 when measured at the same wavelength and in the same cell. What is the nitrate content of the river, in mg/mL, and its precision estimate at the 95% confidence level?

Using (2.13) and (2.14), we obtain the linear least-squares fit of the nine calibration data points:

$$y = 10.229\,x + 0.06363$$

where $y \equiv$ absorbance and $x \equiv$ mg NO_3^-/mL. Also, by (2.17), $s_d = 0.05694$. Then, from (2.19) and (2.20),

$$\hat{x}_u = \frac{0.642 - 0.0636}{10.229} = 0.0565 \qquad \text{and} \qquad s_{\hat{x}_u} = 0.0033$$

From Appendix A, for $\nu = 9 - 2 = 7$, $t_{95,7} = 2.365$; hence,

$$x_u \approx \hat{x}_u \pm t_{95,7} s_{\hat{x}_u} = 0.056_5 \pm 0.007_8$$

13.4   The pH color indicator methyl orange is a weak acid, HIn, with $K_a \approx 10^{-4}$. Both the undissociated form, HIn, and the dissociated form, In^-, of the acid are colored. At the *isosbestic* (equal-absorptivities) wavelength, 470 nm, both HIn and In^- have $\epsilon = 10\,000 \text{ L} \cdot \text{mol}^{-1} \cdot \text{cm}^{-1}$. The absorbance of a solution of HIn, buffered to pH 4.500 and contained in a 1.00 cm cell, was measured at 470 nm and found to be 0.376. What is the formal concentration of methyl orange in the solution?

Since both species absorb to the same extent,

$$A = \epsilon_{HIn} b C_{HIn} + \epsilon_{In^-} b C_{In^-} = \epsilon b C$$

where $C = C_{HIn} + C_{In^-}$ is the formal concentration. Hence,

$$0.376 = (10\,000)(1.00)C \qquad \text{or} \qquad C = 3.76 \times 10^{-5}\,M$$

13.5 The absorbances of three solutions of methyl orange, each at the same (unknown) formal concentration C but at different pH-values, were measured in a 1.00 cm cell at 510 nm; data are:

pH	strongly acidic	4.500	strongly basic
A	0.470	0.176	0.129

What is K_a for methyl orange, if it is known that pK_a is about 4?

In the strongly acidic solution, almost all of the indicator should be in the form of HIn. Therefore,

$$0.470 = \epsilon_{HIn}(1.00)C$$

In the strongly basic solution, almost all of the indicator should be in the form of In^-. Therefore,

$$0.129 = \epsilon_{In^-}(1.00)C$$

At pH 4.500 ($[H^+] = 3.16 \times 10^{-5}$), (13.6) gives

$$0.176 = \epsilon_{HIn}(1.00)(1-f)C + \epsilon_{In^-}(1.00)fC$$

where f is the fraction of methyl orange in the dissociated form. Thus,

$$0.176 = 0.470(1-f) + 0.129f \qquad \text{or} \qquad f = 0.862$$

and
$$K_a = \frac{[H^+][In^-]}{[HIn]} = \frac{(3.16 \times 10^{-5})(0.862\,C)}{0.138\,C} = 1.97 \times 10^{-4}$$

13.6 Titanium and vanadium form colored peroxide complexes, and their concentrations can be determined spectrophotometrically. Absorption data for these two species, measured in a 1.00 cm cell, are:

Complex	Metal Concentration, $\mu g/mL$	Absorbance	
		410 nm	460 nm
Ti	40.0	0.608	0.410
V	120.0	0.444	0.600

The absorbance of a solution containing both titanium and vanadium peroxide complexes was determined using a 1.00 cm cell and found to be 0.849 at 410 nm and 0.755 at 460 nm. Determine the Ti and V contents of the solution.

From Beer's law and the above data: at 410 nm, $\epsilon_{Ti} = 0.0152$ and $\epsilon_V = 0.00370$; at 460 nm, $\epsilon_{Ti} = 0.0102_5$ and $\epsilon_V = 0.00500$—all in units of $mL \cdot \mu g^{-1} \cdot cm^{-1}$. Define x as $\mu g\ Ti/mL$ and y as $\mu g\ V/mL$ in the unknown solution. From (13.6),

$$A_{410} = 0.849 = (0.0152)(1.00)x + (0.00370)(1.00)y$$
$$A_{460} = 0.755 = (0.0102_5)(1.00)x + (0.00500)(1.00)y$$

Solving the equations simultaneously yields $x = 38.1$ and $y = 72.9$.

13.7 1.00 mmol of X in 1.00 L of water has $A = 0.368$ at 500 nm when measured in a 1.00 cm cell. In water, X undergoes the reaction $X \rightarrow 2Y$. At 500 nm, $\epsilon_X = 600$ and $\epsilon_Y = 10.0\ L \cdot mol^{-1} \cdot cm^{-1}$. Find K for $X \rightarrow 2Y$.

If x mol/L of X reacts, then at equilibrium the concentrations are $[X] = 10^{-3} - x$, $[Y] = 2x$. Therefore, from (13.6),

$$0.368 = (600)(1.00)(10^{-3} - x) + (10.0)(1.00)(2x) \qquad \text{or} \qquad x = 4.00 \times 10^{-4}$$

whence $[X] = 6.00 \times 10^{-4}$, $[Y] = 8.00 \times 10^{-4}$, and

$$K = \frac{(8.00 \times 10^{-4})^2}{6.00 \times 10^{-4}} = 1.07 \times 10^{-3}$$

13.8 A metal ion M^{2+} reacts in water with the chelating agent Y^{4-} to form MY^{2-}, which has an absorption maximum at 240 nm; neither M^{2+} nor Y^{4-} absorbs at 240 nm. Two solutions, I and II, were prepared and their absorptions measured at 240 nm:

Solution I contained 2.00 mmol of Na_4Y in 10.00 mL of a $2.00 \times 10^{-3}\ M$ MCl_2 solution. The absorbance was $A_I = 1.000$ when measured in a 1.000 cm cell. No detectable increase in absorbance occurred when additional Na_4Y was added to the solution.

Solution II contained 80.0 μmol of Na_4Y in 10.00 mL of a $2.00 \times 10^{-3}\ M$ MCl_2 solution. The absorbance of this solution was $A_{II} = 1.200$ when measured in a 2.000 cm cell. A detectable increase in absorbance occurred when additional Na_4Y was added to the solution.

Give the stability constant, K, for the reaction $MY^{2-} \to M^{2+} + Y^{4-}$.

Prior to reaction with M^{2+}, the concentrations of the chelating agent were

$$[Y^{4-}]_I = \frac{2.00 \times 10^{-3}}{0.0100} = 0.200 \qquad \text{and} \qquad [Y^{4-}]_{II} = \frac{8.00 \times 10^{-5}}{0.0100} = 0.00800$$

Since added Y^{4-} did *not* increase the absorptivity in solution I, almost all M^{2+} in that solution must have been in the complexed form. Therefore, $[MY^{2-}]_I = 2.00 \times 10^{-3}$, and Beer's law gives

$$A_I = 1.000 = \epsilon_{MY^{2-}}(1.000)(2.00 \times 10^{-3}) \qquad \text{or} \qquad \epsilon_{MY^{2-}} = 500\ \text{L} \cdot \text{mol}^{-1} \cdot \text{cm}^{-1}$$

Since the absorptivity of solution II *did* increase when additional Y^{4-} was added, not all of the M^{2+} is complexed in the original solution, and

$$A_{II} = 1.200 = (500)(2.000)[MY^{2-}]_{II} \qquad \text{or} \qquad [MY^{2-}]_{II} = 1.20 \times 10^{-3}$$

Therefore, in this equilibrium solution,

$$[M^{2+}]_{II} = 2.00 \times 10^{-3} - 1.20 \times 10^{-3} = 0.80 \times 10^{-3} \qquad \text{and} \qquad [Y^{4-}]_{II} = 8.00 \times 10^{-3} - 1.20 \times 10^{-3} = 6.80 \times 10^{-3}$$

which imply

$$K = \frac{[M^{2+}]_{II}[Y^{4-}]_{II}}{[MY^{2-}]_{II}} = \frac{(0.00080)(0.00680)}{0.00120} = 4.5_3 \times 10^{-3}$$

13.9 The amount of contaminant X in a sample of lake water is to be determined by absorption spectrometry. Four solutions are prepared and the transmittance of each determined using the same 1.00 cm cell at the same wavelength; results are given in Table 13-1. What is the concentration of X (ppm by wt) in the lake water? Assume that Beer's law applies to all solutions and that absorbances are additive.

The absorbances of the samples are: $A_1 = 0.000$, $A_2 = 0.140$, $A_3 = 0.907$, $A_4 = 0.485$. Therefore, the absorbance of the 5.00 ppm X is

$$A_3 - A_2 = 0.767 = \epsilon_X(1.00)(5.00) \qquad \text{or} \qquad \epsilon_X = 0.153_4\ \text{ppm}^{-1} \cdot \text{cm}^{-1}$$

The absorbance of X in solution 4 is

$$A_4 - A_2 = 0.345 = (0.153_4)(1.00)C_X \qquad \text{or} \qquad C_X = 2.24_9\ \text{ppm}$$

The concentration of X in the undiluted lake water is 100 times as great, or 225 ppm by wt.

Table 13-1

Solution No.	Composition	T, %
1	Distilled water	100.0
2	Distilled water plus all species (except X) in lake water, each at the same concentration as in lake water	72.5
3	Solution 2 plus 5.00 ppm by wt of X	12.4
4	1.00 mL of lake water diluted to 100.0 mL with solution 2	32.7

13.10 A weak acid, HX, has $K_a = [H^+][X^-]/[HX] = 2.50 \times 10^{-5}$. Only HX absorbs at 450 nm. 0.0100 mmol of HX is added to 100 mL of a pH 4.000 buffer solution; assume no change in volume. The absorbance of this solution at 450 nm measured in a 1.00 cm cell was 0.360. Determine (a) ϵ_{HX} at 450 nm; (b) the absorbance of the solution in a 1.00 cm cell at 450 nm, if the 0.0100 mmol of HX had instead been added to 100 mL of 0.100 M HCl; (c) the absorbance if 20.00 mmol of NaX was added to 100 mL of a pH 9.000 buffer solution.

(a) The concentration of HX prior to equilibration was $0.0100/100 = 1.00 \times 10^{-4}$ M. If x moles per liter of HX dissociates, then, in the buffered equilibrium solution,

$$[HX] = 1.00 \times 10^{-4} - x \qquad [H^+] = 1.00 \times 10^{-4} \qquad [X^-] = x$$

Therefore,

$$2.50 \times 10^{-5} = \frac{(1.00 \times 10^{-4})(x)}{1.00 \times 10^{-4} - x}$$

which gives $x = 2.00 \times 10^{-5}$, $[HX] = 8.0_0 \times 10^{-5}$. Therefore, from (13.5),

$$0.360 = \epsilon_{HX}(1.00)(8.0_0 \times 10^{-5}) \qquad \text{or} \qquad \epsilon_{HX} = 4.5_0 \times 10^3 \text{ L} \cdot \text{mol}^{-1} \cdot \text{cm}^{-1}$$

(b) If x moles per liter of HX dissociates, then, at equilibrium,

$$[HX] = 1.00 \times 10^{-4} - x \qquad [H^+] = 0.100 + x \qquad [X^-] = x$$

However, since $K_a = 2.50 \times 10^{-5}$ is small, x will be small compared with 0.100, and $[H^+] \approx 0.100$. Therefore,

$$2.50 \times 10^{-5} \approx \frac{(0.100)(x)}{1.00 \times 10^{-4} - x}$$

which results in $x = 2.50 \times 10^{-8}$, $[HX] = 1.00 \times 10^{-4}$. From (13.5),

$$A = (4.5_0 \times 10^3)(1.00)(1.00 \times 10^{-4}) = 0.45_0$$

(c) Prior to equilibration, the concentration of X^- is $20.00/100 = 0.200$ M. If x moles per liter of X^- reacts with H^+ to produce HX, then, in the buffered equilibrium solution,

$$[HX] = x \qquad [H^+] = 1.00 \times 10^{-9} \qquad [X^-] = 0.200 - x$$

Thus

$$2.50 \times 10^{-5} = \frac{(1.00 \times 10^{-9})(0.200 - x)}{x}$$

so that $x = 8.00 \times 10^{-6} = [HX]$ and $A = (4.5_0 \times 10^3)(1.00)(8.00 \times 10^{-6}) = 0.036$.

13.11 The reaction of a new, biomedically important compound, B, with compound C is either $B + C \rightarrow BC$ or $B + 2C \rightarrow BC_2$. Only the species B absorbs at 250 nm, with absorptivity 2000 $L \cdot mol^{-1} \cdot cm^{-1}$. Two solutions were prepared by mixing 50.0 mL of B and 50.0 mL of C, and the absorbances of the equilibrated reaction mixtures were then determined at 250 nm using a 1.00 cm spectral cell. The data are:

Solution No.	Premixing [B]	Premixing [C]	Absorbance
1	2.00×10^{-3}	8.00×10^{-3}	0.800
2	2.00×10^{-3}	12.34×10^{-3}	0.400

Determine whether the reaction product is BC or BC_2 and the value of the equilibrium constant, K, for the reaction.

The absorbance data allow calculating the equilibrium [B] using (*13.5*). Thus, for solution 1,

$$0.800 = (2000)(1.00)[B]_1 \qquad \text{or} \qquad [B]_1 = 4.00 \times 10^{-4}$$

Similarly,

$$0.400 = (2000)(1.00)[B]_2 \qquad \text{or} \qquad [B]_2 = 2.00 \times 10^{-4}$$

The concentrations of B and C after mixing the two solutions, but prior to equilibration, were: for solution 1, $[B]_{o1} = 1.00 \times 10^{-3}$ and $[C]_{o1} = 4.00 \times 10^{-3}$; for solution 2, $(B)_{o2} = 1.00 \times 10^{-3}$ and $[C]_{o2} = 6.17 \times 10^{-3}$.

If the reaction was $B + C \rightarrow BC$, then the equilibrium concentrations would be:

Solution 1. $[B]_1 = 4.00 \times 10^{-4}$, as determined from the absorbance. $[BC]_1$ equals the amount of B consumed per L:

$$[BC]_1 = [B]_{o1} - [B]_1 = 6.0 \times 10^{-4}$$

$[C]_1$ equals the original concentration minus the amount of B consumed per L:

$$[C]_1 = [C]_{o1} - [BC]_1 = 3.40 \times 10^{-3}$$

Therefore,

$$K_1 = \frac{[BC]_1}{[B]_1[C]_1} = \frac{6.0 \times 10^{-4}}{(4.00 \times 10^{-4})(3.40 \times 10^{-3})} = 4.4_1 \times 10^2$$

Solution 2. $[B]_2 = 2.00 \times 10^{-4}$; and, calculating as above,

$$[BC]_2 = [B]_{o2} - [B]_2 = 8.0 \times 10^{-4}$$
$$[C]_2 = [C]_{o2} - [BC]_2 = 5.37 \times 10^{-3}$$
$$K_2 = \frac{[BC]_2}{[B]_2[C]_2} = \frac{8.0 \times 10^{-4}}{(2.00 \times 10^{-4})(5.37 \times 10^{-3})} = 7.4_5 \times 10^2$$

Since $K_1 \neq K_2$, the reaction is *not* $B + C \rightarrow BC$.

The reaction then must be $B + 2C \rightarrow BC_2$. This may be verified by computing K_1 and K_2 as above, except that now

$$[C] = [C]_o - 2[BC_2]$$

The result is: $K_1 = K_2 = K = 1.9_1 \times 10^5$.

13.12 A photocell in a spectrophotometer can register T to ± 0.007. A 1.00 cm cell is filled with a solution of R which has $\epsilon = 260 \ L \cdot mol^{-1} \cdot cm^{-1}$ at 600 nm; only R absorbs at this wavelength. The transmittance of the solution at 600 nm was found to be 0.720. Find the concentration, $\bar{C}$, of R and the uncertainty in the concentration, σ_C.

From (13.5),

$$\bar{C} = \frac{A}{\epsilon b} = \frac{-\log 0.720}{(260)(1.00)} = 5.50 \times 10^{-4} \ M$$

By (13.9),

$$\sigma_C = -0.434 \, \frac{\bar{C}\sigma_T}{\bar{T} \log \bar{T}} = -0.434 \, \frac{(5.50 \times 10^{-4})(0.007)}{(0.720)(\log 0.720)}$$

$$= 1._6 \times 10^{-5} \ M$$

Thus, $\bar{C} \pm \sigma_C = 0.55_0 \pm 0.01_6$ mmol/L.

13.13 The iron content of a sample was determined in a photometric titration. First, 0.5993 g of the sample was dissolved in HCl and the solution diluted to 100.0 mL. Next, 10.0 mL of this solution was taken, salicylic acid was added to form the complex $FeSal_3^{3-}$, and the resulting solution was diluted in a separate flask to $V_S = 100.0$ mL. This solution was then transferred to a beaker, in which it was titrated with 0.01496 M EDTA, the reaction being

$$FeSal_3^{3-} + H_3Y^- + 3H^+ \rightarrow FeY^- + 3H_2Sal$$

Samples of the titration mixture were withdrawn periodically during the titration, the absorbance measured, and the sample returned to the titration mixture before adding additional titrant. The experimental data, corrected by use of (13.10), are displayed in Table 13-2. What was the wt % Fe in the original sample?

Table 13-2

V_T, mL	31.22	31.39	31.57	31.82	32.01	32.17	32.38
A_{corr}	0.600	0.537	0.470	0.414	0.353	0.295	0.249
V_T, mL	32.60	32.81	32.97	33.22	33.58	33.97	36.02
A_{corr}	0.187	0.127	0.0751	0.0482	0.0484	0.0485	0.0493

A plot of A_{corr} versus V_T shows two straight-line sets of data, like Fig. 13-3(d), intersecting at approximately $V_{H_3Y^-} = 33$ mL. The experimental data point at $V_T = 32.97$ mL will be disregarded, since it occurs at the bend in the data sets. A linear least-squares fit of the data prior to $V_T = 32.97$ mL ($N = 9$; Section 2.9) is given by

$$A = -0.292_4 V + 9.71_5$$

and of the data after $V_T = 32.97$ mL ($N = 4$) by

$$A = 3.84_9 \times 10^{-4} V + 0.0354_4$$

The two lines intersect (share a common value of A) at the equivalence point:

$$-0.292_4 V + 9.71_5 = 3.84_9 \times 10^{-4} V + 0.0354_4 \qquad \text{or} \qquad V = 33.0_6 \text{ mL}$$

Thus, $FeSal^{3-}$, and therefore Fe^{3+}, was present in the amount $(33.0_6)(0.01496) = 0.494_6$ mmol. Since only 10.0 mL of the original 100.0 mL of solution was used, the mmols Fe in the sample was

$$\frac{100.0}{10.0} (0.494_6) = 4.94_6$$

giving, finally, as the iron content of the original sample:

$$\frac{(4.94_6 \text{ mmol Fe})(55.85 \text{ mg Fe/mmol Fe})}{599.3 \text{ mg sample}} (100\%) = 46.0_9 \text{ wt \% Fe}$$

13.14 Sketch the type of photometric titration curve to be expected if (a) the substance being titrated absorbs markedly, but the titrant and products absorb only slightly; (b) the total absorption of the products is 30% that of the substance being titrated, and the titrant absorbs only slightly; (c) species A and B are titrated in sequence, with only A and titrant T absorbing.

(a)   Initial A will decrease until the equivalence point, whereafter it will slowly increase. See Fig. 13-4(a).

(b)   The absorbance will decrease until the equivalence point, where A will be 30% of the original A; the absorbance will then slowly increase. See Fig. 13-4(b).

(c)   The absorbance will decrease to zero at the equivalence point for species A. It will then remain at zero while species B is titrated, and will increase after the equivalence point for B. See Fig. 13-4(c).

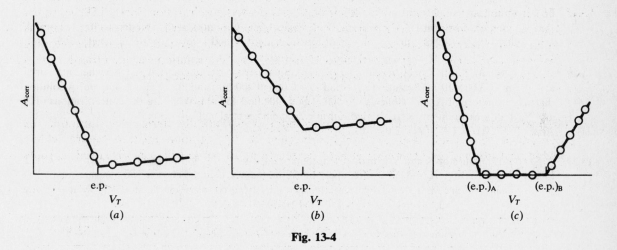

Fig. 13-4

Supplementary Problems

13.15 Express the "energy" $\bar{\nu} = 5000$ cm^{-1} in units of (a) nm, (b) Å, (c) THz (see Table 1-2), (d) pJ (see Table 1-2), (e) eV. (f) What is the energy, in kilocalories, of one mole of such photons?
Ans. ·(a) 2000 nm, (b) 2.000×10^4 Å, (c) 149.9 THz, (d) 9.932×10^{-8} pJ, (e) 0.6200 eV, (f) 14.29 kcal/mol

13.16 A 6.44×10^{-4} M solution of Ti-peroxide complex had $T = 0.340$ when measured in a 1.00 cm cell at 410 nm. What is the concentration of Ti-peroxide complex in a solution that had $T = 0.516$ when measured at 410 nm in a 0.500 cm cell?   *Ans.* 7.89×10^{-4} M

13.17 A weak acid, HZ, is analyzed at 600 nm; both HZ and Z$^-$ absorb at this wavelength. A series of buffer solutions, of 4.00×10^{-3} M formal concentration in HZ, were prepared and the absorbance of each solution determined at 600 nm using a 1.00 cm cell. Assume that the buffer solution components do not absorb at 600 nm. The data are:

Buffer pH	1.00	2.00	3.00	4.00	5.00	6.00	7.00	8.00	9.00	10.00
Absorbance	0.272	0.272	0.274	0.289	0.395	0.606	0.674	0.683	0.684	0.684

What are the values of ϵ_{HZ} and ϵ_{Z^-} at 600 nm and what is the value of K_a for HZ?
Ans. $\epsilon_{HZ} = 68.0$ L·mol^{-1}·cm^{-1}, $\epsilon_{Z^-} = 171$ L·mol^{-1}·cm^{-1}; $K_a = 4.24 \times 10^{-6}$

13.18 Samples of blood, urine, and acetone solutions are analyzed using the same cell and at the same wavelength. Except for an instrumental "blank" correction of $A = 0.045$ for all solutions, only acetone absorbs at the wavelength chosen for measurement. Data for acetone solutions are:

$C_{acetone}$, mg/100 mL	0.500	1.000	2.000	4.000	6.000	8.000
A	0.057	0.068	0.091	0.137	0.183	0.229

Samples of blood and urine were obtained from a normal subject and from a patient suffering from ketosis (abnormally high ketone levels). Data are: for the normal, $A_{blood} = 0.068$, $A_{urine} = 0.097$; for the patient, $A_{blood} = 0.189$, A_{urine} (1/25 dilution) $= 0.198$. Calculate the acetone levels, in mg acetone/100 mL fluid, for the four samples.

Ans.   Based on the linear least-squares fit of the corrected absorbances of the standards: for the normal, blood level $= 0.992$, urine level $= 2.26$; for the patient, blood level $= 6.26$, urine level $= 166$.

13.19 A 2.00×10^{-4} M solution of X had $A = 1.000$ when measured at 400 nm, and $A = 0.125$ when measured at 450 nm. A 3.00×10^{-3} M solution of Y had $A = 0.060$ at 400 nm, and $A = 0.636$ at 450 nm. A solution that was a mixture of X and Y had $A = 0.500$ at both 400 and 450 nm. What are the concentrations of X and Y, if all measurements were carried out in a 1.00 cm spectral cell?

Ans. $[X] = 9.16 \times 10^{-5}$, $[Y] = 2.09 \times 10^{-4}$

13.20 A 50.00 mL mixture of X and Y is titrated photometrically with 0.02538 M titrant T. Each reacts with T in a 1:1 ratio. Only X and T absorb appreciably at the wavelength chosen for measurement. The products and species Y absorb only slightly. For the titration data of Table 13-3, what are the equivalence points and the concentrations of X and Y?

Ans.   Based on linear least-square fits of the first nine, the next seven, and the last eight sets of data, equivalence points are found at $V_T = 22.3_3$ mL and 38.3_8 mL; $[X] = 1.13 \times 10^{-2}$, $[Y] = 8.15 \times 10^{-3}$.

Table 13-3

V_T, mL	2.05	5.12	6.94	9.56	12.03	14.27	16.48	18.43
A_{corr}	0.740	0.632	0.571	0.493	0.399	0.322	0.238	0.165
V_T, mL	20.37	23.05	26.13	28.84	31.62	34.10	35.93	37.36
A_{corr}	0.103	0.057	0.016	0.068	0.075	0.081	0.091	0.095
V_T, mL	39.45	40.44	41.77	43.13	44.47	45.34	46.51	47.78
A_{corr}	0.141	0.201	0.258	0.327	0.380	0.430	0.483	0.548

Chapter 14

Radioactivity

14.1 TYPES OF DECAY

Three different types of decay are exhibited by radioactive species. The most common is β^- *decay*, wherein an electron (*beta particle*) is emitted by the nucleus and, simultaneously, a neutron in the nucleus converts to a proton. This nuclear process can be indicated as:

$$\,_0^1n \to \,_1^1p + \,_{-1}^0e \tag{14.1}$$

where the superscripts are the nearest-whole-number atomic weights (in u; see Table 1-1) and the subscripts are the charges (in units of e; see Table 1-3).

In β^+ *decay*, the nucleus emits a positron, $\,_{+1}^0e$, and a proton in the nucleus is converted to a neutron:

$$\,_1^1p \to \,_0^1n + \,_{+1}^0e \tag{14.2a}$$

A related process, *K-electron capture*, leads to the same change in the nucleus: the nucleus absorbs one of the electrons from the inmost shell and a proton in the nucleus converts to a neutron. Thus,

$$\,_1^1p + \,_{-1}^0e \to \,_0^1n \tag{14.2b}$$

As more energy is released in (*14.2b*) than in (*14.2a*), if proton conversion occurs by β^+ decay, then K-electron capture is also possible. For instance, 60% of the decays of $\,_{29}^{61}Cu$ (half-life, 3.32 h) involve β^+ emission and 40% involve K-electron capture. [To be strictly correct, the right-hand sides of (*14.1*) and (*14.2*) should involve another particle, a *neutrino*. However, since this particle has zero charge and zero (rest) mass, it may be omitted as chemically insignificant.]

Many of the very heavy elements undergo α *decay*, emitting an alpha particle (a helium nucleus, $\,_2^4He$):

$$2\,_0^1n + 2\,_1^1p \to \,_2^4He \tag{14.3}$$

Isotopes that have undetectable decay rates are called *stable* isotopes. A plot of the number of neutrons N versus the number of protons (*atomic number*) Z for all stable isotopes is shown in Fig. 14-1. All elements above bismuth ($Z = 83$) have detectable decay rates; for each element, the isotope with the longest half-life is indicated on the figure.

The observed decay process is that which brings the resultant isotope closer to the band of stable isotopes shown in Fig. 14-1. As seen in that figure, β^- decay corresponds to a 45° downward shift, since it involves a unit decrease in the number of neutrons and a unit increase in the number of protons. As a result, any radioactive isotope whose N/Z ratio places it above the band of stability will decay by β^- emission.

If the radioactive isotope is below the stable band and also above atomic number 83, then α decay is the most probable decay mode. If the radioactive isotope is below the stable band and also below atomic number 84, then β^+ decay and/or K-electron capture is the most probable decay mode; the only exceptions in this group are a few isotopes of Bi, Au, Pt, and a few other large-Z elements, where decay is by α emission.

EXAMPLE 14.1   The atomic weight of arsenic is 74.9. Predict the decay mode of $\,_{33}^{77}As$.

The *mass number* of an isotope, $A = N + Z$, is the number of particles in the nucleus. Because the mass of each particle is very nearly 1 u, the atomic weight of a chemical element, which is the abundance-weighted average of the atomic weights of all stable isotopes of that element, is essentially equal to $\bar{A} = \bar{N} + Z$. Thus, $N \gtrless \bar{N}$ (i.e., an isotope is above or is below the band of stability in Fig. 14-1) is equivalent to $A \gtrless$ atomic weight.

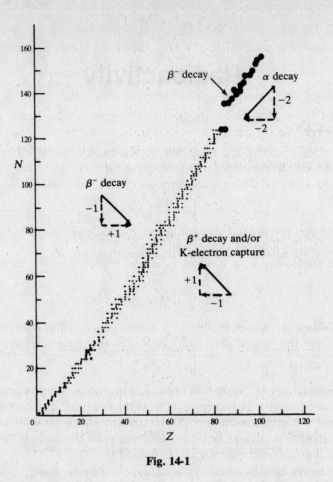

Fig. 14-1

Here, since $77 > 74.9$, the arsenic isotope should decay by β^- emission. [It does so, with a half-life of 38.7 h.]

One other general observation can be based on Fig. 14-1: The more distant an isotope is from the band of stability, the less stable will be the isotope and the shorter its half-life.

14.2 DECAY RATE EQUATIONS

Radioactive decay is a *first-order* process in that the number of atoms undergoing decay per unit time is proportional to the total number present. Thus, if A^* radioactive atoms are present at time t,

$$\frac{dA^*}{dt} = -\lambda A^* \qquad\qquad (14.4)$$

in which $\lambda > 0$ is the *decay constant* of the species. By integration of (*14.4*),

$$\ln A^* = \ln A_0^* - \lambda t \qquad \text{or} \qquad A^* = A_0^* e^{-\lambda t} \qquad\qquad (14.5)$$

where $A_0^* \equiv A^*(0)$ is the number of atoms present at $t = 0$. The half-life, $t_{1/2}$, of the species is defined as the time required for a collection A^* to be reduced to $\frac{1}{2}A^*$ by decay. From (*14.5*),

$$\ln A^* = \ln A_0^* - \lambda t$$
$$\ln \tfrac{1}{2}A^* = \ln A_0^* - \lambda(t + t_{1/2})$$

whence, by subtraction,

$$\lambda t_{1/2} = \ln 2 = 0.69315 \cdots \qquad (14.6)$$

By (14.6), if time is measured in half-lives, $t = nt_{1/2}$, then (14.5) takes the simple form

$$A^* = \frac{A_0^*}{2^n} \qquad (14.7)$$

According to (14.5), a plot of $\ln A^*$ or $\log A^*$ versus time should yield a straight line. Experimentally, however, we do not directly measure A^*; rather, we make a count of the decay particles (β^-, β^+, or α) or of the photons (X-ray, γ-ray) accompanying the decays. Let $P(t)$ be the number of these decay products. Then, if there is one product per decay event (as is generally the case—see below),

$$D_s \equiv \frac{dP}{dt} = -\frac{dA^*}{dt} = \lambda A^* \qquad (14.8)$$

Thus, A^* is directly proportional to the *activity* (the disintegration rate), D_s, of the radioactive sample, which is in turn directly proportional to the *observed counting rate*, R_s:

$$R_s = \begin{cases} ED_s & \text{for particles} \\ F_\gamma ED_s & \text{for } \gamma\text{-ray photons} \end{cases} \qquad (14.9)$$

Table 14-1

Stable Species X	Natural Isotopic Abundance, %	σ, barns	Product Isotope X*	$t_{1/2}$	E_γ, MeV	F_γ, %
$^2_1$H	0.0148	0.0005	$^3_1$H	12.26 y	β^- only	
$^{23}_{11}$Na	100.0	0.53	$^{24}_{11}$Na	14.96 h	1.369 2.754	100
$^{27}_{13}$Al	100.0	0.235	$^{28}_{13}$Al	2.31 min	1.780	100
$^{37}_{17}$Cl	24.47	0.40	$^{38}_{17}$Cl	37.29 min	1.642 2.170	38 47
$^{51}_{23}$V	99.75	4.9	$^{52}_{23}$V	3.75 min	1.434	100
$^{79}_{35}$Br	50.52	2.9 8.5	$^{80m}_{35}$Br $^{80}_{35}$Br	4.38 h 17.6 min	— 0.511 0.618	5 7
$^{121}_{51}$Sb	57.25	6.1	$^{122}_{51}$Sb	2.80 d	0.564 0.686	66 3.4
$^{127}_{53}$I	100.0	6.4	$^{128}_{53}$I	24.99 min	0.441	14
$^{139}_{57}$La	99.911	8.9	$^{140}_{57}$La	40.22 h	0.487 1.596	40 96
$^{197}_{79}$Au	100.0	98.8	$^{198}_{79}$Au	64.73 h	0.412	95
$^{196}_{80}$Hg	0.146	880	$^{197}_{80}$Hg	65.0 h	0.191	2

The factor $0 < E < 1$ measures the efficiency of the counting device being used: an additional factor is required for γ-radiation because only a fraction F_γ of the decays may result in the particular photon that is being counted (see Table 14-1). The official units for D_s or R_s are as given in Table 1-1; in practice, activity is often measured in decays/min. Because D_s or R_s is proportional to A^*, it too obeys (14.5), and the graph of $\log D_s$ or $\log R_s$ versus t is also linear.

Some further notation connected with the experimental measurement of radioactivity follows:

$C_T \equiv$ total count (from sample and background)

$C_b \equiv$ background count

$\tau_T \equiv$ duration of total count

$\tau_b \equiv$ duration of background count

$R_T \equiv C_T/\tau_T \equiv$ average total count-rate

$R_b \equiv C_b/\tau_b \equiv$ average background count-rate

On the assumption that $\tau_T \ll t_{1/2}$, instantaneous count-rates will be constant throughout the interval τ_T and so will coincide with average count-rates; hence,

$$R_T - R_b = R_s \qquad\qquad (14.10)$$

14.3 NEUTRON ACTIVATION

If atoms of a stable species X are irradiated with neutrons, the most common resultant nuclear reaction is one whereby a certain proportion of the X-nuclei absorb neutrons to become heavier by one atomic mass unit. Since only a gamma ray (or rays) is emitted, there is no change in atomic number, and the product species is an isotope, X^*, of X. However, the X^*-nuclei are subject to first-order decay, as described in Section 14.2, during the period of neutron irradiation. Consequently, the number A^* of X^*-nuclei is simultaneously increasing at a rate proportional to the number A of X-nuclei and decreasing at a rate proportional to itself, A^*:

$$\frac{dA^*}{dt} = \phi\sigma A - \lambda A^* \qquad\qquad (14.11)$$

Here ϕ is the incident neutron flux, usually given in neutrons$\cdot$cm$^{-2}\cdot$s^{-1}; σ is the *absorption cross-section* of X (the effective area presented by an X-nucleus to the neutron stream), commonly measured in *barns* (1 barn $= 1.00 \times 10^{-24}$ cm^2); and λ is the decay constant of X^*. Some values of σ and of $t_{1/2} = (\ln 2)/\lambda$ are given in Table 14-1.

During irradiation, ϕ will be essentially constant, as will A, since only a tiny fraction of X-nuclei can be expected to absorb a neutron. Thus, (14.11) may be integrated by standard means, using the initial condition $A^*(0) = 0$, to give:

$$A^* = \frac{\phi\sigma A}{\lambda}(1 - e^{-\lambda t}) \qquad\qquad (14.12)$$

The corresponding equation for R_s is, from (14.8) and (14.9),

$$R_s = E\phi\sigma A(1 - e^{-\lambda t}) \qquad \text{or} \qquad EF\gamma\phi\sigma A(1 - e^{-\lambda t}) \qquad\qquad (14.13)$$

14.4 RADIOACTIVITY STATISTICS

The number of counts, C, recorded during the experimental period obeys *Poisson statistics*, as do many counting data (the number of faults in a mile of cable, the number of cars per hour arriving at a given intersection, etc.). The theoretical ($N \to \infty$; see Section 2.3) Poisson distribution depends on only one parameter, the mean $\mu = \bar{C}$; for the standard deviation we have:

$$\sigma_C = \sqrt{\mu} \qquad\qquad (14.14)$$

This is in contrast to the Gaussian distribution (Section 2.3), in which the mean and standard deviation are independent parameters. Using (14.14), we may apply the propagation-of-error formulas of Section 2.7 to combinations of count-data.

EXAMPLE 14.2 A radioactive sample is monitored for $\tau_T = 10.0$ min and registers $\bar{C}_T = 20\,000$ counts; a $\tau_b = 60.0$ min count of the background registers $\bar{C}_b = 600$ counts. Determine the expected count rate, $\bar{R}_s$, of the sample and the standard deviation σ_{R_s}, assuming that the time measurements are precise.

By (14.10),

$$R_s = \frac{1}{\tau_T} C_T - \frac{1}{\tau_b} C_b$$

Hence, following Example 2.9 and using (14.14),

$$\bar{R}_s = \frac{1}{\tau_T}\bar{C}_T - \frac{1}{\tau_b}\bar{C}_b = \frac{1}{10.0}(20\,000) - \frac{1}{60.0}(600) = 1990 \text{ counts/min}$$

$$\sigma_{R_s} = \sqrt{\left(\frac{1}{\tau_T}\right)^2 \sigma_{C_T}{}^2 + \left(\frac{1}{\tau_b}\right)^2 \sigma_{C_b}{}^2} = \sqrt{\frac{\bar{C}_T}{\tau_T{}^2} + \frac{\bar{C}_b}{\tau_b{}^2}}$$

$$= \sqrt{\frac{20\,000}{100} + \frac{600}{3600}} = 14.1 \text{ counts/min}$$

Thus, the experimental result would be cited as $R_s = 1990 \pm 14$ counts/min.

It can be shown that σ_{R_s} is minimized if the total available counting time, $\tau_T + \tau_b$, is distributed in the ratio

$$\frac{\tau_T}{\tau_b} = \sqrt{\frac{\bar{R}_T}{\bar{R}_b}} \qquad\qquad (14.15)$$

Solved Problems

TYPES OF DECAY

14.1   Predict how $^{43}_{21}$Sc should decay.

Since the atomic weight of Sc is 44.9, this isotope has fewer neutrons than the average stable Sc isotope, and β^+ emission and/or K-electron capture would be the expected mode of decay. [It decays by β^+ emission.]

14.2   Predict how $^{21}_{9}$F should decay.

Since the atomic weight of F is 19.0, this isotope has a greater number of neutrons than the average stable F isotope, and β^- emission would be the expected mode of decay.

14.3   Predict how $^{239}_{92}$U should decay.

Since the average atomic weight of the long-lived isotopes of U is 238.0, this isotope has a neutron surplus, and β^- emission would be the expected mode of decay.

14.4   Predict how $^{234}_{93}$Np should decay.

Since the average atomic weight of the long-lived isotopes of Np is 237.1, this isotope has a neutron deficit, and α radiation would be the expected mode of decay.

14.5 $^{238}_{92}$U undergoes a series of decays emitting α and β^- particles; eventually the decay chain ends in stable $^{206}_{82}$Pb. Evaluate x and y in the overall decay equation

$$^{238}_{92}\text{U} \to ^{206}_{82}\text{Pb} + x\,^{0}_{-1}e + y\,^{4}_{2}\text{He}$$

Since the superscript for the electron is zero, it is easiest first to balance the equation on atomic mass number:

$$y(4) + 206 = 238 \qquad \text{or} \qquad y = 8$$

Then, by a balance of the subscripts (the nuclear charges),

$$92 = 82 + x(-1) + 8(2) \qquad \text{or} \qquad x = 6$$

DECAY RATE EQUATIONS

14.6 The tritium ($^3_1$H) decay rate of a sample is $15\,260$ min^{-1}. Use data from Table 14-1 to calculate the tritium content of the sample.

From (14.6) and (14.8),

$$A^* = \frac{D_s}{\lambda} = \frac{D_s}{0.693}\, t_{1/2}$$
$$= \frac{15\,260 \text{ min}^{-1}}{0.693}[(12.26 \text{ y})(365 \text{ d/y})(24 \text{ h/d})(60 \text{ min/h})]$$
$$= 1.419 \times 10^{11} \text{ tritium atoms}$$

14.7 The average human adult has 0.18 lbm (see Table 1-1) of potassium distributed in his body. One isotope of potassium, $^{40}_{19}$K, is radioactive, with $t_{1/2} = 1.26 \times 10^9$ y. In nature, 0.0118% of potassium is $^{40}_{19}$K. How many decays per minute of $^{40}_{19}$K occur in the average adult body?

From (14.4) and (14.6),

$$-\frac{dA^*}{dt} = \frac{0.693}{t_{1/2}}A^* = \frac{(0.693)(0.18 \text{ lbm K})(453.6 \text{ g/lbm})(0.000118 \text{ g }^{40}\text{K/g K})(6.022 \times 10^{23} \text{ mol}^{-1})}{(1.26 \times 10^9 \text{ y})(365 \text{ d/y})(24 \text{ h/d})(60 \text{ min/h})(40 \text{ g }^{40}\text{K/mol})}$$
$$= 1.5 \times 10^5 \text{ decays/min}$$

[*Note*: Potassium is necessary in human nutrition. Since the half-life is of the order of the age of the universe, radioactive potassium has always been part of the human body.]

14.8 Mass spectrometry shows that a sample of uranium is 99.28 wt % $^{238}_{92}$U. 1.26 g of U_2O_3 of the same isotopic composition has a $^{238}_{92}$U disintegration rate of 8.41×10^5 min^{-1}. If the average atomic weight of U in nature is 238.03, whereas the atomic weight of $^{238}_{92}$U is 238.05, what is the half-life of $^{238}_{92}$U?

From (14.6) and (14.8),

$$t_{1/2} = \frac{0.693}{D_s}A^*$$
$$= \frac{0.693}{8.41 \times 10^5 \text{ min}^{-1}}(1.26 \text{ g U}_2\text{O}_3)\left(\frac{476.06 \text{ g U}}{524.06 \text{ g U}_2\text{O}_3}\right)\left(\frac{0.9928 \text{ g }^{238}\text{U}}{1 \text{ g U}}\right)$$
$$\times \left(\frac{1 \text{ mol }^{238}\text{U}}{238.05 \text{ g }^{238}\text{U}}\right)(6.022 \times 10^{23} \text{ mol}^{-1})$$
$$= 2.37 \times 10^{15} \text{ min} = 4.51 \times 10^9 \text{ y}$$

14.9 A 1.00×10^{-3} M solution of HCl includes $^{36}_{17}$Cl ($t_{1/2} = 3.09 \times 10^5$ y). The *specific activity* (the activity per unit mass or unit volume) of the solution is 5000 decays $\cdot$ min$^{-1}\cdot$ mL^{-1}. What fraction of the Cl atoms in solution are ^{36}Cl atoms?

In 1 mL of solution, there are

$$A^* = \frac{D_s}{0.693}\, t_{1/2} = \frac{5000 \text{ min}^{-1}}{0.693}(3.09 \times 10^5 \text{ y})(365 \text{ d/y})(24 \text{ h/d})(60 \text{ min/h})$$

$$= 1.17 \times 10^{15}\ ^{36}\text{Cl atoms}$$

and there are altogether

$$(1.00 \times 10^{-6} \text{ mol Cl})(6.022 \times 10^{23} \text{ mol}^{-1}) = 6.02 \times 10^{17} \text{ Cl atoms}$$

The desired fraction is thus

$$\frac{1.17 \times 10^{15}}{6.02 \times 10^{17}} = 0.00194$$

14.10 A sample contains $^{24}_{11}$Na. At noon on Monday, a 2.00 min count registered 2625 gamma rays at $E_\gamma = 1.369$ MeV. How many 1.369 MeV gamma rays would be expected during a 1.00 min count at noon on Wednesday of the same week? Refer to Table 14-1.

By (*14.5*) for R_s, with $t = 2$ d $= 48.00$ h,

$$R_s = \left(\frac{2625 \text{ counts}}{2.00 \text{ min}}\right)e^{-(0.693)(48.00 \text{ h})/(14.96 \text{ h})} = 142 \text{ counts/min}$$

14.11 A sample containing some $^{140}_{57}$La registered 3667 gamma rays at $E_\gamma = 0.487$ MeV during a 5.00 min count starting at 9:00 a.m. today. At this energy, the detector has a counting efficiency of 15.0%. What was the activity of this sample at 9:02:30 a.m. a week ago? Refer to Table 14-1.

From (*14.9*), the activity at 9:02:30 a.m. *today* (the midpoint of the counting period) was

$$D_s(0) = \frac{R_s}{F_\gamma E} = \frac{3667/5.00}{(0.40)(0.150)} = 1.2_2 \times 10^4 \text{ min}^{-1}$$

Taking this back one week—i.e., using

$$t = -(7.00 \text{ d})(24.00 \text{ h/d}) = -168.00 \text{ h}$$

$$\lambda = \frac{0.693}{40.22}\text{ h}^{-1}$$

in (*14.5*) for D_s—we find:

$$D_s = (1.2_2 \times 10^4)e^{(0.693)(168.00)/40.22} = 2.2_1 \times 10^5 \text{ min}^{-1}$$

14.12 $^{14}_{6}$C is continually being produced in the upper atmosphere as a result of cosmic ray bombardment of nitrogen. Living matter, in equilibrium with this source of carbon, shows a specific activity of 15.8 decays $\cdot$ min$^{-1}\cdot$ (g carbon)$^{-1}$. A sample of charred wood from an archeological site gave a count-rate of 6.38 min$^{-1}\cdot$ (g carbon)$^{-1}$, the detector being 82.2% efficient. What was the age of the sample, if the $^{14}_{6}$C half-life is 5568 y?

For the 1 g of carbon,

$$D_s = \frac{R_s}{E} = \frac{6.38}{0.822} = 7.76 \text{ min}^{-1}$$

today, whereas, when the wood was cut t years ago, the activity was presumably $D_s(0) = 15.8 \text{ min}^{-1}$. Thus, from (14.5) for D_s,

$$t = \frac{\ln D_s(0) - \ln D_s}{\lambda} = \frac{\ln 15.8 - \ln 7.76}{0.693/5568} = 5.71_3 \times 10^3 \text{ y}$$

14.13 A sample contains $^{38}_{17}\text{Cl}$ ($t_{1/2} = 37.5$ min) and $^{128}_{53}\text{I}$ ($t_{1/2} = 25.0$ min). At $t = 0$, the sample registered 4000 counts/min (corrected for background); at $t = 75.0$ min, 875 counts/min (corrected for background). What were the individual count-rates at $t = 0$?

We are given that

$$R_{Cl}(0) + R_I(0) = 4000 \qquad (1)$$

$$R_{Cl} + R_I = 875 \qquad (2)$$

Further, since $75.0 = 2(37.5) = 3(25.0)$, (14.7) implies that

$$R_{Cl} = \frac{R_{Cl}(0)}{2^2} \qquad (3)$$

$$R_I = \frac{R_I(0)}{2^3} \qquad (4)$$

Simultaneous solution of these four equations yields

$$R_{Cl}(0) = 3000 \text{ min}^{-1} \qquad R_I(0) = 1000 \text{ min}^{-1}$$

NEUTRON ACTIVATION

14.14 A 1.00 g sample of NH_4I was irradiated with neutrons for 25.0 min, producing $^{128}_{53}\text{I}$ ($t_{1/2} = 25.0$ min). Exactly 25.0 min after the end of the irradiation, the activity of the sample was found to be D_1. A second 1.00 g sample of NH_4I was irradiated for 50.0 min; 50.0 min after the end of this irradiation, an activity D_2 was found. Calculate D_1/D_2.

At the end of an irradiation lasting ν half-lives, (14.12) gives

$$D_0 = \phi\sigma A(1 - 2^{-\nu})$$

Then, by (14.5), the activity n half-lives later is

$$D = D_0 2^{-n} = \phi\sigma A(1 - 2^{-\nu})2^{-n} \qquad (1)$$

The two samples have identical ϕ, σ, and A; therefore,

$$\frac{D_1}{D_2} = \frac{(1 - 2^{-1})2^{-1}}{(1 - 2^{-2})2^{-2}} = \frac{4}{3}$$

14.15 A 2.000 g sample containing an unknown amount of chloride was irradiated with neutrons for 37.5 min together with a 0.500 g sample of a standard containing 300 ppm by wt (μg/g) of chloride. The half-life of the radioactive product isotope, $^{38}_{17}\text{Cl}$, is 37.5 min. When the standard was counted for 1.00 min, beginning 37.5 min after the end of the irradiation, 2500 counts ($^{38}_{17}\text{Cl}$ + background) were recorded. Using the same counting equipment, the unknown was counted for 1.00 min, beginning 187.5 min (five half-lives) after the end of the irradiation; 1900 counts ($^{38}_{17}\text{Cl}$ + background) were recorded. A separate 10.00 min count of the background resulted in 1000 counts. Determine the weight of Cl in the sample.

Both the sample and the standard were irradiated together in the same neutron flux, for the same period of time; both were counted with the same equipment. The ratio of counting rates is then, from (1) of Problem 14.14,

$$\frac{R_s}{R'_s} = \frac{D}{D'} = \frac{A2^{-n}}{A'2^{-n'}} = \frac{W2^{-n}}{W'2^{-n'}} \qquad (1)$$

where a prime refers to the standard, and where we have used the proportionality between the weight (mass) W of Cl and the number A of Cl atoms. Substituting in (1) the values

$$R_s = R_T - R_b = \frac{1900}{1.00} - \frac{1000}{10.00} = 1800 \text{ min}^{-1}$$

$$R'_s = R'_T - R_b = \frac{2500}{1.00} - \frac{1000}{10.00} = 2400 \text{ min}^{-1}$$

$n = 5.00$, $n' = 1.00$, and $W' = (300 \times 10^{-6})(0.500) = 150 \ \mu g$, we solve to find:

$$W = 1.80 \times 10^3 \ \mu g = 1.80 \text{ mg Cl}$$

14.16 A chemical standard containing iodine produces $^{128}_{53}I$ ($t_{1/2} = 25.0$ min) when irradiated with neutrons. Data are found in Table 14-2 for an irradiation and activity analysis of sample X of this standard. The count-rate of this sample turned out to be quite low (100 min^{-1}); better statistics would be obtained from a rate of 2000 min^{-1}. What must be the weight of sample X' to yield the data indicated in Table 14-2?

Table 14-2

	ppm I	Weight, mg	Neutron Flux ϕ, $cm^{-2} \cdot s^{-1}$	Duration of Irradiation, $t_{1/2}$	Delay before Count, $t_{1/2}$	Corrected Counts in 5.00 min
X	1.00	10.0	2×10^{12}	1	3	500
X'	1.00	?	2×10^{12}	2	1	10 000

For the two samples, the values of ϕ, σ, E, the iodine concentration, and the counting time are the same; hence, reasoning as in Problems 14.14 and 14.15, we have:

$$\frac{10\ 000}{500} = \frac{W'(1 - 2^{-2})2^{-1}}{(10.0)(1 - 2^{-1})2^{-3}} \qquad \text{or} \qquad W' = 33.3 \text{ mg}$$

14.17 A 1.00 g sample containing an unknown amount of aluminum was irradiated in a nuclear reactor, resulting in the production of $^{28}_{13}Al$ ($t_{1/2} = 2.30$ min); the irradiation time was $2t_{1/2}$. The neutron flux was $2.15 \times 10^{12} \ cm^{-2} \cdot s^{-1}$. The sample was counted a time $3t_{1/2}$ after the irradiation and found to have an activity of $4.80 \times 10^5 \text{ min}^{-1}$. What was the aluminum content (in μg) of the sample?

From Table 14-1, $\sigma_{Al} = 0.235 \times 10^{-24} \ cm^2$. Substituting the data in (1) of Problem 14.14,

$$(4.80 \times 10^5 \text{ min}^{-1})\left(\frac{1 \text{ min}}{60 \text{ s}}\right) = (2.15 \times 10^{12} \ cm^{-2} \cdot s^{-1})(0.235 \times 10^{-24} \ cm^2)A(1 - 2^{-2})2^{-3}$$

whence $A = 1.65 \times 10^{17}$. Thus the Al content was

$$\frac{1.65 \times 10^{17} \text{ atoms}}{6.02 \times 10^{23} \text{ atoms/mol}} (26.98 \text{ g/mol}) = 7.40 \times 10^{-6} \text{ g} = 7.40 \ \mu g$$

ANALYTICAL USES OF RADIOACTIVITY

14.18 It is desired to know the amount (in mg) of a certain organic product, P, of a reaction. Because recovery of pure P is only partial, only a lower limit for "mgs P" can be determined. However, using a radioactive tracer, an *isotopic dilution* was performed as follows: 2.00 mg of pure P was tagged with sufficient $^{14}_{6}C$ ($t_{1/2} = 5568$ y) to give a specific activity of 1500 min^{-1}·mg^{-1}. This tagged sample was added to the reaction mixture, and the resultant mixture stirred until homogeneous. It was possible to recover 5.25 mg of pure P, which was now a mixture of the reaction product and the added tagged P; as before, the percent recovery was less than 100%. This 5.25 mg of P showed a specific activity of 40.0 min^{-1}·mg^{-1}. How many mg of P were produced originally?

The total activity added to the mixture was $(1500)(2.00) = 3000$ min^{-1}. Since only $(40.0)(5.25) = 210$ min^{-1} was recovered, only

$$\frac{210}{3000} = 0.0700$$

of the total P (reaction product + tagged 2.00 mg) was recovered. Therefore,

$$\frac{5.25}{0.0700} = 75.0 \text{ mg total P}$$

must have been present, and the product yield was $75.0 - 2.00 = 73.0$ mg.

14.19 When 500.0 mL of 0.100 M Pb(NO$_3$)$_2$ is mixed with 500.0 mL of 0.100 M NaCl, some PbCl$_2(s)$ precipitates. 10.0 mL of the original NaCl solution registered 2000 counts/min due to the presence of some $^{36}_{17}Cl$ ($t_{1/2} = 3 \times 10^5$ y); 10.0 mL of the post-precipitation solution registered 200.0 counts/min using the same counting apparatus. Both counts were corrected for background. What is the value of K_{sp} for PbCl$_2(s)$?

Prior to any precipitation reaction, a 10.0 mL portion of the mixture would have registered 1000 counts/min, since the NaCl solution had been diluted by a factor of 2.00. At this stage, [Pb^{2+}] = 0.0500 and [Cl$^-$] = 0.0500. Following precipitation, the [Cl$^-$] is considerably smaller, since the measured count-rate is only 200.0 min^{-1}; in fact, the [Cl$^-$] must now be

$$(0.0500)\frac{200.0}{1000} = 0.0100$$

Therefore, since the total volume is 1.000 L, $0.0500 - 0.0100 = 0.0400$ mol of Cl$^-$ must have precipitated. From the stoichiometry of the precipitate, 0.0200 mol of Pb^{2+} must also have precipitated, so that the final [Pb^{2+}] is

$$0.0500 - 0.0200 = 0.0300$$

Thus, $K_{sp} = [\text{Pb}^{2+}][\text{Cl}^-]^2 = (0.0300)(0.0100)^2 = 3.00 \times 10^{-6}$.

14.20 For MX$_2(s)$, $K_{sp} = 4.00 \times 10^{-9}$. Excess MX$_2(s)$, tagged with radioactive X ($t_{1/2} = 1000$ y), was added to 1.00 L of H$_2$O. After the equilibrium

$$\text{MX}_2(s) \rightleftharpoons \text{M}^{2+} + 2\text{X}^-$$

became established, it was found that the solution had specific activity 60 000 min^{-1}·mL^{-1}. If the tagged MX$_2(s)$ had instead been added to 1.00 L of 1.00 M M(NO$_3$)$_2$, what would have been the specific activity of the solution?

In the first solution, when S moles of MX$_2(s)$ per liter dissolve, there result [M^{2+}] = S and [X$^-$] = $2S$, so that

$$K_{sp} = [\text{M}^{2+}][\text{X}^-]^2 = 4S^3 = 4.00 \times 10^{-9} \qquad \text{or} \qquad S = 1.00 \times 10^{-3}$$

Therefore, $[X^-] = 2.00 \times 10^{-3}$ results in the specific activity $60\,000$ min$^{-1} \cdot$mL^{-1}. For the hypothetical solution, it may be assumed that at equilibrium $[M^{2+}] = 1.00$. Therefore,

$$[X^-] = \sqrt{K_{sp}/[M^{2+}]} = \sqrt{(4.00 \times 10^{-9})/1.00} = 6.33 \times 10^{-5}$$

The following proportion can now be set up:

$$\frac{x}{60\,000} = \frac{6.33 \times 10^{-5}}{2.00 \times 10^{-3}}$$

Solving, $x = 1.90 \times 10^3$ min$^{-1} \cdot$mL^{-1}.

COUNT STATISTICS

14.21 The $_6^{14}$C content of a sample was determined using a liquid scintillation counter. In 10.00 min, 700 counts of $_6^{14}$C + background were recorded. A separate background count gave 300 counts in 15.0 min. Determine the mean and the standard deviation of the ^{14}C count-rate.

Following Example 14.2,

$$\bar{R}_s = \frac{700}{10.00} - \frac{300}{15.0} = 50.0 \text{ counts/min} \qquad \sigma_{R_s} = \sqrt{\frac{700}{(10.00)^2} + \frac{300}{(15.0)^2}} = 2.9 \text{ counts/min}$$

14.22 The two counts of Problem 14.21 took a total of 25.0 min. If 60.0 min was available, what division of that time would minimize the uncertainty in the count-rate?

Solving (*14.15*),

$$\frac{\tau_T}{\tau_b} = \sqrt{\frac{70.0}{20.0}}$$

and $\tau_T + \tau_b = 60.0$ simultaneously, we obtain: $\tau_T = 39.1$ min, $\tau_b = 20.9$ min.

Supplementary Problems

TYPES OF DECAY

14.23 Predict how $_{49}^{122}$In should decay.   *Ans.* β^-

14.24 Predict how $_{88}^{222}$Ra should decay.   *Ans.* α

14.25 Predict how $_{13}^{25}$Al should decay.   *Ans.* β^+ and/or K-electron capture

14.26 Evaluate x and y in the equation $_{94}^{241}$Pu $\rightarrow$ $_{83}^{209}$Bi $+ x_{-1}^0e + y_2^4$He.   *Ans.* $x = 5$, $y = 8$

DECAY RATE EQUATIONS

14.27 If 32.0 mg of pure $_{94}^{239}$PuO$_2$ has an activity of 6.40×10^7 s^{-1}, what is the half-life of $_{94}^{239}$Pu? *Ans.* 2.44×10^4 y

14.28 A piece of silver was irradiated with neutrons for 1.00 min resulting in the production of two radioactive isotopes of silver: $_{47}^{108}$Ag ($t_{1/2} = 147$ s) and $_{47}^{110}$Ag ($t_{1/2} = 24.5$ s). At the instant the irradiation ended ($t = 0$), the activities of these radioisotopes were 2500 s^{-1} and 6000 s^{-1}, respectively. Evaluate the ratio

$$\text{atoms } _{47}^{108}\text{Ag/atoms } _{47}^{110}\text{Ag}$$

(*a*) at $t = 0$, (*b*) at $t = 147$ s.   *Ans.* (*a*) 2.50; (*b*) 80.0

NEUTRON ACTIVATION

14.29 The uranium and thorium contents of a 1.20 g sample of high-purity quartz were determined by simultaneous neutron irradiation of the sample and a chemical standard. The data are:

	Uranium	Thorium
Content of chemical standard	9.60 μg	9.85 μg
Counts/3000 s; chemical standard at $t = 0$	19 602	28 441
Counts/3000 s; 1.20 g quartz at $t = 24.0$ h	148	166
Half-life of isotope measured	2.35 d	27.0 d

Both sample and standard were counted using the same equipment and using the same gamma-ray peaks for analysis. Find the uranium and thorium contents of the sample in ppb (ng/g) by wt.
Ans. 81.2 ppb U, 49.1 ppb Th

14.30 20.00 mg of a fish sample was irradiated with neutrons; in the same irradiation capsule was a 15.00 mg chemical standard containing 4.00 ppm by wt Hg. The fish sample was counted for 10.00 min and registered 500.0 counts/min (corrected for background) in the $^{197}_{80}$Hg gamma-ray peak. 195 h later (one half-life of $^{197}_{80}$Hg), the chemical standard was counted for 10.00 min and registered 100.0 counts/min (corrected for background) in the $^{197}_{80}$Hg gamma-ray peak. What is the Hg content of the fish sample?
Ans. 1.88 ppm

ANALYTICAL USES OF RADIOACTIVITY

14.31 The volume of blood in a laboratory animal is determined *in vivo* by isotopic dilution. 1.00 mL of a tritiated saline solution registered 60 000 counts/min owing to the presence of $^3_1$H ($t_{1/2} = 12.26$ y). After 25.0 μL of this saline solution had been injected into the bloodstream of the animal and allowed to mix, a 1.00 mL sample of blood was drawn; it registered 12.00 counts/min due to $^3_1$H. What is the blood volume of the animal?   *Ans.* 125.0 mL

14.32 50.0 mL of a 6.00×10^{-3} M solution of NaI was added to 50.0 mL of an x M solution of Pb(NO$_3$)$_2$, resulting in the formation of some PbI$_2$(s) precipitate; K_{sp} for PbI$_2$ is 1.50×10^{-9}. The original NaI solution (just prior to mixing with the Pb(NO$_3$)$_2$) registered 60.0 counts · min^{-1} · mL^{-1} owing to the presence of $^{131}_{53}$I ($t_{1/2} = 8.00$ d). Immediately after mixing and precipitation, the resultant solution registered 10.0 counts · min^{-1} · mL^{-1}. Find x.   *Ans.* 5.00×10^{-3}

14.33 Some, but not necessarily all, of the Cl atoms in PCl$_5$ can undergo rapid exchange with either Cl atom in Cl$_2$. The exchange reaction can be examined by starting with one of the compounds tagged with radioactive $^{36}_{17}$Cl ($t_{1/2} = 3.08 \times 10^5$ y). Suppose that 10.0 mmol of untagged PCl$_5$ and 5.00 mmol of Cl$_2$ (tagged with $^{36}_{17}$Cl) are dissolved in an organic solvent, and, after equilibrium, 75.0% of the $^{36}_{17}$Cl activity arises from PCl$_5$ and 25.0% from Cl$_2$. How many Cl atoms in PCl$_5$ participated in the rapid exchange with the Cl$_2$?   *Ans.* 3 of each 5

COUNT STATISTICS

14.34 When using a liquid scintillation counter, you can select the beta-energy regions for which you want to obtain counts. Two regions, I and II, were chosen for the analysis of a sample that contains both tritium and carbon-14. A pure $^3_1$H source registers only in region I; a pure $^{14}_6$C source registers in both regions, with the count-rate ratio $R_I/R_{II} = 0.6500$. If a 10.00 min count of a sample registers 2500 counts in region I and 1600 counts in region II, and there is negligible background in both regions, how many counts in region I are expected from each element and what are the two standard deviations?
Ans. For tritium, $\bar{C} \pm \sigma_C = 1460 \pm 56$; for carbon-14, $\bar{C} \pm \sigma_C = 1040 \pm 26$.

Two-Tailed Confidence Coefficients $t_{CL,\nu}$ for Student's t Distribution with ν Degrees of Freedom

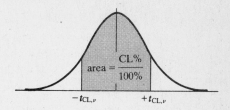

CL, % ν	50.0	68.3	80.0	90.0	95.0	95.5	99.0	99.7	99.9	99.99	99.999	99.9999
1	1.000	1.837	3.078	6.314	12.706	13.998	63.657	244.850	636.619	6 366.198	63 661.977	636 619.772
2	0.816	1.321	1.886	2.920	4.303	4.532	9.925	19.573	31.598	99.992	316.225	999.999
3	0.765	1.197	1.638	2.353	3.182	3.309	5.841	9.338	12.924	28.000	60.397	130.155
4	0.741	1.141	1.533	2.132	2.776	2.872	4.604	6.691	8.610	15.544	27.771	49.459
5	0.727	1.110	1.476	2.015	2.571	2.650	4.032	5.555	6.869	11.178	17.897	28.477
6	0.718	1.090	1.440	1.943	2.447	2.518	3.707	4.943	5.959	9.082	13.555	20.047
7	0.711	1.077	1.415	1.895	2.365	2.430	3.499	4.562	5.408	7.885	11.215	15.764
8	0.706	1.066	1.397	1.860	2.306	2.368	3.355	4.305	5.041	7.120	9.782	13.257
9	0.703	1.058	1.383	1.833	2.262	2.321	3.250	4.120	4.781	6.594	8.827	11.637
10	0.700	1.052	1.372	1.812	2.228	2.285	3.169	3.980	4.587	6.211	8.150	10.516
12	0.694	1.043	1.356	1.782	2.179	2.233	3.055	3.785	4.318	5.694	7.261	9.085
16	0.690	1.032	1.337	1.746	2.120	2.170	2.921	3.562	4.015	5.134	6.330	7.642
20	0.687	1.025	1.325	1.725	2.086	2.134	2.845	3.439	3.850	4.837	5.854	6.927
30	0.683	1.017	1.310	1.697	2.042	2.088	2.750	3.285	3.646	4.482	5.299	6.119
60	0.679	1.008	1.296	1.671	2.000	2.044	2.660	3.143	3.460	4.169	4.825	5.449
120	0.677	1.004	1.289	1.658	1.980	2.022	2.617	3.076	3.373	3.997	4.573	5.102
∞	0.674	1.000	1.282	1.645	1.960	2.000	2.576	3.000	3.291	3.891	4.417	4.892

Index

Catalog

If you are interested in a list of SCHAUM'S
OUTLINE SERIES send your name
and address, requesting your free catalog, to:

SCHAUM'S OUTLINE SERIES, Dept. C
McGRAW-HILL BOOK COMPANY
1221 Avenue of Americas
New York, N.Y. 10020